PRECISION ASSEMBLY TECHNOLOGIES FOR MINI AND MICRO PRODUCTS

IFIP – The International Federation for Information Processing

IFIP was founded in 1960 under the auspices of UNESCO, following the First World Computer Congress held in Paris the previous year. An umbrella organization for societies working in information processing, IFIP's aim is two-fold: to support information processing within its member countries and to encourage technology transfer to developing nations. As its mission statement clearly states,

> *IFIP's mission is to be the leading, truly international, apolitical organization which encourages and assists in the development, exploitation and application of information technology for the benefit of all people.*

IFIP is a non-profitmaking organization, run almost solely by 2500 volunteers. It operates through a number of technical committees, which organize events and publications. IFIP's events range from an international congress to local seminars, but the most important are:

- The IFIP World Computer Congress, held every second year;
- Open conferences;
- Working conferences.

The flagship event is the IFIP World Computer Congress, at which both invited and contributed papers are presented. Contributed papers are rigorously refereed and the rejection rate is high.

As with the Congress, participation in the open conferences is open to all and papers may be invited or submitted. Again, submitted papers are stringently refereed.

The working conferences are structured differently. They are usually run by a working group and attendance is small and by invitation only. Their purpose is to create an atmosphere conducive to innovation and development. Refereeing is less rigorous and papers are subjected to extensive group discussion.

Publications arising from IFIP events vary. The papers presented at the IFIP World Computer Congress and at open conferences are published as conference proceedings, while the results of the working conferences are often published as collections of selected and edited papers.

Any national society whose primary activity is in information may apply to become a full member of IFIP, although full membership is restricted to one society per country. Full members are entitled to vote at the annual General Assembly, National societies preferring a less committed involvement may apply for associate or corresponding membership. Associate members enjoy the same benefits as full members, but without voting rights. Corresponding members are not represented in IFIP bodies. Affiliated membership is open to non-national societies, and individual and honorary membership schemes are also offered.

PRECISION ASSEMBLY TECHNOLOGIES FOR MINI AND MICRO PRODUCTS

Proceedings of the IFIP TC5 WG5.5 Third International Precision Assembly Seminar (IPAS '2006), 19-21 February 2006, Bad Hofgastein, Austria

Edited by

Svetan Ratchev
University of Nottingham
United Kingdom

Precision Assembly Technologies for Mini and Micro Products
Edited by Svetan Ratchev

p. cm. (IFIP International Federation for Information Processing, a Springer Series in Computer Science)

ISSN: 1571-5736 / 1861-2288 (Internet)

Printed on acid-free paper

ISBN 978-1-4419-4063-6 e-ISBN 978-0-387-31277-4

9 8 7 6 5 4 3 2 1
springeronline.com

TABLE OF CONTENTS

Preface ix

International Advisory Committee xi

PART I – Micro Handling and Feeding Techniques 1

1. Design of a Capillary Gripper for a Submillimetric Application 3
   Lambert, P., Seigneur, F., Koelemeijer, S., Jacot, J.

2. Multi-Axes Micro Gripper for the Handling and Alignment of Flexible Micro Parts 11
   Brecher, C., Peschke, C., Freundt, M. Lange, S.

3. Design, Fabrication and Characterization of a Flexible System Based on Thermal Glue for in Air and in SEM Microassembly 21
   Clévy, C., Hubert, A., Fahlbusch, S., Chaillet, N., Michler, J.

4. Design and Experimental Evaluation of an Electrostatic Micro-Gripping System 33
   Lang, D., Tichem, M.

5. A Generic Approach for a Micro Parts Feeding System 43
   Paris, M., Perrard, C., Lutz, P.

6. Pneumatic Contactless Feeder for Microassembly 53
   Turitto, M., Chapius, Y., Ratchev, S.

PART II – Robotics and Robot Applications for Precision Assembly 63

7. "Parvus" A Micro-Parallel-SCARA Robot for Desktop Assembly Lines 65
   Burisch, A., Wrege, J., Soetebier, S., Raatz, A., Hesselbach, J., Slatter, R.

8. Methods for Comparing Servo Grippers for Mini and Micro Assembly Applications 75
Karjalainen, I., Tuokko, R.

9. Compliant Parallel Robots 83
Raatz, A., Wrege, J., Burisch, A., Hesselbach, J.

10. Test Environment for High-Performance Precision Assembly – Development and Preliminary Tests 93
Prusi, T., Heikkilä, R., Uusitalo, J., Tuokko, R.

11. Sensor Guided Micro Assembly by Using Laser-Scanning Technology 101
Rathmann, S., Wrege, J., Schöttler, K., Raatz, A., Hesselbach, J.

12. High Speed and Low Weight Micro Actuators for High Precision Assembly Applications 109
Degen, R., Slatter, R.

PART III – Design and Planning for Microassembly 119

13. Automated Assembly Planning Based on Skeleton Modelling Strategy 121
Bley, H., Bossmann M.

14. Morphological Classification of Hybrid Microsystems Assembly 133
Kurniawan, I., Tichem, M., Bartek, M.

15. First Steps in Integrating Micro-Assembly Features into Industrially used DFA Software 149
Salmi, T., Lempiäinen, J.

16. Tolerance Budgeting in a Novel Coarse-Fine Strategy for Micro-Assembly 155
Henneken, V., Tichem, M.

17. The Importance of Concept and Design Visualisation in the Production of an Automated Assembly and Test Machine 167
Smale, D.

18. Development of Passive Alignment Techniques for the Assembly of Hybrid Microsystems 181
Brecher, C., Weinzierl, M., Lange, S.

PART IV – Modular Assembly Systems and Control Applications **191**

19. Miniature Reconfigurable Assembly Line for Small Products 193
Codourey, A., Perroud, S., Mussard, Y.

20. Conception of a Scalable Production for Micro-Mechatronical Products 201
Fleischer, J., Krahtov, L., Volkmann, T.

21. Towards an Integrated Assembly Process Decomposition and Modular Equipment Configuration 215
Lohse, N., Schäfer, C., Ratchev, S.

22. Evolvable Skills for Assembly Systems 227
Hoppe, G.

23. Toward the Vision Based Supervision of Microfactories Through Images Mosaicing 239
Bert, J., Dembélé, S., Lefort-Piat, N.

24. Precision Multi-Degrees-Of-Freedom Positioning Systems 251
Olea, G., Takamasu, K., Raucent, B.

PART V – Economic Aspects of Microassembly **265**

25. What is the Best Way to Increase Efficiency in Precision Assembly 267
Koelemeijer, S., Bourgeois, F., Jacot, J.

26. Life Cycle and Cost Analysis for Modular Re-Configurable Final Assembly Systems 277
Heilala, J., Helin, K., Montonen, J., Väätäinen, O.

27. Impact of Bad Components on Costs and Productivity in Automatic Assembly 287
Oulevey, M., Koelemeijer, S., Jacot, J.

PART VI – Microassembly – Solutions And Applications **295**

28. The Reliable Application of Average and Highly Viscous Media 297
Lenz, T., Othman, N.

29. Laser Sealed Packaging for Microsystems 307
Seigneur, F., Jacot, J.

30. Modelling and Characterisation of an Ortho-Planar Micro-Valve 315
Smal, O., Dehez, B., Raucent, B., De Volder, M., Peirs, J., Reynaerts, D., Ceyssens, F., Coosemans, J., Puers, R.

31. Microhandling and Assembly: the Project ASSEMIC 327
Almansa, A., Bou, S., Fratila, D.

Author Index 333

Keyword Index 334

Preface

Customers increasingly expect products that are smaller, have improved functionality and reliability and cost less. Miniaturisation and integration of mechanical, sensing and control functions within confined spaces is becoming an important trend in designing new products in industries such as automotive, biomedical, pharmaceutical and telecommunications.

In micromanufacture often manual assembly becomes unfeasible due to the small size of the products. As a result microassembly is becoming a sector of strategic importance in high labour cost areas due to the specific needs of automated fabrication and assembly processes which make outsourcing a less attractive option. It is well recognised that the production of miniaturised products will require radical rethinking and restructuring of the underlying technologies and system engineering approaches in high precision assembly as well as developing unprecedented and unique commercial concepts and infrastructure for delivering new technologies.

In precision assembly there is a clear need for modular and highly customisable miniaturised production systems based on plug and produce assembly units with micro and nano accuracy of operation. From the equipment point of view the key emphasis is on developing new solutions for the automatic handling of large volumes of very small parts; development of multi-process microassembly machines using a smaller mechanical base and incorporating a wide variety of specialised product specific processes, capable of meeting the increased demands on process capability, repeatability and traceability.

The critical enabling technologies currently being developed include: high precision positioning devices; precision tracking and control of applied forces; process monitoring and feedback; miniaturised "super-clean" environments; automated micro and nano assembly, testing and packaging techniques. There is also a need to develop new approaches for microassembly system design which will allow the deployment of reconfigurable systems for volume manufacture of products in close proximity to the customer.

The International Precision Assembly Seminar (IPAS) is a premier international forum for reporting and discussing key technological developments in the field of mini and micro assembly automation. The contributions to the 3rd IPAS'2006 seminar have been grouped into 6 sections. Part 1 deals with new techniques for the handling and feeding of micro parts. Micro-robotics and robot applications for micro assembly are discussed in Part 2. An overview of different design and planning applications for microassembly is provided in Part 3. Part 4 is dedicated to reconfigurable and modular micro assembly systems and control applications. The economic aspects of microassembly including new business models are discussed in Part 5 while Part 6 presents specific technical solutions and microassembly applications.

The seminar is sponsored by the International Federation of Information Processing (IFIP) WG5.5, the International Institution of Production Engineering Research (CIRP), the European Factory Automation Committee (EFAC) and the Robotics and Mechatronics Professional Network of the Institution of Electrical Engineers (IEE).

The organisers should like to express their gratitude to the members of the International Advisory Committee for their support and guidance and to the authors of the papers for their original contributions. My special thanks go to Professor Luis Camarinha-Matos, Chair of the IFIP WG5.5 and Professor Helmut Bley, Chair of the STC A of CIRP for their continuous support and encouragement. And finally my thanks go to Rachel Brereton, Ruth Strickland and Dr Kevin Phuah from the Precision Manufacturing Group of the University of Nottingham for handling the administrative aspects of the seminar, putting the proceedings together and managing the detailed liaison with the publishers.

Svetan M. Ratchev

International Advisory Committee

Seminar Administration

Mrs Rachel Brereton/Miss Ruth Strickland
School of M3, University of Nottingham
Nottingham NG7 2RD, UK
rachel.brereton@nottingham.ac.uk
ruth.strickland@nottingham.ac.uk

PART I

Micro Handling and Feeding Techniques

DESIGN OF A CAPILLARY GRIPPER FOR A SUBMILLIMETRIC APPLICATION

Pierre Lambert
Université libre de Bruxelles CP 165-14
Avenue F.D. Roosevelt, 50
B - 1050 Bruxelles
pierre.lambert@ulb.ac.be

Frank Seigneur, Sandra Koelemeijer and Jacques Jacot
Ecole Polytechnique Fédérale de Lausanne
STI IPR LPM1 - Station 17
CH - 1015 Lausanne
frank.seigneur@epfl, sandra.koelemeijer@epfl.ch, jacques.jacot@epfl.ch

Abstract This paper describes the study of a gripper using the surface tension effects to pick and place the 0.5mm or 0.3mm diameter balls of a millimetric watch bearing. Two liquid supply strategies have been tested (pressure drive and tip dip). The effects of the coating (through the measurement of the contact angles), the presence of an internal channel and the size of the gripper have been studied. Analytical and numerical force models have been developped and validated thanks to a test bed allowing the measurement of the developped force (typically of the order of 100μN) with a resolution of 1μN. A complete pick and place cycle has been performed using the 0.5mm diameter gripper. Such a test has still to be done in the future with the 0.3mm diameter gripper.

Keywords: Microassembly, capillary forces, surface tension, submillimetric application

1. INTRODUCTION

This study aims at using the surface tension effects in the handling of submillimetric parts (the balls of a millimetric bearing). First, a brief state of the art is presented in section 2, highlighting the previous work concerning the surface tension based gripping. The proposed case study is detailed in section 3, showing the design aspects. Then section 4 presents two new models to compute the force (they are adapted from existing ones to this case study). Section 5 summarizes the inputs of the models. Then, in section 6, a functional validation of the gripper is presented and some force measurements confirm the proposed models. Finally, conclusions are drawn in section 7.

2. STATE OF THE ART

Within the framework of microassembly (and in particular in microgripping), the classical way consists in downscaling the two-fingered tweezer principle. This leads to considerable diffficulties therefore a lot of new gripping principles has been proposed these last years: Lambert, 2004 refers to some existing reviews. These new principles take advantage of the scaling laws, because the forces involved decrease slower than the weight of the submillimetric components. They consequently represent an innovative alternative to miniaturized tweezers.

One of these principles, namely the capillary gripping, seems very promising as an alternative to miniaturized tweezers or vacuum pipets: indeed, this kind of gripping is well adpated to flat components with only free accessible surface (typically SMD components) but can also pick up more complex geometries (rings, balls). Its scaling law is very promising since the capillary force is directly proportional to the linear dimension ($F \div l$), i.e. this principle generates forces larger than the weight ($W \div l^3$) and is more efficient than the vacuum gripping ($F \div l^2$) for the handling of submillimetric components. Moreover, as indicated in Grutzeck and Kiesewetter, 1998, there exists a damping effect which prevents mechanical damage due to high contact pressures.

This principle has already been proposed by Grutzeck and Kiesewetter, 1998 and Bark, 1999: they essentially validated the concept by handling millimetric silicon components. Lambert, 2004 deeply studied this gripping principle by modelling, numerical simulation and experimental validation. More recently Biganzoli et al., 2005 and Obata et al., 2004 proposed two different solutions in relation with an applicative framework: Biganzoli suggested to modify the gripper curvature while Obata tuned the volume of liquid.

3. CASE STUDY

The chosen case study deals with the design of a gripper to be used in the insertion of small balls (diameter 500μm) in a hole. One of the requirements is to avoid the conventional tweezers and vacuum grippers, because of the scratches they provoke on the balls. Due to the very small weight of the balls (about 3.8μN), the surface tension based gripper is largely strong enough since it generates forces up to 150μN. The handling scheme is illustrated in figure 1: the picking force is provided by the capillary force and the releasing task is ensured by laterally moving the gripper once the ball is in the hole.

Since the gripper uses capillary forces, a liquid has to be dispensed before each manipulation, but there is no need to eject the liquid (such as for example

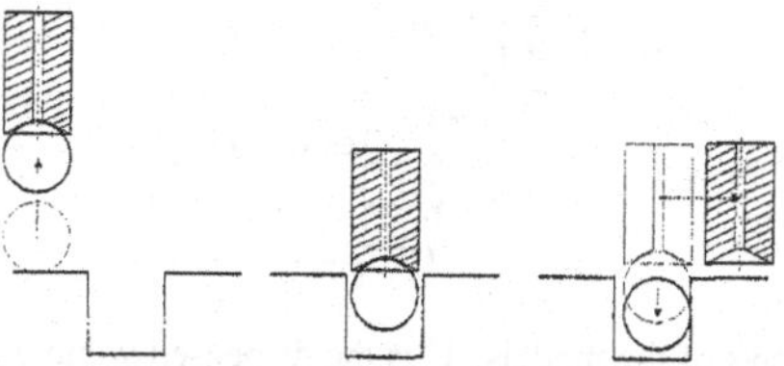

Figure 1. Handling scheme of the capillary gripping for the insertion of a ball in a hole.

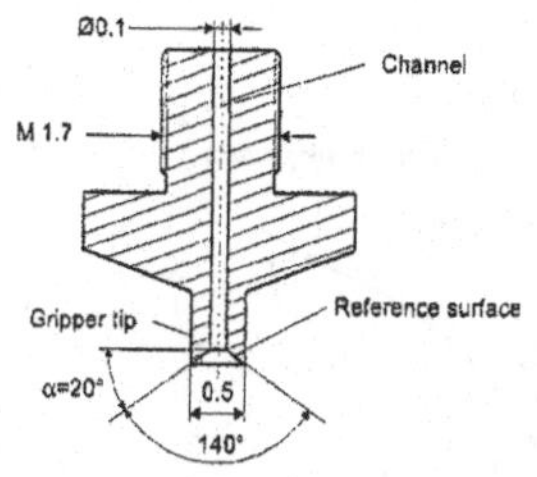

Figure 2a. Schematic view of the designed gripper: the so-called reference surface is the surface which contacts the ball during the handling.

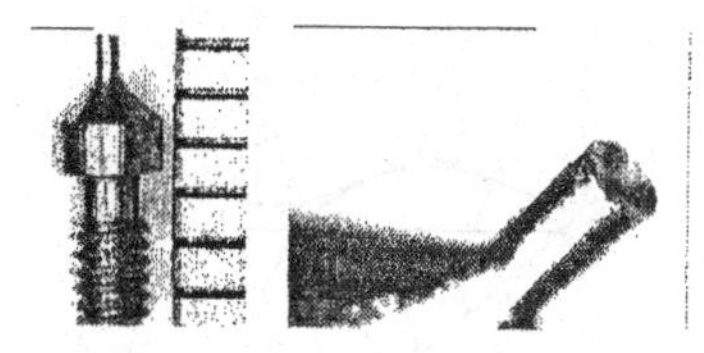

Figure 2b. Prototype (the distance between the graduations is 1mm) and detail of the conical tip.

in ink jet printing): it is sufficient to bring a bit of liquid in contact with the ball to pick up. Beside this dispensing functionnality, the other functions of the developped gripper tip can be summarized as follows: (1) to develop a picking force larger than the weight of the object ($W \approx 3.8\mu N$); (2) to develop a picking force large enough to handle the component with reasonable accelerations (manual handling); (3) to center the ball with respect to the gripper in order to ensure its positioning; (4) to release the ball once it is inserted in the hole. The proposed gripper is shown in figure 2b. It has been machined in stainless steel. Other prototypes have been coated with a hydrophobic silane-based coating.

Two solutions have been tried to supply the gripping liquid: (1) to drive the pressure through the gripper channel; (2) to dip the tip in the liquid.

Finally, the first solution has been discarded because of the instabilities of the droplet when its height approaches a half diameter of the gripper. It has been taken advantage of the good repeatability of the volume transferred to the gripper by dipping it to the liquid. For a large variety of grippers (see table 1), the ratio of the droplet height h to the gripper diameter D has been found to be 0.304, with a standard deviation of 0.020 and a maximal residual of 0.061.

4. MODELS

The force model discussed in this section can be viewed as a black box, as shown in figure 3. The theoretical capillary force developped by the gripper has

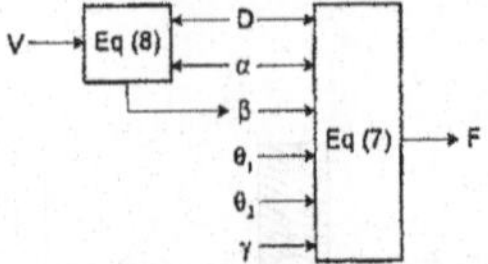

Figure 3. Schematic views of the models: V is the dispensed volume of liquid, D is the ball diameter (R is the radius), α is a geometrical parameter of the gripper, β is the filling angle, θ_1 and θ_2 are the contact angles and γ is the surface tension.

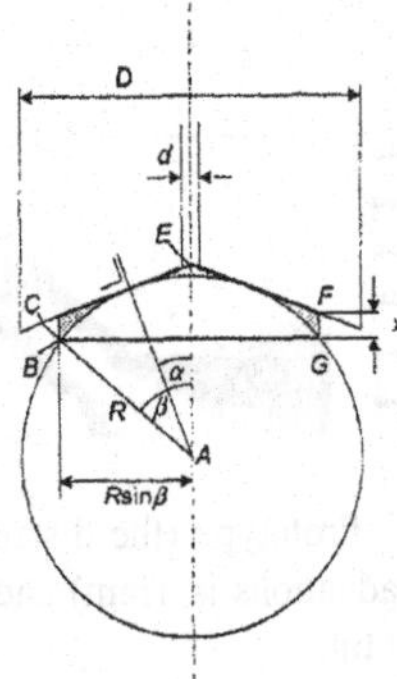

Figure 4a. Geometrical details of the gripper tip.

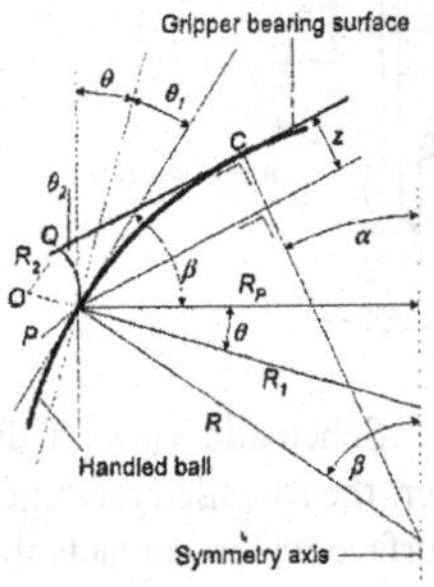

Figure 4b. Geometrical details of the meniscus.

been computed with an analytical model based on the so-called circle approximation of the meniscus geometry (the detailed problem is shown in figure 4b) and with a numerical model based on the Laplace equation, derived from that one presented in Lambert and Delchambre, 2005. The latter model is based on equations that were first proposed by Orr et al., 1975.

The situation depicted in figure 4b shows that the meniscus wets the ball along the circle containing P (the position of this circle, whose radius is equal to R_P, is determined by the filling angle β).

The contact angle θ_1 is the angle between the tangent to the meniscus on the one hand and the tangent to the ball on the other hand: it is determined by the wetting properties of the materials, i.e. it is determined by the triple {handling liquid, material of the ball, surrounding environment}. On the gripper side (point Q), since the gripper and the ball are made of different materials, the contact angle θ_2 can be different from θ_1. The so-called circle approximation assumes a circular shaped meniscus: in the case of figure 4b, this circle is centred in O and has a radius R_2. The capillary force can be written as:

$$F = \pi R \sin^2 \beta \gamma [\frac{\sin(\beta + \theta_1)}{\sin \beta} + \frac{\cos \theta_2 + \cos(\theta_1 + \beta - \alpha)}{1 - \cos(\beta - \alpha)}]$$

Gripper	D (mm)	Channel	Coated	θ_A (°)	σ_{θ_A} (°)	θ_B (°)	σ_{θ_R} (°)	$\frac{h}{D}$
B	0.5	Yes	Yes	22.7	4.2	19.3	3.5	0.287
F	0.5	Yes	No	21.7	4.2	18.8	2.0	0.293
H	0.5	No	No	21.0	2.6	16.8	2.5	0.305
J	0.3	No	Yes	29.8	3.5	24.8	2.9	0.329

Table 1. Properties of the tested grippers: D is the gripper diameter, 'channel' denotes the presence of an internal channel, 'coated' denotes the presence of a silane-based coating, θ_A is the advancing contact angle, θ_R is the receding contact angle and $\frac{h}{d}$ is the ratio between the height h of the hanging droplet after the gripper has been dipped in liquid and D.

$$+2\pi R \sin\beta\gamma\sin(\beta+\theta_1) \qquad (1)$$

(this equation gives the force as a function of the handling liquid (γ), the materials (θ_1,θ_2), the size of the ball (radius R), the volume of liquid through the filling angle β and the gripper geometry (α)). One may notice that the meniscus is not defined for $\beta = \alpha$, leading to non physical result for the force (the force tends towards infinity).

The link between the dispensed volume of liquid V and the filling angle β can be determined as follows: in figure 4a, the dispensed volume of liquid fills the $BCEFG$ area between the ball and the gripper, therefore:

$$V = V_1 - V_2 \qquad (2)$$

where V_1 is the sum of the volumes of the cylindre $BCFG$ and the cone CEF, and V_2 is the volume of is the volume of the spherical cap of the ball limited by the filling angle β.

5. EXPERIMENTAL SET UP

The experimental set up used in this study has already been described in Lambert, 2004. It allows to measure the advancing and receding contact angles and to measure the capillary force with a resolution of the order of 1 μN. Four grippers have been tested whose properties are summarized in table 1 (the liquid which was used was a lubrication oil with a surface tension $\gamma = 34.5\text{mNm}^{-1}$. Note that oil has been used because it does not evaporate during the experiments. Moreover, since it can be the lubrication oil of the assembled device, it is not a drawback to have residual traces of liquid on the component after the manipulation). The balls are made in Z_rO_2.

6. RESULTS

Figure 5 illustrates an assembly cycle which has been achieved with the test bed mentionned in the previous section. On the first image, the ball to handle can be seen on the right side, 'floating' on a droplet of liquid (it 'floats' thanks to the surface tension effect, like the water strider, and not due to the Archimedes principle). On the left side of the image there is the bearing in which the ball has to be placed. On the second image, the gripper is aligned above the ball. The third image illustrates the picking of the ball, which is then moved (images 4, 5 and 6) to its final location inside the bearing. The seventh image illustrates the radial shift of the gripper required to perform the release task. Finally, the gripper is moved away from the bearing.

Concerning the force models, the results of both models (analytical and numerical) are plotted in figures 6a and 6b. On each figure, both the force and the volume of liquid between the gripper and the component are plotted as a function of the filling angle β. Two regimes can be distinguished: for $\beta < \alpha$ (here $\alpha = 20$ °), the force is increasing with β while the volume of liquid remains quite constant. In the case $\beta > \alpha$, the volume of liquid quickly increases with the filling angle while the force is decreasing. The simulation has been run until the liquid overflows, i.e. when $\beta \approx 60$ °. The behaviour of both models is quite similar however there is a small difference between the analytical model (dashed line) and the numerical model (solid line). The force values corresponding to the maximal filling angles are respectively equal to 97μN for the Φ0.3mm gripper and to 168μN for the Φ0.5mm gripper.

Finally, direct force measurements are shown in figure 7, for 4 different grippers (B, F, H, J). In each case, the gripper has been dipped to liquid and put in contact with ball to handle (the ball diameter is equal to that one of the gripper). This first contact discards the main part of liquid so that the volume of liquid becomes smaller than the conical cavity of the grippers (henceforth, it cannot be seen with the camera). Then, the gripper is applied onto the ball n times without refilling it and the capillary force between the gripper and the ball is measured. After each rupture of the meniscus (i.e. after each contact), a bit of liquid is left on the ball, so that the volume of liquid involved in the

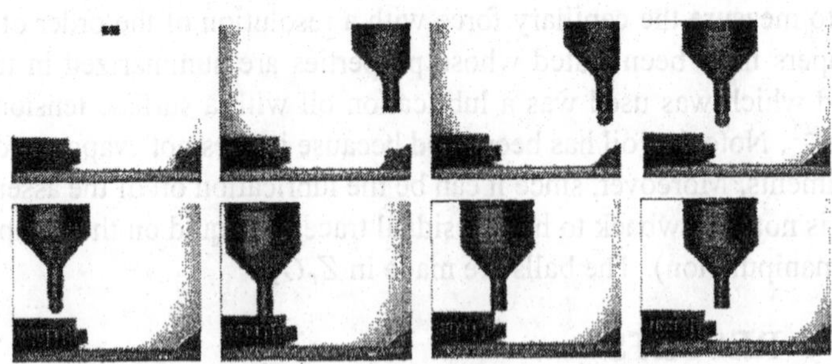

Figure 5. Assembly sequence: the diameter of the handled ball is 0.5mm.

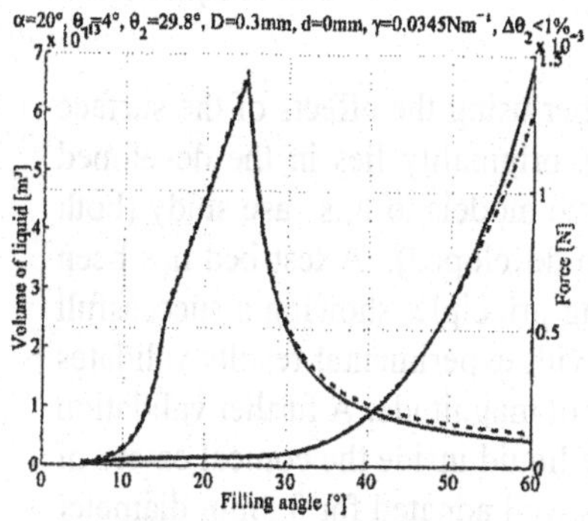

Figure 6a. Volume of liquid and force as a function of the filling angle β (numerical model: solid line, analytical mode: dashed line) for the gripper J.

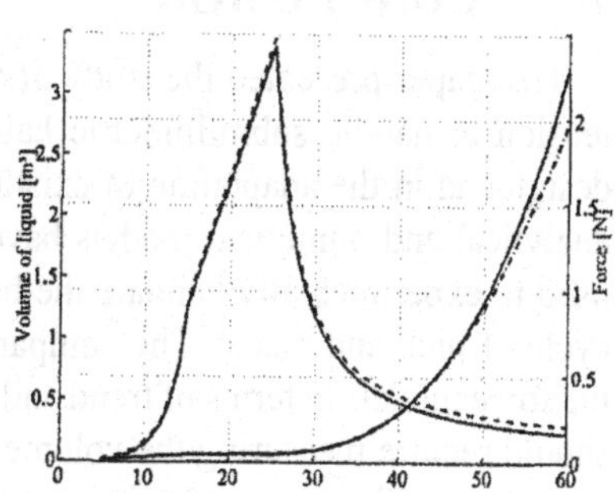

Figure 6b. Volume of liquid and force as a function of the filling angle β (numerical model: solid line, analytical mode: dashed line) for the gripper H.

manipulation is decreasing (however it cannot be measured). Therefore, according to the models, the force should increase, which can be observed in figure 7 for grippers H and J. For grippers B and F, the simulation cannot be run because the effects of the channel have not been modelled. Nevertheless, the effect of this channel is to decrease the force (the force generated by larger grippers B and F is even smaller than the one generated by the smaller gripper J).

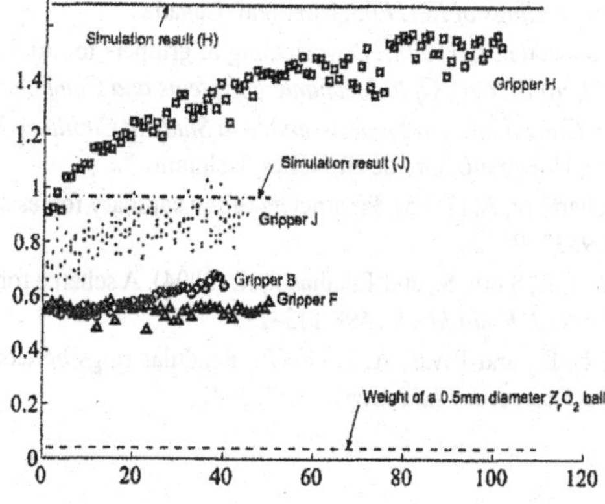

Figure 7. Force measurements.

7. CONCLUSION

This paper presented the study of a gripper using the effects of the surface tension to handle submillimetric balls. Its originality lies in the developed design and in the adaptation of existing force models to this case study (both analytical and numerical models have been developed). A test bed has been used to experimentally validate the handling principle, showing a successfull cycle of pick and place. The comparison with experimental results validates the force models in terms of trend and order of magnitude. A further validation should require measuring the volume of the liquid inside the conical cavity of the gripper. The proposed gripper design is well adpated for 0.5mm diameter balls and develops a force which is large enough too in the case of 0.3mm balls. The liquid supply has been achieved by dipping the gripper tip to the handling liquid. At this step, the presence of an internal channel is not necessary to supply the liquid. Moreover, the channel decreases the force. As a concluding remark, let us note that other applications than the handling of small balls are going to be studied. On a short term, the gripper will be moved by a Sysmelec robot, in order to assess the performances of this gripping principle.

ACKNOWLEDGEMENTS

The authors wish to thank J.-J. Crausaz and P. Zbinden for the manufacturing, G. Kulik and P. Hoffmann for the coating of the prototypes.

REFERENCES

1. Bark, K.-B. (1999). *Adhäsives Greifen von kleinen Teilen mittels niedrigviskoser Flüssigkeiten.* Springer.
2. Biganzoli, F., Fassi, I., and Pagano, C. (2005). Development of a gripping system based on capillary force. In *Proceedings of ISATP05*, Montreal, Canada.
3. Grutzeck, H. and Kiesewetter, L. (1998). Downscaling of grippers for micro assembly. In *Proc. of 6th Int. Conf. on Micro Electro, Opto Mechanical Systems and Components*, Potsdam.
4. Lambert, P. (2004). *A Contribution to Microassembly: a Study of Capillary Forces as a gripping Principle.* PhD thesis, Université libre de Bruxelles, Belgium.
5. Lambert, P. and Delchambre, A. (2005). Parameters ruling capillary forces at the submillimetric scale. *Langmuir*, 21:9537–9543.
6. Obata, K. J., Motokado, T., Saito, S., and Takahashi, K. (2004). A scheme for micro-manipulation based on capillary force. *J. Fluid Mech.*, 498:113–21.
7. Orr, F. M., Scriven, L. E., and Rivas, A. P. (1975). Pendular rings between solids: meniscus properties and capillary force. *J. Fluid Mech.*, 67:723–42.

MULTI-AXES MICRO GRIPPER FOR THE HANDLING AND ALIGNMENT OF FLEXIBLE MICRO PARTS

Development of compact and shock resistant gripper components

Christian Brecher[1], Christian Peschke[2], Martin Freundt[3], Sven Lange[4]

[1, 2, 3, 4] Fraunhofer Institute for Production Technology IPT, Steinbachstrasse. 17, 52072 Aachen, Germany

Abstract: Optical micro parts, such as glass fibres, require handling and alignment accuracies down to the sub micrometer range. Addressing this task, one aim of the Fraunhofer IPT is the development of new concepts of active gripper systems. In this context a highly integrated, adaptive, rugged and economical gripper system particularly for accurate handling and alignment of flexible micro parts down to the sub-micron level has been developed. This gripper system can be used on conventional robot systems for carrying out micro-assembly operations. The robot system does the pre-positioning, the tolerances necessary for the micro-assembly are subsequently realized directly at the tip of the gripper with the gripper integrated multi-axes system. Positioning systems that achieve the required positioning increments in the sub micron range are already existent. However, the problem of such systems is that they are normally highly sensitive against mechanical impact and extremely cost intensive. In this paper the development of a highly robust gripper-integrable axes system and its integration in a novel gripper design with a multi-axis adjustment system is presented.

Key words: Assembly; Flexible Micro Parts; Handling; Micro Gripper; Active Alignment; Gripper-Integrable Multi-Axes System; Gripper System.

1. INTRODUCTION

Assembly plays an important role in the manufacture of hybrid micro systems. In the process of micro assembly up to 80% of all arising expenses in the manufacturing process[1] of a micro system are generated. Therefor flexible automated solutions are needed to reduce the costs on the one hand

and to satisfy the continuously increasing quality demand concerning assembly accuracy in the sub-micron range[2,3,4] on the other hand. It is the handling and assembly technology which provides the most valuable opportunities for a more variable approach, while joining techniques[5,6] generally have to be selected function-specific.

Especially in the assembly of hybrid micro systems for tele-communications such as switches, star couplers, modulators, and power splitters the demands on accuracy are especially high. Even misalignments of less than 1 µm between the optical components and the glass fibres can lead to significant losses in the transmission coupling of light into the fibres.

The current state of the research and technology in this field is that in many cases glass fibres are still being assembled manually. It is problematical is that accuracies in the sub-micrometer range can not be guaranteed by using manual assembly. In these cases, expensive assembly systems with active alignment strategies have to be used. On behalf of this background more research has to be conducted, particularly on the behaviour of the flexible glass fibres in robotic grippers, before automated assembly is possible. For this reason, experimental work on the optimisation of gripper design is currently being conducted at Fraunhofer IPT, to attain the highest possible repeatable positional accuracy when gripping flexible optical fibres.

In addressing this trend the Fraunhofer IPT developed an all-purpose gripper, suitable to be integrated into a conventional robot in order to reduce cost by performing micro position and alignment tasks in conventional positioning systems. Therefore the gripper is equipped with an integrated stage system allowing positioning and alignment with an accuracy of less then one micron. In order to run processes with short secondary process times the grippers design is shock resistant and compact in size by using a hydrostatic transmission. In this paper, the development of the hydrostatical transmission system and the novel gripper design with a multi-axes system will be described.

2. DEVELOPMENT OF MINIATURISED, HIGHLY-PRECISE AND ROBUST STAGES

For the design of the described multi-axis system compact and shock resistant stages are needed. The application requires that all components of these stages such as transmission devices and guides have to be shock resistant and compact in their dimension. In this context, the Fraunhofer IPT has developed a highly precise and robust, miniaturised stage. Based on a

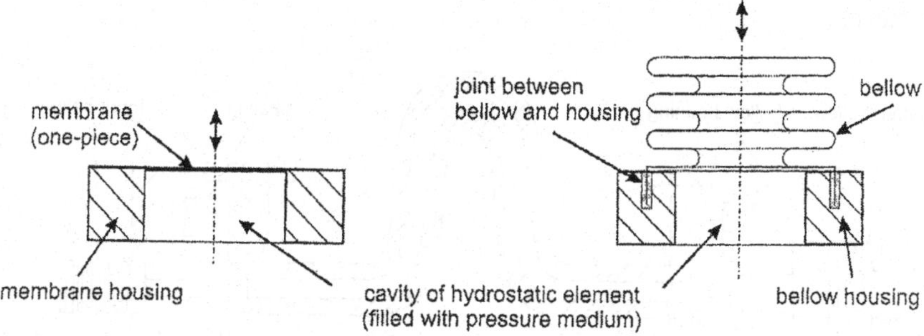

Figure 1. Membrane and bellows as movable piston elements

conventional piezomotor a hydrostatic transmission principle allows extremely precise movements down to 0.1 μm. Currently most micro positioning systems use levers for the transmission. Comparing the principal of a lever, the hydrostatical principle is building more compact in dimension and allows for larger slip-stick free transmission of displacements. Furthermore the hydrostatical principle is more robust against high acceleration rates. This innovative principle – hydrostatic transmission – has not been used in connection with micro-positioning technology before.

The system developed at the Fraunhofer IPT consists of a reservoir filled with a hydrostatic fluid and two movable piston elements for actuation and output. As a result, these elements can be freely positioned in relation to each other, allowing an extremely compact stage design.

The development of the hydrostatic transmission element involved fundamental research into the design of the movable piston elements and the seal for these elements. Investigations on hydrostatic fluids were also necessary in order to determine, which is best dedicated for a highly precise transmission system. Solid state joints were used to allow for movement without play, hysteresis or friction. Two possible design approaches of moveable pistons as solid state elements were investigated: membrane systems and bellows (figure 1). The use of bellows allows an extremely compact design of the transmission element – compared to membranes of the same diameter, the bellows are significantly smaller and more elastic in the direction of motion. Such a system must be leak-proof in order to generate a reproducible transmission behaviour. Furthermore the element must be filled with a fluid, therefore a detachable seal is necessary. Adhesive bonds (two-component adhesive) were successful tested and put into use as non-detachable seals. O-sealings were tested as detachable seals but as they displaced too much volume, thereby critically increasing initial load in the transmission element, rectangular sealing rings were used in the final design instead.

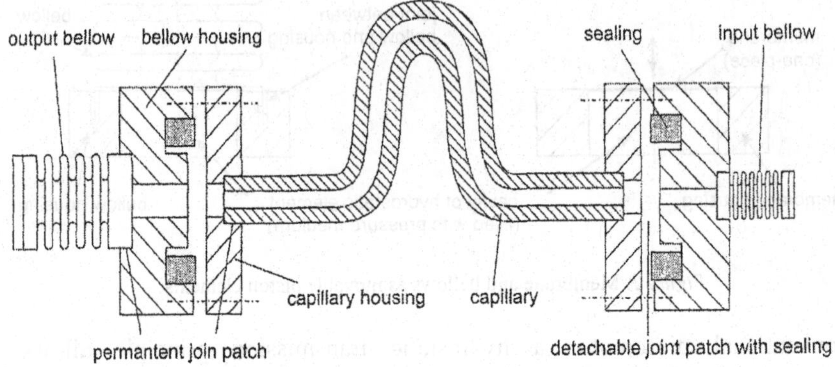

Figure 2. Design of hydrostatic transmission element for integration into gripper

Highly precise transmission through a capillary only works when friction between the fluid and the capillary is low – viscosity therefore plays an important role. Glycerine, water and pressure oils (HLP) were tested in the system. Glycerine's high viscosity (1500 times more viscous than water) makes it impossible to fill the transmission element without air bubbles. Because of their low viscosity air bubbles are not a problem when using water and HLP oils. Further tests proved that both, water and HLP oils, can generate reproducible and consistent motional behaviour. In the end, water was selected to transmit the pressure because it is easier to handle. The ratio of the effective diameters of the bellows results in a theoretical reduction ratio of 1:4.4. Figure 2 shows the design of the final hydrostatic transmission device.

As the actuator, a 20 g commercially available piezo-drive was chosen which uses the friction principle to move the piston rod. The minimum increments vary slightly from motor to motor between 0.6 ± 0.1 µm and 1.0 ± 0.1 µm.

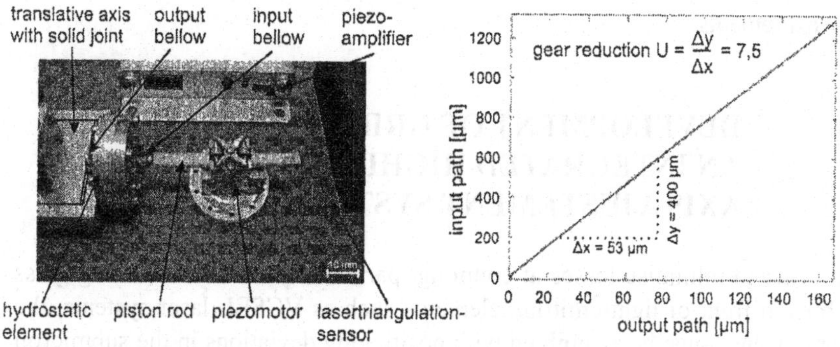

Figure 3. Prototype setup of a single-axis high precise positioning unit

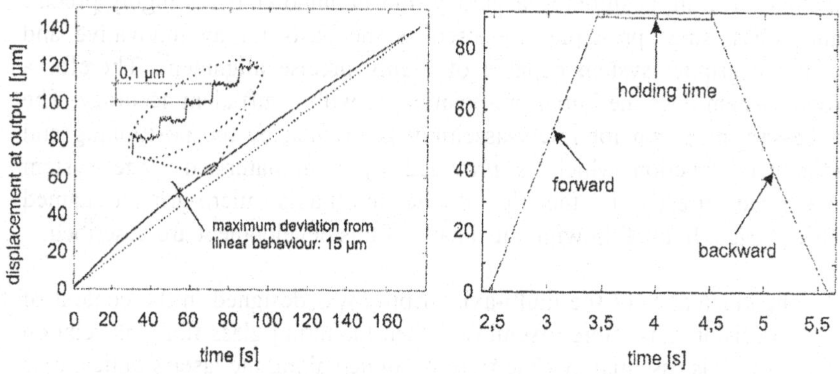

Figure 4. Motional properties of the hydrostatic transmission element

Using the described piezomotor an arithmetical reduction ratio of at least 3.5 is needed to achieve the required precision. Having moved the motor one step along, frictional losses fix the system in its position without the need of additional power. The prototype design of this unit, consisting of the piezomotor and the hydrostatic drive and output bellows is shown in figure 3 (left). The result of measurements for transition behaviour of the drive is also shown in figure 3 (right). A reduction ration of 1:7.5 ± 0.2 was measured in the prototype single-axis system. This deviation from the theoretical value is caused by the internal elasticity of the hydrostatic element but it has a positive effect on the minimum achievable increment.

The output bellow is coupled with a stage equipped with solid state joints, resulting in minimum increments of 0.1 ± 0.02 µm. This configuration can travel up to a distance of 140 µm with a maximum linear deviation of 15 µm along the entire path of motion. In this configuration, the maximum output speed is 90 µm/s. Figure 4 shows the results of these measurements.

3. DEVELOPMENT OF GRIPPER SYSTEM WITH AN INTEGRATED HIGHLY PRCISE MULTI-AXIS AJUSTEMENT SYSTEM

In order to minimise losses at coupling, particularly when assembling glass fibres in front of light-emitting elements such as VCSEL laser systems, the components must be assembled with positioning deviations in the submicron range. Initial investigations have shown that strategies are needed to compensate for positioning inaccuracies during the highly precise micro-assembly of flexible micro parts. In order to compensate this and other factors affecting positioning accuracy, the Fraunhofer IPT's highly precise and robust stage principle was used as the basis for an innovative and adaptive gripper system capable of highly precise alignment. Therefor a robot system does the initial pre-positioning with insufficient accuracy. The necessary precision for micro-assembly is reached by the positioning and alignment function which is provided by a miniaturised stage system integrated directly in the tip of the multi-axis micro-gripper named »Flexogrip«. In the following the details of this development are described.

The kinematics of the multi-axis gripper was designed in the context of high precision glass fibre assembly. When mounting glass fibres in relation to a laser, it is essential that the fibre is aligned along the laser's optical axis with a precision in the submicron range. The kinematics of the gripper must therefor include four highly precise axes: two translational stages perpendicular to the fibre axis and two rotation stages around each of the two translational stages. The rotational movement around the fibre itself and the translational movement along to the fibre axis are less important as the demands on precision are significantly lower (a few micrometers). Serial kinematics were used in order to make the four-axis design as compact as possible. The general gripper concept is shown in figure 5. Such an

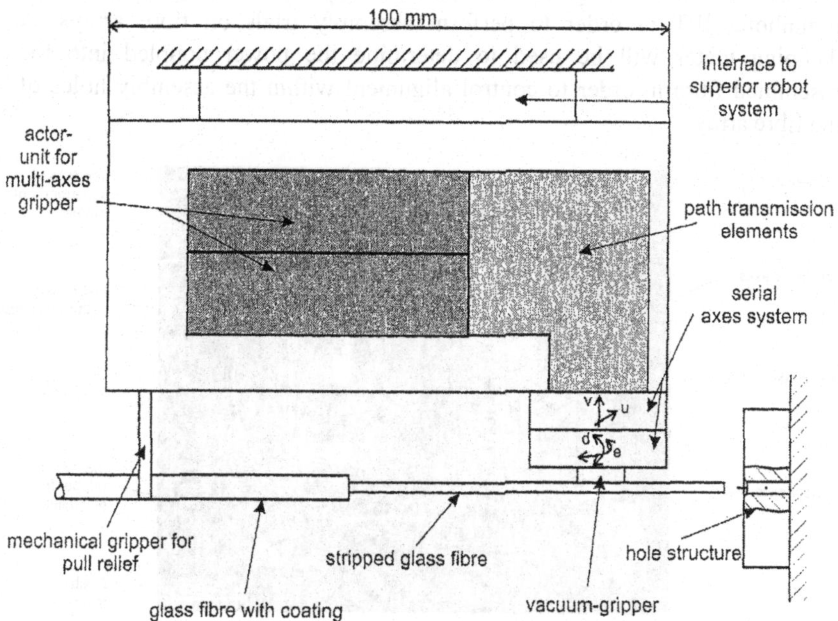

Figure 5. General concept of the multi-axis gripper

innovative gripper system, capable of moving in increments of 0.1 - 0.3 µm by having integrated a miniaturised multi-axis system directly in the tip of the gripper, was made possible by the hydraulic transmission system described above. The minimum travel distance and angle of the gripper integrated stage system utilized to compensate for positioning errors, were determined by testing the reproducible gripping precision, the behaviour of the flexible components and positioning errors: a maximum angle of ± 0.2° and a travel range of at least 100 µm were realized.

This concept was integrated in the »Flexogrip« which is designed to handle and precisely align flexible micro-components. The gripper weighs approx. 350 g, is 115 x 75 x 40 mm³ in size and is equipped with pneumatically driven mechanical gripper used as pull relief. The movable, alignable gripper includes a high precision V-grooved vacuum gripper to ensure that flexible micro parts are gripped accurately. Non-friction and low maintenance solid state joints act as the linear guides and bearings for rotational movements.

Figure 6 shows the final gripper without its housing. Exact measurements are still needed to determine the abilities of the combined stage system. In order to be able to monitor the process, a pressure sensor will be integrated into the vacuum-gripper and its behaviour will be characterised. The gripper system will also be integrated into an existing pre-positioning system at the Fraunhofer IPT in order to perform assembly trials on fibre arrays. A damping meter will be used to determine the power coupled into the assembled fibre in order to control alignment within the assembly holes of the fibre array.

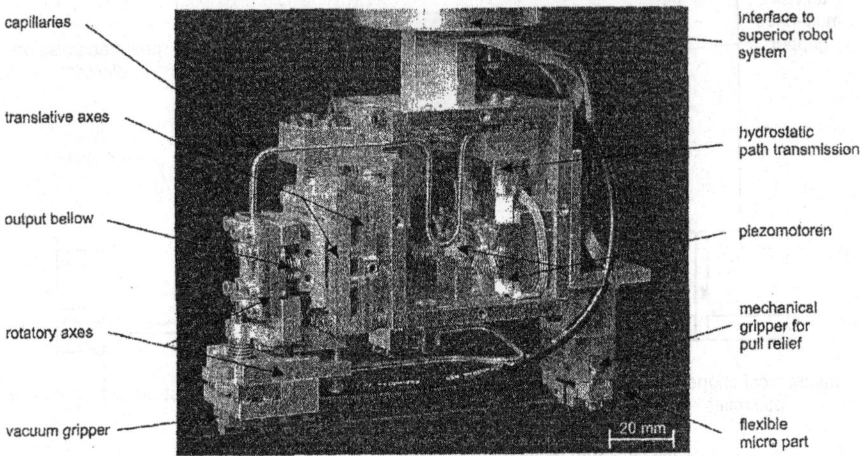

Figure 6. Multi-axis micro-gripper for handling and positioning flexible micro-components

4. CONCLUSION AND OUTLOOK

This paper describes the development of a unique, highly precise, robust and compact stage with a resolution of 0.1 µm. The stage is based on the principle of a hydrostatic transmission principle, which has not been used in micro-assembly before. The hydrostatic transmission consists of a capillary tube and metal bellows, making it very robust. The stage was accelerated up to 90 m/s² in experiments, without being damaged.

This highly precise and at the same time robust stage was put to use in the development of a gripper for the precise assembly of flexible micro-parts. By integrating four of these highly precise and robust stages into a gripper, it is now possible to finely align the micro-components with four degrees of freedom.

ACKNOWLEDGMENT

The authors thank the Deutsche Forschungsgemeinschaft (DFG) for the encouragement during the accomplishment of the research project (SFB 440

"Assembly of Hybrid Micro Systems"). Furthermore the authors would like to thank the European Commission. Within the participation of the Fraunhofer IPT in the framework of the EC Network of Excellence »Multi-Material Micro Manufacture: Technologies and Applications (4M)« parts of the results were presented to and discussed with European research and development partners.

REFERENCES

1. C. Ossmann, Back-End and Assembly Production of Cost Sensitive Microsystems, mst news 2005, Nr. 1/05, pp. 40-41
2. K. Heuer, J. Hesselbach, M. Berndt and R. Tutsch, Sensorgeführtes Montagesystem für die Mikromontage, Robotik 2004, VDI-Berichte Nr. 1841, ISBN 3-18-091841-1, München, 17.-18. June 2004, pp. 39-46
3. B. Petersen, Flexible Handhabungstechnik für die automatisierte Mikromontage. Dissertation, RWTH Aachen, 2003
4. G. Pokar, Untersuchung zum Einsatz von ebenen Parallelrobotern in der Mikromontage, Diss. TU Braunschweig, Vulkan, Essen, 2004
5. M. Weck and C. Peschke, Assembling hybrid microsystems – challenges and solutions, IPAS'2003, ISBN 0-85358-117-7, Q3 Digital/Litho, Loughborough, 2003
6. M. Weck and C. Peschke: Handhabung von Mikrobauteilen – Herausforderungen und Lösungen. Tagungsband: Kolloquium Mikroproduktion, ISBN 3-8027-8670-X, Vulkan, pp. 119-128, Essen, 2003

DESIGN, FABRICATION AND CHARACTERIZATION OF A FLEXIBLE SYSTEM BASED ON THERMAL GLUE FOR IN AIR AND IN SEM MICROASSEMBLY

Cédric Clévy[1], Arnaud Hubert[1], Stephan Fahlbusch[2], Nicolas Chaillet[1] and Johann Michler[2]

[1]*Laboratoire d'Automatique de Besançon -UMR CNRS 6596 - ENSMM - UFC 24 rue Alain Savary, 25000 Besançon, FRANCE*

[2]*Swiss Federal Laboratories for Materials Testing and Research (EMPA), Feuerwerkerstrasse 39, 3602 Thun, SWITZERLAND*

Abstract This paper presents the design, fabrication and characterization of a device able to exchange the tip part (so-called the tools) of a two fingered microgripper. The principle of this tool changer is based on the use of a thermal glue whose state (liquid or solid) is changed by heating or cooling. Several kinds of pairs of tools have been designed. The suitable pair of tools can be chosen according to the size, shape and material of the object to manipulate. The tool changer enables one to perform a sequence of elementary micromanipulation tasks (i.e. an assembly sequence) by using only one gripper mounted on only one manipulator. The tool changer has been automated and successfully tested in air and in the vacuum chamber of a Scanning Electron Microscope (SEM). It brings flexibility to the micromanipulation cell and contributes to reduce the costs, the used space and experimentations time for micromanipulations in the SEM. The assembly of a ball bearing (the balls are 200 μm in diameter) has been successfully tested using the microgripper equipped with the tool changer in a SEM. This tool changer has been designed for a microgripper but can be easily adapted to lots of other kinds of systems.

Keywords: Micromanipulation cell, Tool Changer, Micromanipulation, Microassembly, Flexibility, Microfactory, Scanning Electron Microscope, Automation.

1. Introduction

Great developments have been done in the field of microrobotics for the past few years. Several very efficient actuators have been developed

1 Introduction

Great developments have been done in the field of microrobotics for the past few years. Several very efficient actuators have been developed to be used in microassembly workcells ([3] [1]). Nevertheless, in the field of micromanipulation, two main challenging topics are still growing.

First of all, some research works are done to automate microassembly cells ([7]) ([18]). Very small and precise sensors must be developed and a lot of work is also done to integrate vision capabilities ([8]) ([17]). These researches may lead toward the development of microfactories ([2]) ([15]).

Secondly, performing micromanipulation tasks inside the chamber of a Scanning Electron Microscope (SEM) is very useful. Indeed, the large depth of focus, the high magnification and the clean environment of a SEM provide very good conditions to micromechanical studies or to microassembly. These researches led toward the development of very compact and SEM compatible devices ([11]) ([14]).

These application fields require flexible micromanipulation cells. To perform micromanipulation tasks with flexibility, we have designed a microgripper and its tool changer. Up to now, few works have been done to bring flexibility ([10]) ([9]) ([21]). Several systems have been developed but are closer to a miniaturization approach ([11]) ([19]) using for example revolver turrets ([20]) ([6]). Few devices are adapted to the microworld. They allow temporary mechanical fixation ([16]) ([12]). Our solution was developed to close a part of this gap. The principle of our tool changer will be presented in section 2 and its characterization will be studied in section 3. Finally, this system will be used to perform the assembly of a ball bearing in a SEM (section 4).

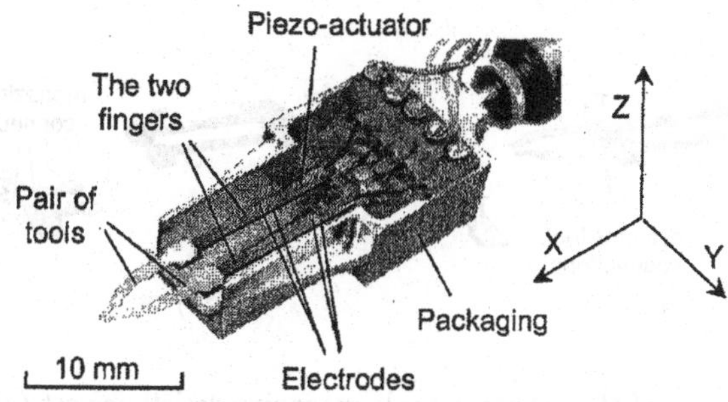

Figure 1: Overview of the microgripper.

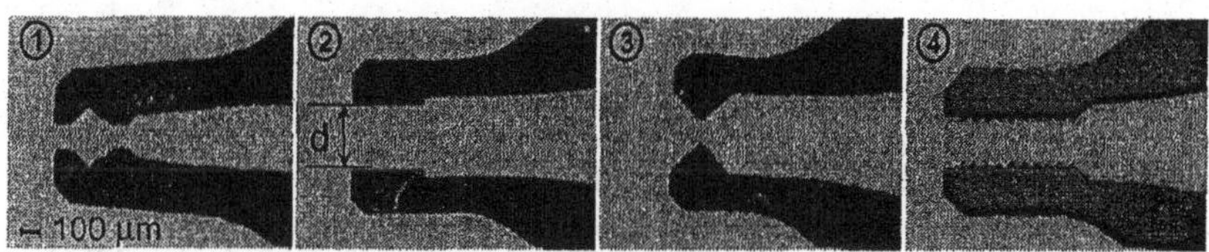

Figure 2: Different kinds of pair of available tools. These tools are made of Nickel, are 200 μm thick and have been fabricated by UV LIGA. *d* is the initial gap between both tools.

2 Working principle of the tool changer

To perform micromanipulation tasks of any kind of objects from 20 to 500 μm in size, we have developed a microgripper (figure 1) that can be fixed on nearly any kind of microrobot (the suitable mechanical interface must be done to connect them together). In our case, the microgripper has been fixed on a X-Y-Z table actuated by linear stages (M-112 1DG from Physik Instrumente) for air application and by piezoelectric stick and slip actuators from the LSRO (Laboratoire des Systèmes RObotiques, Ecole Polytechnique Fédérale de Lausanne, Switzerland) for in SEM applications. The piezoelectric actuator of the microgripper has got four degrees of freedom that combine an in-plane (Y axis) and an out-of-plane motion (Z axis). At the tip of this actuator a pair of tools that are made of Nickel is fixed. The strokes of this microgripper are +/- 80 μm along the Y axis and +/- 200 μm along the Z axis for +/- 100 V supply voltage. The blocking forces are 55 mN and 10 mN along Y and Z respectively. More details about this microgripper are given in ([13]).

Several kinds of pairs of tools have been designed with different tip shapes and different initial gaps between the tools *d* (figure 2). The suitable pair of tools can be chosen depending on the size, shape and material of the object to manipulate.

To perform a microassembly sequence or micromanipulation in confined

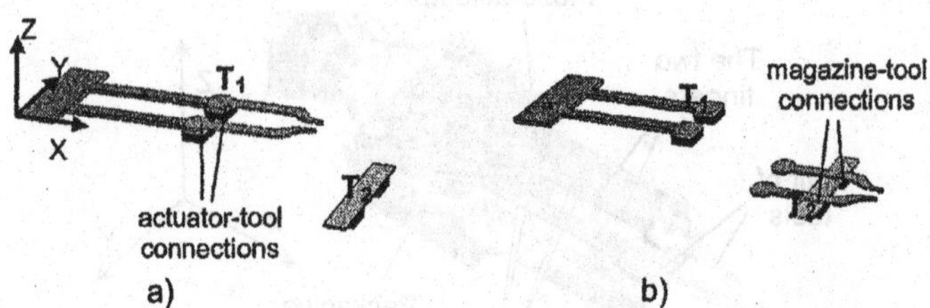

Figure 3: Diagram of the actuator-tools-magazine-resistance set (a) in manipulation configuration, i.e. when a pair of tools is fixed at the tip of the actuator (b) in tool exchange configuration, i.e. when a pair of tools is fixed on the magazine.

spaces with flexibility, a tool changer adapted to the microgripper has been designed. This system allows to fix a pair of tools alternatively at the tip of the actuator or in a magazine (figure 3). Several pairs of tools are available in the magazine with different characteristics. The temporary fixation is possible due to the use of a thermal glue (Crystalbond 555-HMP belongs to Crystalbond^{TM} series made by Aremco Products, Inc., USA) that is liquid at 65 °C (in the air), 62 °C (in the vacuum) and solid at room temperature. Cycles of liquefaction and solidification can be performed without that the glue looses its properties. Very small amounts of glue (about 4 nL per contact) are placed at the contacts tools-actuator and tools-magazine (figure 4). Surface Mounted Devices resistances of 6 Ω each have been placed under these contacts and can be heated up when supplied with a current (liquefaction of the glue). To solidify the glue, the supply of the resistances is switched off. Cycles of tool exchanges can be performed in air environment as well as in the vacuum chamber of a SEM. Figure 5 details the successive steps to do in order to perform a tool exchange. A user interface has been developed with Borland Builder C++ to control both the microgripper and the tool changer allowing automatic tool exchanges for in air applications. Two minutes are necessary to exchange the pair of tools. This time is short compared to the one required to open the chamber of the SEM, change the gripper (for example), calibrate the new position of the tools, close the door of the SEM and obtain the required vacuum (15 minutes at least).

To improve the flexibility of the tool changer both resistances of the actuator can be supplied separately (R_1 on figure 4). Both resistances of each place of the magazine can also be supplied separately (R_2 on figure 4). By combination, it is so possible to fix one tool at the tip of the actuator whereas the second one is fixed on the magazine allowing the correction of the relative position between both tools (gap d). Figure 6 displays the successive steps to perform in order to change the gap between two tools.

This system is also used to set up the tools in the magazine. The first time, the tools are approximately placed in the magazine by hand using tweezers. The

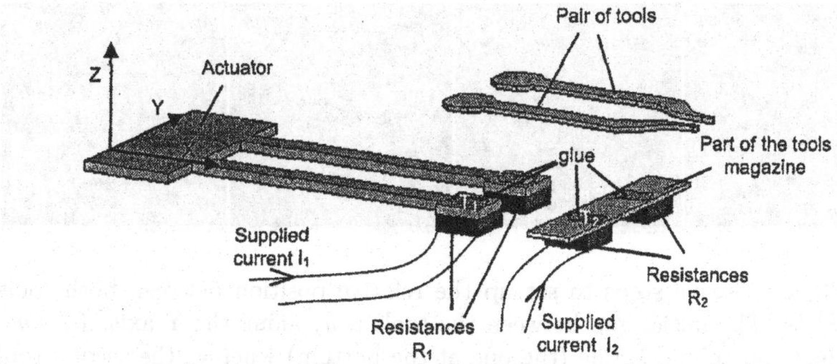

Figure 4: Diagram of the actuator-tools-magazine-resistances-glue set. Small amounts of glue are placed at the actuator-tools and tools-magazine contacts.

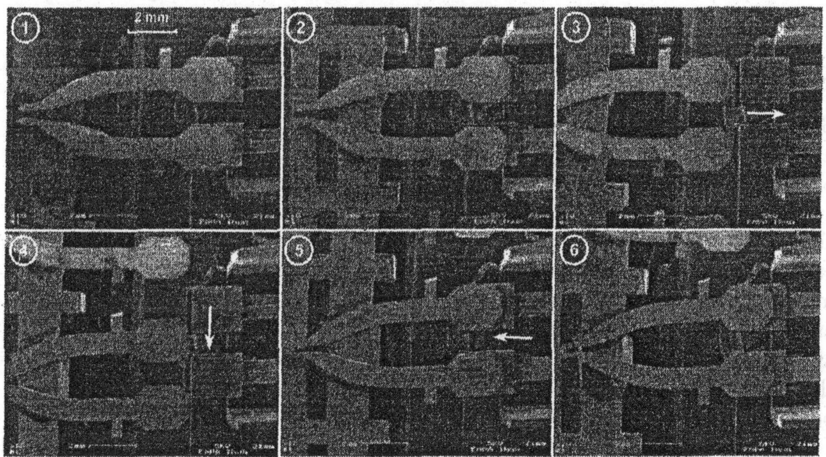

Figure 5: Successive steps for exchanging a pair of tools in the SEM: (1) a pair of tools is fixed at the tip of the actuator to perform the first micromanipulation task - (2) the first pair of tools is fixed both at the tip of the actuator and on the magazine - (3) the first pair of tools is released in the magazine - (4) the actuator alone reaches the position of the second pair of tools - (5) the second pair of tools is fixed both at the tip of the actuator and on the magazine - (6) the second pair of tools is fixed at the tip of the actuator to perform the second micromanipulation task.

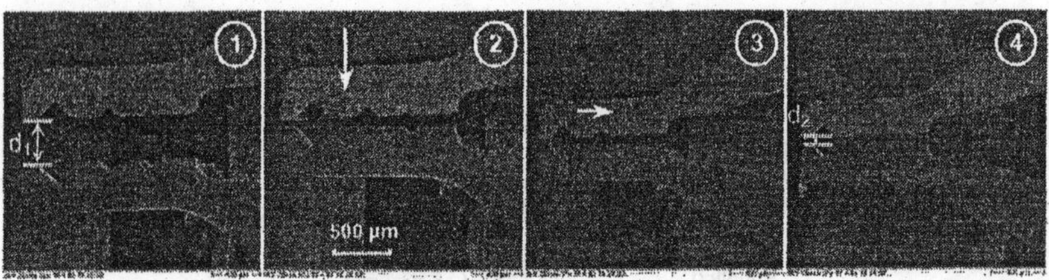

Figure 6: Successive steps to set up the relative position between both tools in the SEM. The initial gap between the tools is d_1 along the Y axis. (1) One tool is fixed on the magazine (the one at the bottom) whereas the second tool is fixed at the tip of the actuator (the one at the top) - (2) motion in the Y direction to set the gap between the tools to d_2 - (3) correction of the relative position of the tools along the X axis (if necessary) - (4) taking the pair of tools out of the magazine. The gap between the tools is now d_2.

relative position of the tools can then be corrected precisely using this system.

3 Characterization of the tool changer

Several studies have been performed to characterize the tool changer. First of all the tool positioning accuracy has been studied in air environment. Hundreds of cycles have been performed including one reference measurement, one measurement at the tip of one tool, the deposition of one tool, a displacement of the actuator alone, the removal of the same tool of the magazine. The measurements use a laser sensor (LC 2420 from Keyence) and are relative. The maximum positioning errors between two tool exchanges are 3.2 μm, 2.3 μm and 2.8 μm along the X, Y and Z axes respectively. The standard deviations are 0.73, 0.47 and 1.16 μm and the averages are 0.74, 0.62 and 0.03 μm along the X, Y and Z axes respectively. When a deviation is too large, the relative position of the tools can be modified using the sequence defined in figure 6.

The mechanical performances of the glue film have also been measured. These measurements showed that a force of 300 mN has to be applied at the tip of one tool along the Y axis to break the film of glue between tool and actuator. 400 mN are necessary to break the actuator and the blocking force of the actuator during micromanipulation tasks is 110 mN.. Thus, the film of glue acts as a fuse. As a consequence, micromanipulation tasks are performed safely.

Finally, during the liquefaction, the glue generates gas. This is not a problem in the air but can cause damage in the chamber of a SEM. Indeed, the pressure inside the chamber of a SEM must stay lower than 1.5×10^{-5} millibars to allow a good working of the electron beam. So, the degassing process of the glue must be quantified to know whether it can prevent the good working of the electron

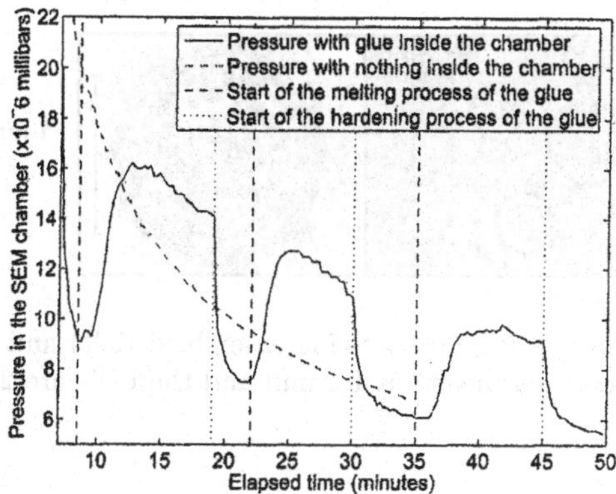

Figure 7: Pressure variation versus time during when a part of glue is successively liquefied and solidified several times inside the chamber of a SEM. This measurement can be compared to the pressure variation versus time given when there is nothing inside the chamber of the SEM (reference).

beam by affecting the vacuum. Pressure measurements have been performed to compare the pressure evolution versus time when there is nothing inside the SEM chamber and when there is an amount of glue undergoing cycles of liquefaction and solidification. The results of these measurements are given in figure 7 showing that there is a degassing process but low enough to allow the good working of the electron beam of the SEM. More details about the caracterization of the tool changer are available in ([5]) and ([4]).

4 Microassembly of a ball bearing using the tool changer

The assembly of a ball bearing has been tested in the chamber of a SEM (High vacuum SEM, Carl Zeiss DSM 962) to demonstrate both the microgripper capabilities and the effectiveness of the tool changer in vacuum environment. The external diameter of this bearing measures 1.6 mm and the diameter of the balls of this bearing is 200 μm (picture 8). Figure 9 shows the successive steps that were performed to assemble the bearing. Several kinds of pairs of tools were necessary requiring the use of the tool changer. The first pair of tools was used to manipulate the first three parts. These parts have different sizes, so once again, the tool changer was used to correct the gap between the tools.

Figure 8: Five balls bearing before being assembled (left) and once assembled (right). The external ring measures 1.6 mm and the balls are 200 μm in diameter.

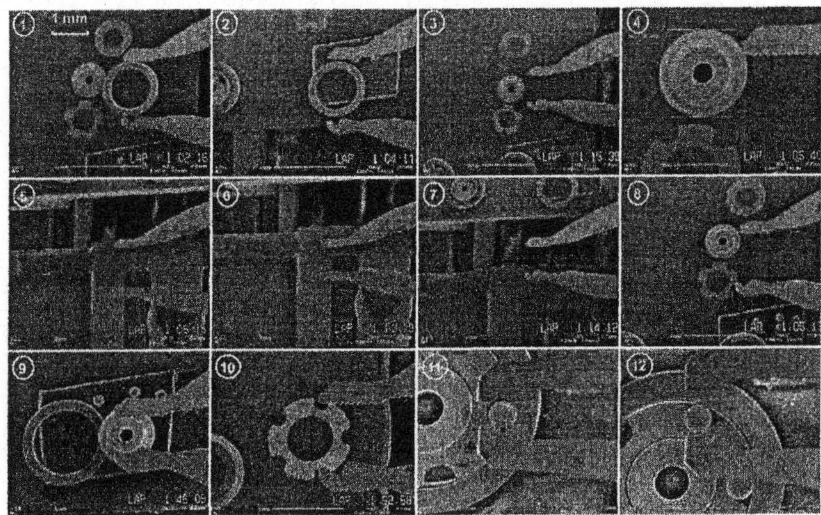

Figure 9: Assembly of a ball bearing in the SEM: sequence of elementary operations (1) taking of the external ring of the bearing. The gap between the tools measures 1.6 mm and corresponds to the diameter of the ring - (2) release of this ring on the workplane - (3) the gap between the tools is too large to take the axle of the bearing - (4) measurement of the size of the part to take - (5) reaching the magazine - (6) correction of the gap between the tools. This gap now measures 1 mm - (7) taking the pair of tools of the magazine - (8) approach to the second part to manipulate - (9) manipulation of the axle of the bearing - (10) pick and place of the ball bearing casing using the same pair of tools and the same gap than before - (10-11) sequence of tool exchange to take a second pair of tools - (11) pick and place of the first ball - (12) pick and place of the second ball.

5 Conclusion

To perform an assembly process or more generally micromanipulation tasks in air environment or inside the vacuum chamber of a SEM, we have designed a four degrees of freedom microgripper and a tool changer. This tool changer enables one to perform micromanipulation tasks using the pair of tools that is adapted to the object to manipulate. It is also possible to correct the relative position of the tools. Hundreds of automatic tool exchanges can be performed. The SEM compatibility of the tool changer has been successfully tested. This device brings flexibility and compactness to the micromanipulation cell in which it is used. As an example, the assembly of a ball bearing has been successfully performed in the vacuum chamber of a SEM. The principle of the tool changer is based on the use of a thermal glue and could be adapted on other kinds of microgripper or even on other kinds of devices.

Acknowledgment

The authors would like to thank the MPS company for the donation of microbearings, the AMiR institute (Oldenburg University, Germany) for the pressure measurements in their SEM and the LEO company (Oberkochen, Germany) for the thermal measurements inside their SEM. This work has notably been supported by the ROBOSEM project (European Project FP5 G1RD-CT2002-00675).

References

[1] K. F. Bohringer, R. S. Fearing, and K. Y. Goldberg. *Handbook of industrial robotics*. Wiley and sons, 1998. Chapter Microassembly.

[2] J. M. Breguet and A. Bergander. Toward the personal factory? *SPIE*, 4568:293–303, 2001.

[3] H. Van Brussel, J. Peirs, D. Reynaerts, A. Delchambre, G. Reinhart, N. Roth, M. Weck, and E. Zussman. Assembly of microsystems. *Annals of the CIRP*, 49(2):451–472, 2000.

[4] C. Clévy, A. Hubert, J. Agnus, and N. Chaillet. A micromanipulation cell including a tool changer. *Journal of Micromechanics and Microengineering*, 15:292–301, July 2005.

[5] C. Clévy, A. Hubert, and N. Chaillet. A new micro-tools exchange principle for micromanipulation. In *IROS*, Sendai, Japan, September 2004.

[6] R. Eberhardt, T. Scheller, G. Tittelbach, and V. Guyenot. Automated assembly of micro-optical components. *SPIE*, 3202:117–127, 1998.

[7] S. Fatikow, A. Kortschack, H. Hudsen, T. Sievers, and T. Wich. Towards fully automated microhandling. In *IWMF*, pages 34–39, Shanghai, China, 2004.

[8] T. Kasaya, H. Miyazaki, S. Saito, and T. Sato. Micro object handling under sem by vision-based automatic control. In *ICRA*, pages 2189–2196, Detroit, USA, 1999.

[9] B. Kim, H. Kang, D.H. Kim, G.T. Park, and J.O. Park. Flexible microassembly system based on hybrid manipulation scheme. In *International Conference on Intelligent Robots and Systems*, pages 2091–2066, Las-Vegas, USA, October 2003.

[10] B.E. Kratochvil, K.B. Yesin, V. Hess, and B.J. Nelson. Design of a visually guided 6 dof micromanipulator system for 3d assembly of hybrid mems. In *International Workshop on Microfactories*, pages 128–133, Shanghai, China, 2004.

[11] H. Miyazaki and T. Sato. Mechanical assembly of three-dimensional microstructures from fine particles. *Advanced robotics*, 11(2):139–185, 1997.

[12] M. Nienhaus, W. Ehrfeld, F. Michel, V. Graeff, and A. Wolf. Handling and bonding of millimeterwave monolithic integrated circuits with high density interconnections for automotive and it applications. In *3rd workshop on "Area array packaging technologies"*, Berlin, Germany, 1999.

[13] R. Perez, J. Agnus, C. Clévy, A. Hubert, and N. Chaillet. Modelling, fabrication and validation of a high performance 2 dof microgripper. *ASME/IEEE Transaction on Mechatronics*, 10(2), April 2005.

[14] S. Saito, H. Miyazaki, and T. Sato. Pick and place operation of a micro object with high reliability and precision based on micro physics under sem. In *ICRA*, pages 2736–2743, Detroit Michigan, USA, May 1999.

[15] E. Shimada, J.A. Thompson, J. Yan, R. Wood, and R.S. Fearing. Prototyping millirobots using dextrous microassembly and folding. In *ASME IMECE/DSCD*, pages 1–8, Orlando, USA, November 2000.

[16] G. D. Skidmore, M. Ellis, E. Parker, N. Sarkar, and R. Merkle. Micro assembly for top down nanotechnology. In *Int symposium on Mechatronics and human science*, pages 3–9, Nagoya, Japan, 2000.

[17] T. Tanikawa, M. Kawai, N. Koyachi, T. Arai, T. Ide, S. Kaneko, R. Ohta, and T. Hirose. Force control system for autonomous micro manipulation. In *ICRA*, pages 610–615, Seoul, Korea, May 2001.

[18] J.A Thompson and R.S. Fearing. Automating microassembly with orthotweezers and force sensing. In *IROS*, Maui HI, 2001.

[19] M. Weck and C. Peschke. Equipment technology for flexible and automated micro-assembly. *Microsystem technologies*, 10(3):241–246, 2004.

[20] B. Winzek, S. Schmitz, and T. Sterzl. Microgrippers with shape memory thin film actuators. In *IPAS*, pages 77–84, Bad Hofgastein, Austria, February 2004.

[21] G. Yang, J. A. Gaines, and B. J. Nelson. A flexible experimental workcell for efficient and reliable wafer-level 3d microassembly. In *ICRA*, pages 133–138, Seoul, Korea, May 2001.

DESIGN AND EXPERIMENTAL EVALUATION OF AN ELECTROSTATIC MICRO-GRIPPING SYSTEM

Defeng Lang, Marcel Tichem
3mE, dep. PME, section PMA, Delft University of Technology

Abstract: The paper describes the modeling and experimental investigation of the application issues of electrostatic based micro-gripping. The design of an electrostatic gripping system for both grip force measure and pickup and place experiment is presented. A finite element model is made to study the gripping environment and process related features. The design and validation of the model are provided. Investigation of the influences that the gripping process gives out to and may receive from the operating environment is discussed. The preliminary result shows that grounded objects that come into the gripping area do not influence the gripping action significantly. Some real gripping actions are performed. Discussions on the scope of gripping environment and process related features are raised on basis of gripping experiments and observations. The phenomena of charging and discharging on the gripper isolator may add difficulties to gripping control. The investigation is concluded in a form of Process Data Sheet. The research draws an insight view on the application criteria of electrostatic gripping technology. Both advantages of the gripping principle *e.g.* flexible in terms of part dimension and geometry, and restrictions of the application are illustrated.

Key words: Micro-gripping; Electrostatic; Grip force; Modeling.

1. INTRODUCTION

The development trend of products miniaturization requests the assembly process to be extended from conventional scale into micro-scale. Micro-gripping, as the entire process which brings the downscaled parts from a loose state to a connected state, has different characteristics and raises new technological challenges compared to the conventional gripping process. Micro-gripping operations deal with parts with typical dimensions in the range of sub-millimetres up to a few millimetres; part features may be in the micron range. The typical accuracy in parts relation is in the range of 0.1 to 10.0 μm. Micro-objects react differently to forces compared to macro-objects. The surface related forces start to dominate over the volume related forces, which opens new possibilities to handle small objects. Electrostatic gripping is a method that is based on one of these possibilities.

A charge difference between two particles causes a force of repulsion or attraction. In electrostatic gripping this electric force field is used to operate target parts. The gripping method has been investigated and practically demonstrated by other researchers. An optically transparent bipolar gripper has been demonstrated[1]. Grip force higher than 2mN has been measured under 200 Volts bias voltage. The method is also implemented under a scanning electron microscope. By this means, the electrostatic field is generated by the microscope[2]. The tool can quickly pick up a 100 microns diameter metal ball.

However, to gain more understanding on the features of the gripping method is still very meaningful. This paper presents a computer model and test bench, which are made to investigate the gripping environment and process related features. Some results are addressed and presented in a form of the process data sheet[3] of the operation method. The research provides more understanding on the electrostatic gripping principle; it also contributes to build up the knowledge of micro-gripping technology in general[4].

2. STRUCTURE OF THE GRIPPING SYSTEM

A gripping and force measurement system has been developed which structure is shown in Figure 1 together with its equivalent circuit. The gripper is made as a cylinder with a diameter of 5.7 mm. The material is aluminum. a piece of wire attached to its top. Via that different electrical potential can be added to the gripper. A piece of glass with thickness of 0.18 mm is used as the isolation layer, which is stuck to the bottom of the gripper by conductive glue. The gripper is attached to a robot. The two entities are electrically isolated from each other. The force measurement part includes a cantilever beam with well-defined spring constant, and a laser displacement sensor that measure the deflection of the cantilever. The smallest increment that can be measured (sensitivity = ration I think) of the force measurement is 8 micron Newtons. In the setup, the cantilever, the displacement sensor and the robot are all grounded and isolated from the gripper.

When enough potential difference is added between the gripper and the cantilever, the cantilever can be pulled up. The displacement sensor measures the deflection value, which is converted to force value.

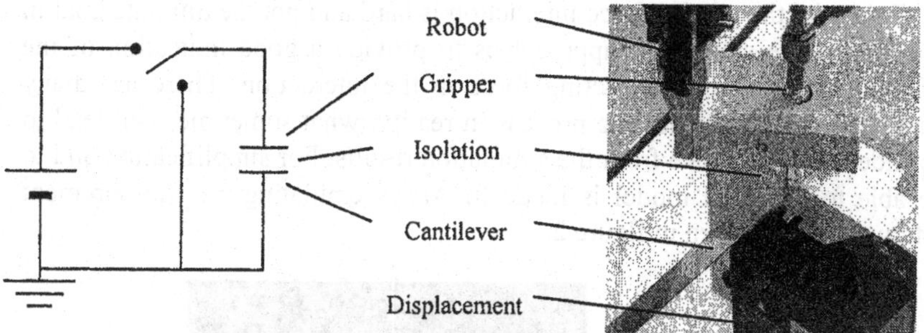

Figure 1. The structure of the gripping system and its equivalent circuit

3. MODELING OF THE GRIPPING PROCESS

A computer model is made on basis of the physical principle of the electrostatic interaction, which is used to study gripping environment related features.

3.1 Electrostatic interaction

By Coulomb's Law, Electrostatic force F is defined as the electrical force of repulsion or attraction induced by an electric field E. The electrostatic force F between two point charges q at point Q and q' at point Q' (with a distance r between them) can be stated as in equation 1. The constant 4π is included because it simplifies some calculations. The ε_0, is the permittivity of free space.

$$F = \frac{1}{4\pi\varepsilon_0} \frac{qq'}{r^2} \tag{1}$$

Electrostatic forces in parallel plates can be described as in equation 2. Where V is the applied voltage; S is the cross-sectional area; d is the separation distance:

$$F_z = \frac{1}{2} \frac{\varepsilon_0 \varepsilon S V^2}{d^2} \tag{2}$$

3.2 Finite element model

To model the precise force interaction is hard and not the ultimate goal in this work. The simulation approach is to provide a good indication of the gripping process, and the acting force in the interaction. There are many parameters that influence the process in reality, what sometimes can lead to considerable difference from the simulation results. For simplification and to be able to verify, a model is made in Ansys emulating the measurement setup, which is shown in figure 2.

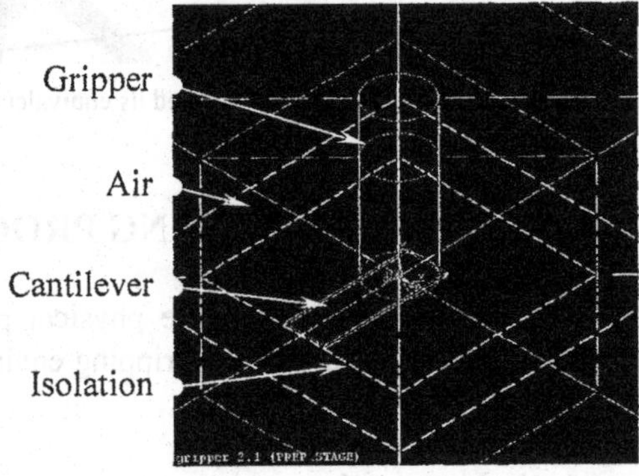

Figure 2. The frame structure of the gripping system model

Three different boundary conditions exist in the model. The first is the potential. The areas that form the gripper has to be assigned with a potential value, for instance 800 Volts; the areas that form the cantilever is set to 0 Volts. The second boundary condition is infinitive areas, which is applied to all the outer areas. The infinitive area is essential for this simulation to run correctly and produce good result. The third boundary condition is Maxwell surface, which is to calculate the nodal forces at the specified areas. After all the boundary conditions are set, the solution can be calculated.

4. VALIDATON AND MODELING RESULTS

To make use of the model, first the simulation results of interaction forces are compared with measuring value on the test bench. Further this model is applied to study the grip force under different geometries, voltages, and gripping situations.

4.1 Validation

On the test bench, the force is measured and calculated by the deflection of the cantilever under certain potential difference and with certain distance away from the gripper.

First, to know the distance between the gripper and cantilever is crucial in this measurement. The gripper is moved away from the cantilever and the displacement is set to zero on the deflection sensor. This will be the point of reference. Then the gripper is moved in and brought closer to the cantilever; meanwhile 830 Volts is added to gripper. When the distance between the gripper and cantilever becomes small enough, the cantilever jumps towards the gripper and makes contact with isolation layer. The position of the gripper compared to the reference point is then known by reading the deflection. The gripper is then switched off and grounded.

Next, a smaller potential is set between the gripper and the cantilever. The cantilever will bend but not touch the gripper. From the difference between the position of the gripper and the reference point the air gap is known; from the displacement of the cantilever the force is calculated. With the distance of the air gap and the known potential, the force can be calculated and compared with the model.

Using the model, thirty different combinations of distance and potential where calculated. To make it easier to compare the model to the measurements, a graph is created and shown in figure 3. A power curve is fitted to the data. Two of the curves are shown in the graph and the corresponding equations are in the bottom left and bottom right of the graph area. The left equation corresponds to the 500 Volts curve; the right equation fits the 830 Volts curve.

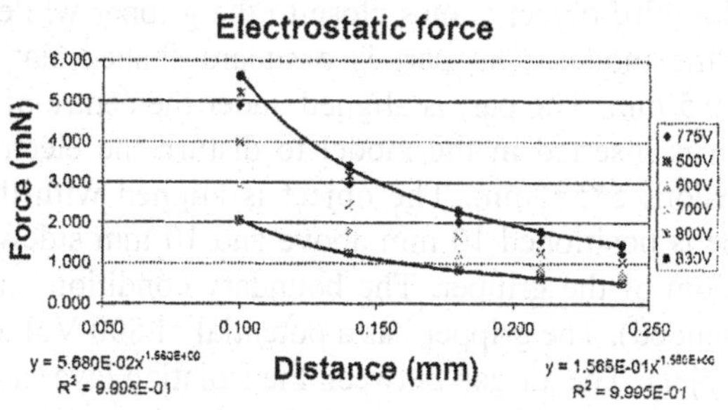

Figure 3. Graph of calculated forces

Figure 4 shows a diagram of a series of measurements, which were performed under potential differences of 500 Volts and 700 Volts. The vertical axis shows the electrostatic force, the horizontal axis shows the distance between the cantilever and the isolation layer. As shown by the figure, the predicted values correspond well to the measured values; although the measured electrostatic force is influenced by humidity of the air and the quality of the bonding with glue. This proves that the model is valid to be applied in further investigation.

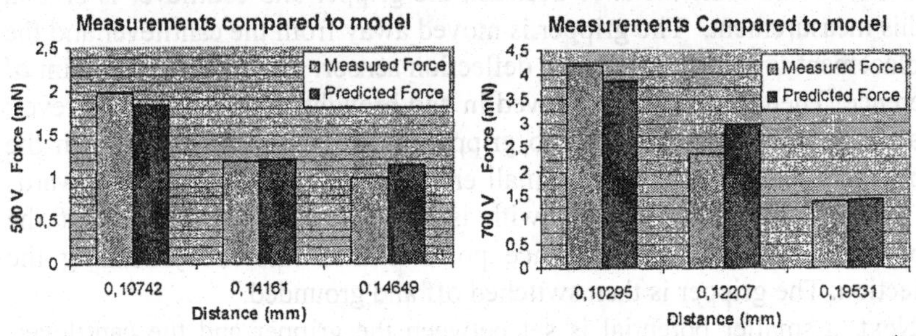

Figure 4. Measurements compared to model

4.2 Prediction model of the gripping operation

The argument has long existed as to whether the electrostatic gripping is reliable enough for industrial application. In more detail, the part is operated in an electrostatic field; any object that presents in this field may receive influence due to induction. Even the to be operated part has a risk of damage. Meanwhile, objects that come into this electrostatic field may give influence to the gripping operation. Thus it is interesting to know the magnitude of the influence.

The above evaluated model is modified to investigate the influence on the grip forces if a third object comes close to the gripper while it operates. Figure 5 shows the model. The part is a square shaped flat object with dimensions 8*8*0.5 mm. The part is aligned under the centre of the gripper. The object which is inserted in the model to disturb the electric field is a cube with dimensions 5*5*5mm. The object is aligned with the part. The centre of the cube is positioned 10 mm above and 10 mm sideway from the centre of the bottom of the gripper. The boundary condition on the cube is set to 0 Volt (grounded). The gripper has a potential of 800 Volt compared to the part and the object. The air gap between the isolation layer and the part is set to 0.4 mm. This is much larger than that occurs in reality. It should be noticed that the Y direction is upward in the model.

The change of the potential field can be seen clearly by comparing the two graphs. The influence can be interpreted. The potential field becomes unsymmetrical around the gripper and underneath of the part. The change of the overall grip force is very limited in both perpendicular and horizontal direction.

Forces without object: Y= 5.86E-04 N, X= 2.68E-07 N, Z= 3.56E-08 N.

Forces with object: Y= 5.82E-04 N, X= -5.31E-07 N, Z= -3.75E-09 N.

The difference is within the range of the calculation errors that are inherent to the model. The simulation suggests that a grounded object in the vicinity of the gripper do not give large influence on the gripping contact.

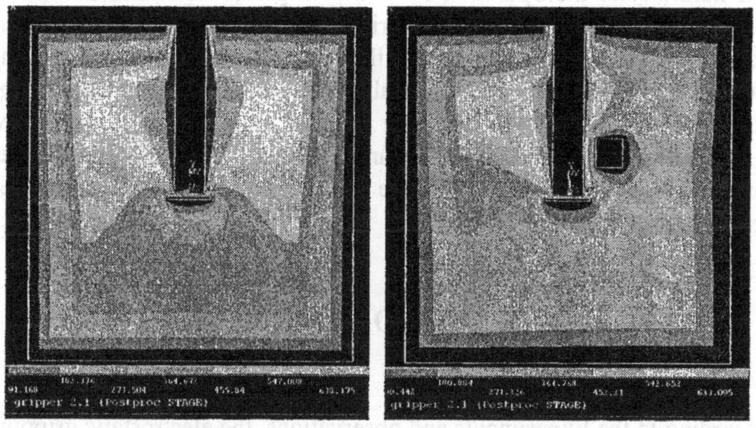

Figure 5. Potential field and forces without and with a grounded object

5. EXPERIMENTS AND OBSERVATION

Some pickup and place experiments were conducted to evaluate the actual performance of the gripping system. Parts made from different conductive and nonconductive materials were addressed. A test that was done on copper parts is presented here. The gripper used in this test has a diameter of 400 microns, isolated by 100 microns thick PTFE layer instead of glass; the target object was measured as 1500 microns diameter and 100 microns thick copper disk. Different voltages were set between the gripper and the part. To release the part, the strategy was to reverse charge the gripper. With 300 Volt added, the probability of pickup was 90% and probability of release was 100%. With 200 Volt added, the probabilities were 35% and 57%, respectively. With 150 Volt added, the probabilities were 0. The gripper is able to pickup plastic and paper parts as well. The probabilities to pickup nonconductive or conductive parts are comparable, but to release a nonconductive part is more difficult.

The charge accumulates over time on the gripper and on the other side of the isolation layer caused some phenomena during the experiments on the cantilever. The reasons behind some of these phenomena cannot be thoroughly explained at the current stage by the authors. Set the gripper and the cantilever away for some distance, the electrostatic force can keep on rising for few tens seconds if keep the electrical potential at constant. It suspects that charge is accumulating in this period, which results the force increasing.

The humidity of the operation environment gives large influence to the reaction of the cantilever in the electric field. Under a humid condition, humidity of the air over 60%, the cantilever was seen to jump toward and contact the isolation layer then after about two seconds be released. The process can repeat itself with constant voltage added to the gripper. Few times under humidity below 40% the cantilever was attracted by the gripper and bumped on the isolation layer at the moment the voltage being switched off (gripper is set to grounded).

6. PROCESS WINDOW OF ELECTROSTATIC GRIPPING

According to the Experiments and observations, the electrostatic gripping can provide adequate force to pickup small objects. The dimension of the gripper is possible to be designed smaller than the target components, which is an advantage, means that it is more flexible in a micro-assembly environment. Meanwhile the same gripper can be applied to handle different parts; the flexibility concerning the part geometry is high.

However, comparing with other means of gripping, the grip force of this method is rather limited (for sure applying higher voltage is an option to raise the force, but it might hoist other limitations.). The Grip force may drop due to the gripped part having contact with conductive objects during the assembly process. In case that the gripping contact has to be assured while assembly takes place, the applicability of the principle can be questioned. Observing the formula of electrostatic force, a thinner isolation layer is beneficial. A gentle contact or a small gap is required in the handling to assure higher force while keeping the voltage low.

Further the gripping method can be concluded in a form of the Process Data Sheet PDS[3]. The purpose of PDS is to capturing the known data on issues related to micro-assembly processes. The data is considered to be important for training, education and engineering purposes in industry and universities. Main concerns of the electrostatic based gripping process are listed in table 1. The data captured in the PDS is based on a literature survey and on own research that addressed in this context.

Table 1. Process Data Sheet of Electrostatic based gripping

Category	Description
Process name	Electrostatic Based Gripping
Process category	Gripping
Description	A charge difference between two particles causes a force of repulsion or attraction. Electrostatic gripping is to utilize this force to operate target parts. According to the number of electrodes, there are monopolar, bipolar grippers.
Process window	Flexible in terms of part geometry and dimension. Few tens microns till few tens millimeters sized parts can be gripped. One contact surface is sufficient. Grip force depends on the contact area between the gripper and part, Permittivity of the isolation material and its thickness, the applied voltage difference. Typical value of the specific grip force is in the rang of mN/mm2. Connection stiffness is limited. Grip force may drop during assembly process due to having contact with conductive objects. Part may have residue charge after being released.
Cycle time	The cycle time is reasonable short, in about sub-second to few seconds. The time for charge accumulation and for residue charge dissipation is main concern.
Material type	Both conductive and non-conductive material can be handled, but not suitable for electric and magnetic sensitive materials and IC chips.
Environment requirements	Gripping action is not greatly influenced by surrounded conductive object if grounded. Surrounded equipments may be affected due to the radiation. Low humidity is preferred.

7. CONCLUSION

An electrostatic gripping system has been presented, which in one way is used to measure the grip force and another way is for real gripping operation. A computer model is built based on the gripping system. The model is verified by comparing the simulation result of the electrostatic force and the measuring result of the force value. Experiment of gripping small copper part shows the application possibility of this principle.

According to the simulation, influences given by a grounded object to the overall grip force is limited. Experiment and observation show that the principle is sensitive the humidity changing of the gripping environment. The phenomena of charging, discharging and residue charge on the gripper, isolator and part raise some uncertainty in the handling process. It gives extra difficulty to make the gripping process reliable and efficient.

In further research, effort will be spent to further investigate how the gripping operation influences o the operation environment, and possibilities to reduce. A gripping system together operation strategy that uses lower electrical potential yet generates sufficient grip force will be the technological direction.

ACKNOWLEDGMENT

The authors wish to thank MSc. student H.J. Hendriks for his contribution to the experiment and modeling work, Mr. H.F.L. Jansen for building the experimental setup, and special thanks to Eng. F. Biganzoli for all the nice discussion and help.

REFERENCES

1. Eniko T. Enikov, Kalin V. Lazarov, An optically transparent gripper for micro assembly, *Journal of Micromechatronics*, Vol. 2, No. 2, pp. 121~140 (2004)
2. K. Tsuchiya, A. Murakami, G. Fortmann, M. Nakano, Y. Hatamura, Micro assembly and micro bonding in Nano Manufacturing World, *SPIE Proceedings* Vol. 3834 (Microrobotics and Microassembly) p.132~140, 1999
3. M. Tichem, F. Bourgeois, P. Lambert, V. Vandaele, D. Lang, Capturing Micro-Assembly Process Windows on Process Data Sheets, *ISATP Proceeding*, 2005
4. M. Tichem, D. Lang, B. Karpuschewski, A classification scheme for quantitative analysis of micro-grip principles, *IPAS2003 Proceeding*, p. 71-78, 2003

A GENERIC APPROACH FOR A MICRO PARTS FEEDING SYSTEM

M. Paris[1], C. Perrard[1], P. Lutz[1]
[1]*Laboratoire d'Automatique de Besançon - UMR CNRS 6596 - ENSMM - UFC*
24 rue Alain Savary, 25000 Besançon, France

Abstract The paper propose a new approach in order to design a generic microparts feeder. The method based on a classification scheme allows to emerge the principal characteristics of each studies solutions. The different criteria take into account the specifities of the micro world and moreover the main characteristics for the feeding functions. Thus, we analyse three systems and confront them to find the generic and flexibility aspects.

Keywords: Feeding, classification scheme, microfactory.

1. INTRODUCTION

It is within the framework of European project EUPASS (Evolvable Ultra-Precision Assembly SystemS) that the whole of work presented here were carried out. The main objectives are to realize a new generation of modular high-precision manipulators, grippers and feeders which will work inside the microworld. As in any system of production, it is essential to take into account the constraints of production which include the criterion of flexibility as well as the intrinsic constraints of micro-objects (i.e roughness, geometry, physicochemical characteristics...). Moreover, it is essential to consider the microscopic forces that interact between the micro objects. These forces also depend on the environment of the micro world (i.e dust, moisture, temperature.). Located between the storage and the assembly system, feeding systems can have a profond effect on the effeciency of an assembly station (Whitney, 2004). The ideal situation is one in which of the part are ready, in the correct orientation, at the moment the ressources need them. No assembly time is wasted (Whitney, 2004). But, rhythm and quantity are dubious variables. Also, a Pick & Place system of production (like the micro-factory developped inside EUPASS) must be placed in the good context of production. This approach is valid only for small and average series with changes of frequent and fast flows. This is the reason why, the micro feeders musn't be dedicated to a particular product but must be as flexible as possible while preserving requested rates. The state of the art carried out allows to emphasize the micro parts feeding systems existing as well as the connected principles with the various functions of feeding. From there and from stated previously constraints, emerges a strategy for the choice of such micro parts feeding systems. This choice is also done according to existing technologies. From explored technologies will be especially exposed the one that were retained.

2. STATE OF THE ART

The classification of the whole of micro feeders met is drawn from the thesis of T. Ebefors (Ebefors, 2000). There are two categories according to the contact. The first describes the systems without contact between the micro feeder and the micro product whose acronymis CF (Contact Free). The second category gathers this times the micro feeders indicated by C (with Contact).

2.1 Contact Free: CF

One counts for the processes without contacts with the micro objects three great classes: **magnetic** (Nakasawa et al., 1999; Kim et al., 1990; Iizuka et al., 1994), **electrostatic** (Moesner et al., 1996; Moesner and Higuchi, 1997; Gengenbach and Boole, 2000; Desai et al., 1999; Fantoni and Santochi, 2004) and **pneumatic levitation** (Konishi and Fujita, 1994; Hirata et al., 1998; Fukuta et al., 2003; Chapuis et al., 2003; Sin and Stephanou, 2003).

The magnetic systems allow to translate in one or two directions a mover. This palet is generally a permanent magnet. The different systems elaborated are classified as micro conveyors. The electrostatics systems are based on the effects of electrophoresis and dielectrophoresis. This principles can separete different components. An Ac electric field induces a dipole moment, which in the presence of a field gradient experiences a force towards a hight or low field intensity region (positive/negative dielectrophoresis)(Zheng et al., 2003). The electrostatic approaches are used in the micro and macro world. U. Gengenbach (Gengenbach and Boole, 2000) realized a palette in levitation by a pneumatic flow and can move it with a electrostatic field. The pneumatic levitation approaches are planar systems of micro conveyances based on the distributed micro motion systems (DMMS). The pneumatic conveyance have some advantages, like: no friction problem, no particle generation, generation of force large enough to convey objects (Fukuta et al., 2003).

2.2 With Contact: C

These methods are gathered in three main categories: the **ciliary motion systems** (Ataka et al., 1993b; Benecke and Riethmller, 1989; Böhringer et al., 1994; Suh et al., 1997; Suh et al., 1999; Tabata et al., 2002), the **ultrasonic feeders** (Haake and Dual, 2003, Kim et al., 2004) and the **micro robots** (Ebefors, 2000, Ferreira, 2000, Ferreira et al., 2004).

Like the contactless pneumatic DMMS micromotion systems, the ciliary micromotion use arrays of tiny simple actuators that co-operate to move objects over relatively large distances and offers possibility in different directions and orientations. Today, the cilia can be moved by the electrostatic forces, magnetic forces, thermic effect and an original concept: a chemical reaction. The first ultrasonics feeders elaborate by (Böhringer et al., 1998), is a table put in vibration by a piezo. The frequency is as, the ultrasonic waves that are generated are able to break surfacic forces. So the components are insulated by a electrostatic field trap (four electrods placed under the table). The last familly is the micro robots. We can found here walker micro robots, pallets. In one hand, the problem is here the energy supply, but in the other hand the main advantage is their height accuracy.

The different processes we saw are assimilated to conveyances. However some works present innovative concepts like the transfer without contact. The next part describe a new approach in order to design a micro parts feeder. The main difference with these feeders or conveyances is to garantee the maintain function, to feed a lot of familly of product with a good orientation.

3. A GENERIC APPROACH

A micro part feeding system must fill two objectives: first, to position the micro objects correctly and in the second place, to orient them. Moreover, it must be able to preserve these two parameters without damage the micro objects. Lastly, if several components have to be treated at the same time, it must be able to insulate them. Entering in the micro feeding system, the objects can be conditionneed, by various manners: bulk, palet or fixed on their support. This is the raison why it can be necessary to order these components before their arrival in the micro part feeding system or inside even of it. After, the feeder separate this components. At the end, the components must be positioned and oriented at the right time required by the assembly cell.

It is essential from the very beginning of the design, to build a sufficiently flexible feeding system that isn't dedicated to criteria relating to a particular component or a particular family of components. As previously noted: the selected strategy is influenced by the product. But in the micro world, the negligible physical phenomena from macro scale become dominating (Van der Waals forces, electrostatic forces, capillary forces). From equations of these physical phenomena, the significant characteristics of the micro objects were retained (Regnier et al., 2004, Huang et al., 2004, Zhou and Nelson, 2000):

- Lifshitz-Van Der Waals constant,
- distance between the surface,
- some geometrical characteristics (like the radius of a sphere),
- surface charge density,
- contact angle between a liquid and a micro component,
- surface tension.

Moreover these adhesions forces are also influenced by the environmental conditions (Zhou et al., 2004):

- dust,
- moisture,
- temperature,
- vibrations,
- operation inside air, inert gaz, liquid. . . ,
- pressure,
- permittivity of the environment.

It becomes significant in these scales by including these parameters in the choice of a strategy. The obtained results show that the strategy for the choice of a micro parts feeding system does not only depends on three criteria: weight, size and form. Thus it is important to take into account the plasticity, the elasticity, the surface roughness of the component, the hydrophobic properties of the surfaces and the physicochemical properties of the components. All this parameters influence the way to grip a component inside the microworld (Tichem et al., 2004). So they are considered at the beginning of their design. Designing a micro parts feeder, will induce the same problem. Now, it is necessary to find the generic aspects among various technological solutions. A micro feeding system is considered quite generic, according to its independance regarding a maximun of criteria, such conventional (weight, size...) and such micro world criteria (plasticity, hydrophobic surfaces, dust, temperature...). After establishing the correlation between the technological criteria and solutions, it is possible to emphasize the limits of the generic aspect of each solution.

4. SOME TECHNOLOGIES EXPLORED

In the framework of the EUPASS project, where objects go several hundreds of microns and few millimeters, the accuracy degree of position is defined according to the size of the objects. This value relates to dimensions of the treated objects. Using the proposed method, according to the constraints of the micro world, and those of european project, various technological solutions were approached. The method is represented by a classification scheme inside which we evaluate the technologies according to the criteria previously quoted. The first row of the baord reopresent the differents technologies and the first columns the differents criteria. This scheme is inspered by (Tichem et al., 2004).

	Housing Packaging	Gel-Pak	Droplet Freezing
Positioning	Dedicated to the shape	Depends to the characteristics of the glue so depends to the height and weight	No limits
Maintain	Contact may lead to damage	Sensivity to the vibration, one great surface	If the droplet is maintain frozen: no limits.
Environment		Dust	Moisture (white frost)
Mechanical properties of the objects	Robust if no force feedback		Accept water.
....	....	....	....

Figure 1. Classification scheme

The figure 1 shows only the most important technologies explored inside the EUPASS project with some criteria. We explored here tree systems (housing packaging, Gel-Pak and freezing) and we retain only two; the Gel-Pak and the freezing. The exemples of the analyse show that the housing packaging is not flexible enough and can't be taken into account. Gel-Pak and the frozen droplet offer more possibilities and can cover more different components. Since a feeding system can fulfill several functions, the first one to satisfy is the transport of the components inside an assembly cell. With this intention, the micro components are not treated directly from the assembly cell. It is more preferable to disconnect storage and load of components of assembly cell by using an intermediate support (see Fig. 2). By this way, the filling of this inter-

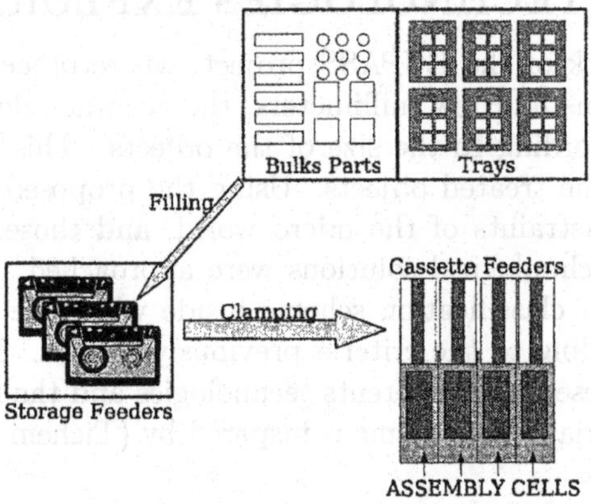

Figure 2. Main frame layout of feeding

mediate support can be adapted in ressources outside and independantly the assembly cell. This is a "Plug & Produce" solution. The constraints of position, orientation and maintain of objects can thus be guaranteed. Two solutions have been retained. They are sufficiently generic to cover a broad various and varied line of products. They are based on the technology of an adhesive gel (Gel-Pak) and freezing. The final shape of these solutions corresponds to a storage stage, like a video or audio cassette which will be placed thereafter in the assembly cell. A plugin interface realize the positioning. Nicely called "Magic-Carpet", it realises the liaison between stock and cell. These two solutions allows to transport and maintain some differents micro objects. The most ineresting system is the freezing. The Gel-Pak system cannot cover a great familly of components if theirs shapes is between some micrometers to few millimeters. In this case, the adhesive band is dedicated to the height. Moreover, the stability is garanted only if the contact aera is large enough. On the contrary, a frozen droplet can maintain any types of micro component without takes into account the shape and the height (more droplets can be used in order to maintain a meso/milli-component).

5. SYNTHESIS

Enumerating all the advanges and disavantages of each technologies allows to emerge the principal caracteristics of each. In fact, we characterize a micro parts feeding system according the three criteria of the GFM pyramid: Genericity, Flexibility, Modularity. The first step to design a micro parts feeder correspond to the base of this pyramid: the generic aspect. Then, we go up each stage in order to find the limits of the flexibility and modularity. When the classification scheme is finished, we know the limits of range of the feeders. Finally, this allows to emphasize the degree of flexibility as well as the necessary modularity of each solution. The applied method, used inside the EUPASS project, for the design of a micro part feeding system, could be represented and summarized by this pyramid.

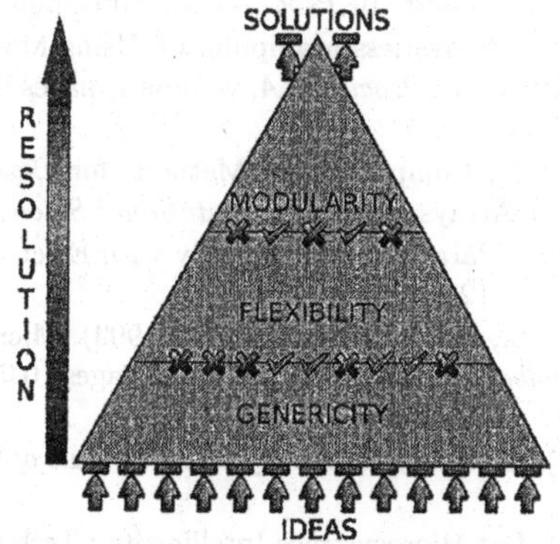

Figure 3. GFM Pyramid

6. CONCLUSION

To conclude, this micro feeder is the base of futurs works to develop a module enabling rapid configuration and deployement of flexible precision assembly systems with minimum investment cost. Thus, the "Magic-Carpet" will evolve in order to be integrated in the future cells of micro assembly, with a strong standardization between the various modules. Our approach permits to cover a great number of component kinds because it does not necessary realizes a design feedback when the production changes. Several analyses will be necessary in order to test and validate this approach. However, the principles of this solution gives a reusable and generic micro feeder that allows to decrease the design time and costs. It gives too an answer to the most important problem of classical feeders: to be dedicated to one product design.

References

Ataka, M. et al. (1993a). Fabrication an Operation of Polyimide Bimorph Actuors for a Ciliary Motion System. *Journal of Microelectromechanical Systems*, 2(4):146–150.

Ataka, M., Omodaka, A., and Fujita, H. (1993b). A Biometric Micro Motion System ~A Ciliary Motion System~. In *Sensors and Actuators*, pages 38–41, Yokohama, Japan.

Benecke, W. and Riethmller, W. (1989). Application of Silicon-Microactuators Based on Bimorph Structures. In *Proc. of the IEEE MEMS Workshop*, An Investigation of Micro Structures, Sensors, Actuators, Machines and Robots, pages 116–120, Salt Lake City, USA.

Böhringer, K. F., Donald, Bruce R., and MacDonald, N. C. (1996). What Programmable Vertor Fields Can (and Cannot) Do: Force Field Algorithms for MEMS and Vibratory Plate Parts Feeders. In *Proc. ICRA*, Mineapolis, USA.

Böhringer, K. F. et al. (1994). Sensorless Manipulation Using Massively Parallel Micofabricated Actuator Arrays. In *Proc. ICRA*, volume 1, pages 826–833, San Diego, USA.

Böhringer, K. F. et al. (1997). Computational Methods for Design and Control of MEMS Micromanipulator Arrays. *IEEE Computational Science & engineering*.

Böhringer, K. F. et al. (1998). Parallel microassembly with Electrostatic Force Field. In *Proc. ICRA*, pages 1204–1211, Leuven, Belgium.

Böhringer, K.F., Fearing, R.S., and Goldberg, K. Y. (1999). Microassembly. In Nof, Shimon, editor, *The Handbook of Industrial Robotics*, pages 1045–1066. John Wiley & Sons, seconde edition.

Brussel, H. Van et al. (2000). Assembly of microsystems. *Annals of the CIRP*, 49(2):451–472.

Chapuis, Y-A. et al. (2003). Les Microsystmes Intelligents : Technologies et Applications. In *Journées Scientifiques Francophones: JSF*.

Desai, A., Lee, S-W., , and Tai, Y-C. (1999). A MEMS ELECTROSTATIC PARTICLE TRANSPORTATION SYSTEM. *Sensors and Actuators, A: PHYSICAL*.

Donald, B. R. et al. (2003). Power Delivery and Locomotion Unthered Microactuators. *Journal of Microelectromechanical Systems*, 12(6):947–959.

Ebefors, T. (2000). *POLYIMIDE V-GROOVE JOINTS FOR THREE - DIMENSIONAL SILICON TRANSDUCERS*. PhD thesis, ROYAL INSTITUTE OF TECHNOLOGY (KTH), Stockholm, Sweden.

Fantoni, G. and Santochi, M. (2004). A contactless linear movement of mini and microparts. In *Proc. of the IMG04*, Genova, Italy.

Ferreira, A. (2000). Design of a Flexible Conveyer Microrobot with Electromagnetic FieldBased Friction Drive Control for Microfactory Stations. *Journal of Micromechatronics*, 1(1):49–66.

Ferreira, A., Cassier, C., and Hirai, S. (2004). Automatic Microassembly System Assisted by Vision Servoing and Virtual Reality. *IEEE/ASME TRANSACTIONS ON MECHATRONICS*, 9(2):321–333.

Fukuta, Y. et al. (2003). Pneumatic Two-Dimesional Conveyance System for autonomous Distributed MEMS. In *The 12th International Conference on Solid-State Sensors, Actuors and Microsystems*, pages 1019–1022, Boston, USA.

Gengenbach, U. and Boole, J. (2000). electrostatic feeder for contactless transport of miniature and microparts. In *Proceeding of SPIE Microrobotics and Microassembly II*, volume 4194, pages 75–81.

Goosen, J.F.L and Wlffenbuttel, R.F. (1995). Object Positionning Using a Surface Micromachined Distributed System. In *The 8th International Conference on Solid-State Sensors and Actuors, and Eurosensors IX*, volume 2, pages 396–399, Stockholm, Sweden.

Haake, A. and Dual, J. (2003). Paricle positionning by a two- or three-dimensional ultrasound field excited by surface waves. In *WCU*, Paris, France.

Hill, M. et al. (2003). A microfabricated ultrasonic particle manipulator with frequency selectable nodal planes. In *WCU*, Paris, France.

Hirata, T. et al. (1998). A Novel Pneumatic Actuator System Realised by Microelectro-Discharge Machining. In *Proceeding of the 11th International Workshop on Micro Electro Mechanical Systems (MEMS'98)*, pages 160–165, Heildelberg, Germany.

Huang, J-T. et al. (2004). Separate and Manipulate Different Kind of Particles by Dielectrophoresis. *Tamkang Journal of Science and Engineering*, 7(2):87–90.

Iizuka, T. et al. (1994). A Micro X-Y-theta Conveyor by Superconducting Magnetic Levitation. In *IEEE Symposium on Emerging Technologies and Factory Automation, ETFA'94*, pages 62–67, IIS The University of Tokyo.

Kim, D-H. et al. (2004). High-Troughput Cell Manipulation Using Ultrasound Fields. In *Proceedings of the 26th Annual International Conference of the IEEE EMBS*, pages 2571–2574, San Francisco, USA.

Kim, Y-K., Katsurai, M., and Fujita, H. (1990). Fabrication and testing of a micro supraconducting actuator using the Meissner effect. In *Proceedings of IEEE 3rd International Workshop on Micro Electro Mechanical Sytems (MEMS'90)*, pages 61–66, Napa Valley, USA.

Konishi, S. and Fujita, H. (1994). A Conveyance System Using Air Flow Based on tne Concept of Distributed Micro Motion Systems. *Journal of Microelectromechanical Systems*, 3(2):54–58.

Liu, C. (1995). A micromachined permalloy magnetic actuor array for micro-robotics assembly systems. In *International Conference of Solid-Sate Sensors and Actuors*, Stockholm, Sweden.

Masudo, T. and Okada, T. (2001). Ultrasonic radiation - novel principle for microparticle separation. In *ANALYTICAL SCIENCES*, volume 17 SUPPLEMENT, pages il341–il344. The Japan Society for Analytical Chemistry.

Mitani, A., Sugano, N., and Hirai, S. (2005). Micro-parts feeding by a saw-tooth surface. In *Proc. ICRA*, Barcelona, Spain.

Moesner, F. M. and Higuchi, T. (1997). Contactless manipulation of microparts by electric field traps. In *Proceedings of the SPIE's International Symposium on Microrobotics and Microsystem Fabrication*, volume 3202, pages 168–175, Pittsburgh.

Moesner, F. M., Higuchi, T., and Tanii, Y. (1996). New considerations for traveling wave particle handling. In *IEEE Industry Application Society Conference Record, 31st Annual Meeting*, pages 1986–1993, San Diego.

Nakasawa, H. et al. (1997). The two-dimensional micro conveyor. In *International Conference of Solid-State Sensors and Actuators*, Transducer'97, pages 33–36, Chicago, USA.

Nakasawa, H. et al. (1999). Electromagnetic Micro-Parts Conveyer with Coil-Diode Modules. In *The 10th International Conference on Solid-State Sensors and Actuators*, volume 2 of *Transducer'99*, pages 1192–1195, Sendai, Japan.

Regnier, S., Rougeot, P., and Chaillet, N. (2004). Modélisation des effets microscopiques pour la micro-manipulation. In *Troisième journée du réseau thématique pluridisciplinaire microrobotique*.

Sin, J. and Stephanou, H. (2003). A parallel micromanipulation method for microassembly. In *Proceeding of SPIE : Micromachining and Microfabrication Process Technology VII*, volume 4557, pages 157–164.

Smela, E. and Kallenbach, M. (1999). Electrochemically Driven Polypyrrole Bilayers for Moving and Positionging Bulk Micromachined Silicon Plates. *Journal of microelectromechanical systems*, 8(4).

Suh, J. W. et al. (1997). Organic thermal and ectrostatic ciliary microactuor array for object manipulation. *Sensor and Actuors: A*, 58:51–60.

Suh, J. W. et al. (1999). CMOS Integrated Ciliary Actuato Array as a General-Purpose Micromanipulation Tool for Small Objects. *Journal of Microelectromechanical Systems*, 8(4):483–496.

Tabata, O. et al. (2002). Ciliary motion actuator using self-oscillating gel. *Sensors and Actuators: A 95*, pages 234–238.

Tichem, M. et al. (2004). A classification scheme for quantitative analisys of micro-frip principles. *Assembly Automation*, 24(1):88–93.

Varstenbosch, J-M. et al. (2004). Theory and experiments on vibration feeding off small parts in the presence of adhesive forces. In *IPAS 2004*, Bad hofgastein, Austria.

Wautelet, M. and Guisbiers, G. (2003). Macro, micro et nanosystèmes: des physiques différentes? *SFT: Congrès franais de thermique.*

Whitney, D. E. (2004). *Mechanical Assemblies.* Oxford University Press.

Zheng, L. et al. (2003). Towards Single Molecule Manipulation with Dielectrophoresis using Nanoelectrodes. In *Proceedings of the Third IEEE Conference on Nanotechnology*, volume 1, pages 437–440.

Zheng, W., Buhlmann, P., and Jacons, H. O. (2004). Sequential shape-and-solder-directed self-assembly of functional microsystems. *PNAS*, 101(35):12814–12817.

Zhou, Q. et al. (2004). Microassembly system with controlled environment. *Journal of Micromechatronics*, 2, Issue 3:227–248.

Zhou, Y. and Nelson, B. J. (2000). The Effect of Material properties and Gripping Force on Micrograsping. In *ICRA*, pages 1115–1120, San Francisco, USA.

PNEUMATIC CONTACTLESS FEEDER FOR MICROASSEMBLY

Michele Turitto[1], Dr. Yves-André Chapius[2] and Dr. Svetan Ratchev[3]
[1,3] School of Mechanical, Materials and Manufacturing Engineering, University of Nottingham, University Park, Nottingham, NG7 2RD, UK [1]epxmt2@nottingham.ac.uk, [3] svetan.ratchev@nottingham.ac.uk
[2] Institute of Industrial Science, The University of Tokyo, 4-6-1 Komaba, Meguro-ku, Tokyo 153-8505, Japan, chapuis@fujita3.iis.u-tokyo.ac.jp

Abstract: The need of fully automated microassembly systems place specific functional requirements on the design and fabrication of critical elements such as grippers, sensors, manipulators and feeders. A new microfeeder design is proposed based on contactless pneumatic distributed manipulation. By cooperation of dynamically programmable microactuators, a number of feeding functions and even some elementary assembly operations can be achieved.

Key words: Microfeeding, Distributed manipulation, Pneumatic, Contactless, Microassembly

1. INTRODUCTION

The semiconductor industry over the last few decades has followed Moore's law: Dr. Gordon Moore in 1965 predicted the doubling of transistors in the same size of an IC every two years. The projections extend to the year 2012, at which time the smallest component of a device would have a linear dimension of 50 nm (1). The requirements in the IC industry have driven development of silicon-based micro and nano products. Silicon products are characterised by monolithic integration as all the process steps are integrated on a single substrate. This is an efficient but, at the same time, quite limiting approach as several different materials have to be combined with a silicon substrate to provide the high versatility and level of functionality needed in a variety of emergent microproducts (e.g. ink jet cartridges for printers, implantable drug delivery systems, chemical and biological sensors). Hence it is necessary to move from monolithic to hybrid integration so that the most appropriate material and fabrication techniques can be utilised.

Such new micromanufacturing requirements demand for an equally innovative approach to microassembly. If microproducts are to be produced in large volumes, novel materials, processes and production technologies have to be developed (2). This need cannot be addressed by simply transferring know-how from the macrodomain no matter how well established. Such a scaling down approach, although plausible, presents a number of flaws as it does not take into account the physical phenomena that emerge at the microscale (where surface-related forces become dominant over mass-related forces) and ignores issues like part fragility or part contamination.

In today's industries, the only alternative to automatic assembly, at least for the most critical operations, is manual assembly even if this can compromise the output in terms of time, cost and quality assurance. If high production volumes of microproducts have to be delivered in an efficient and reliable way, automatic assembly is the ultimate goal to which the efforts have to be aimed at. Figure 1 provides an economical justification for this last statement. It compares the product assembly cost for manual assembly and assembly on a highly flexible assembly cell for different batch sizes. The latter is more cost effective in all cases. The difference in total assembly cost is really significant for microsystems where manual assembly becomes too expensive due to the difficulty faced by a human operator when performing this type of operation. The same is true for the assembly of bigger products such as watches but in this case the advantage is less evident (3). Microassembly systems entail all the elements of an equivalent macroassembly system: grippers, sensors, manipulators, feeders and so forth.

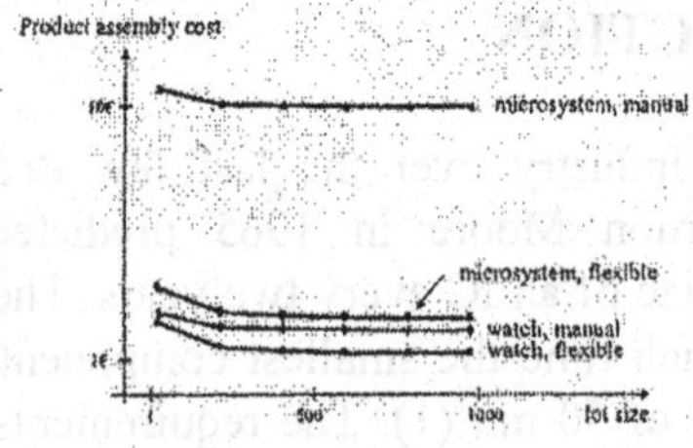

Figure 1. Flexible and manual assembly cost versus lot size for microsystems and watch components (3)

There is a significant body of knowledge focused on microassembly gripping and actuating solutions whereas a crucial element such as feeding has not been sufficiently developed. In fact, it can be argued that part feeding is one of the most restrictive elements in achieving production flexibility and reconfigurability in microassembly systems. Flexibility in part feeding refers to the possibility of introducing new parts into the assembly system with minimal reconfiguration (4). A robust microfeeder has to fulfil this requirement and meet the constraints imposed by the level of precision.

This paper introduces a pneumatic microfeeder based on distributed manipulation fabricated by IC-compatible micromachining processes.

2. MICROFEEDING

Feeders have the function of presenting parts that were previously randomly oriented to an assembly station at the same position, with the correct orientation and the correct speed. In microassembly, distributed manipulation (figure 3) is a quite common approach for conveying microparts.

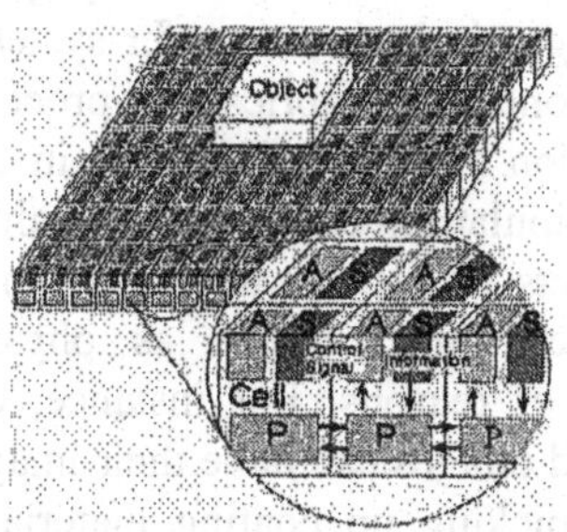

Figure 3. Distributed manipulation

It is based on arrays of tiny actuators where each is able to provide a simple motion. Even though the motion imparted by a single element is within a small range, it is possible to move objects over relatively long distances through the cooperation of a large number of microactuators. In this way a complex motion task is decomposed into many simpler tasks each of which is allocated to the individual units of the array. The actuators are capable of mechanical feats far larger than their size would imply. For example, researchers have located small microactuators on the leading edge of airfoils of an aircraft and have been capable to steer it using only these miniaturised devices (5).

Several physical principles have been exploited in distributed manipulation. Thermal expansion (6), electrostatic actuation (7) and electrostatic torsional resonators (8) are used in "cilia" arrays so called from their biological counterparts that can be found in the human respiratory tract. Through cilia's synchronous vibration small parts can be moved. Cilia arrays are an example of contact manipulation as they rely on friction for moving microparts. This can become an issue when dealing with fragile microparts or parts that cannot be contaminated. In microassembly contactless manipulation is a feasible alternative because of the small size and lightweight of the objects to be moved.

Contactless manipulation is advantageous as (9):

- Surface forces can be completely neglected
- It is suitable for handling fragile, freshly painted, sensitive micron-sized structured surfaces
- It allows the handling of non-rigid microparts
- There's no contamination of and from the end effector

Electrostatic actuation is used in (10) where particles are moved over an array of parallel field electrodes. The application of balanced multi phased voltages creates a travelling field-wave that conveys particles from electrode to electrode. In (11) a magnetic levitated wafer system transport system is developed for the semiconductor fabrication process to deal with the problem of oil contamination that normally exists in conventional transport systems. Contactless manipulation can also be realised using air jets to levitate and transport objects. In (12) an air paper mover is proposed as an alternative to motor-driven pinch rollers as physical contact can be undesirable for a variety of reasons. For instance, coated sheets can be damaged, sheets with toned images can have their images smeared or rollers can be contaminated by picking up coating material. Moreover, moving sheets along non straight trajectories is complex for cylindrical roller systems and the energy required to accelerate them is far greater than the energy effectively transferred to the paper. However this application is not suitable for microassembly as it is not able to provide a high level of precision.

3. FOUR DIRECTIONS MICROACTUATOR

A pneumatic contactless microfeeder based on the principle of distributed manipulation is proposed. The microfeeder consists of an array of micronozzles. Air is used for keeping the parts suspended. The parts are moved through the control of the micronozzles. As can be seen in figure 3, a single microactuator is made up of four nozzles formed by a central electrode and four walls around it.

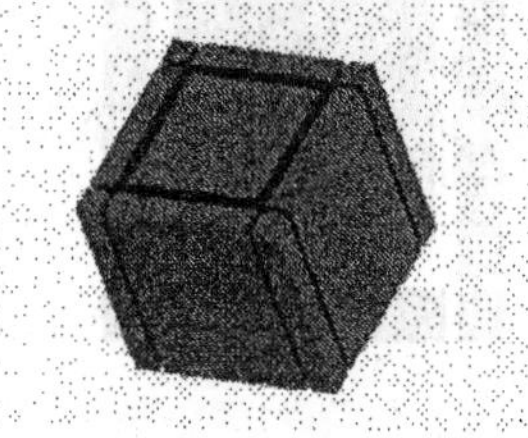

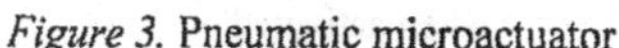

Figure 3. Pneumatic microactuator

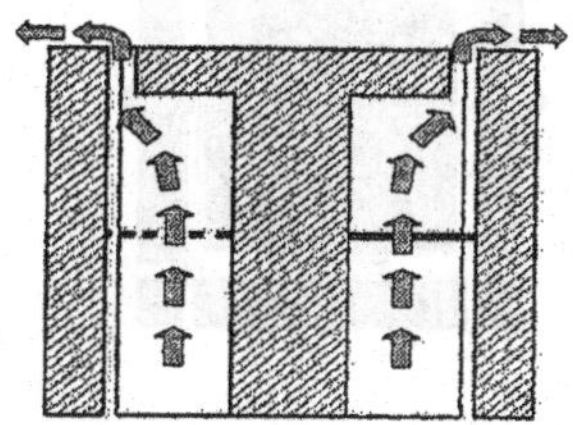

Figure 4. Cross section

The nozzles are opened or closed by electrostatic actuation. In the neutral position the four nozzles are all open (figure 5): the airflow, coming from the bottom of the microactuator, is equally divided among the four nozzles because of the symmetry of the structure. The outcoming airflows are such that the resulting force field causes the micropart to hover above the microactuator. For moving the object the central cursor is attracted towards one of the walls and the corresponding nozzle is closed. In figure 6 the rightwards and leftwards jets compensate each other hence there's a net force that pushes the micropart downwards. A similar working principle was presented in (13). The proposed design is advantageous because the microactuator is more compact as it keeps the dimensions of the airflow channel constant. Moreover, movement in four orthogonal directions is achieved with a single microactuator as opposed to the combination of four different microactuators capable of conveying objects in two directions only. This feature is of paramount importance as distributed manipulation becomes more effective if two conditions are satisfied: the microactuators have to be as small as possible as their size directly affects the minimum size of the parts that can be moved and the density of microactuators has to be high because this directly influences the position resolution that can be achieved.

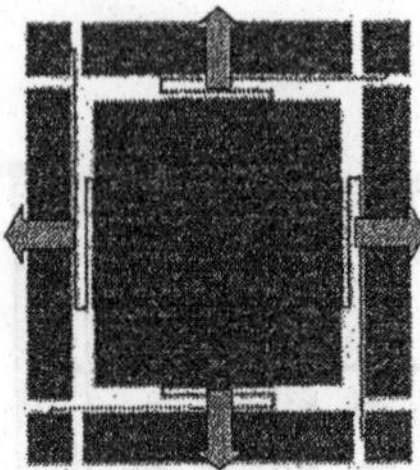

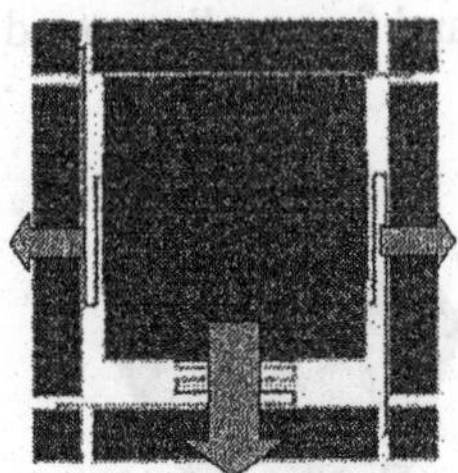

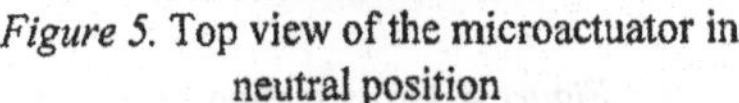

Figure 5. Top view of the microactuator in neutral position

Figure 6. Top view of the microactuator when activated

The microfeeder manufacturing sequence is based on IC-compatible fabrication process so that it is possible to obtain a high number of microactuators all the same time. Details about the process are reported in chapter 4. The array is then mounted and electrically connected to a printed circuit board. This means that the movable central electrode is fixed at its base and bends slightly for closing the nozzles. For this reason, as can be seen in figure 7, its lower part is connected with four springs to four "pillars" placed between the electrodes. The springs increase the robustness of the structure and help the electrode to return to its central position. This task can be accomplished also through the control of the electric field that acts upon the cursor.

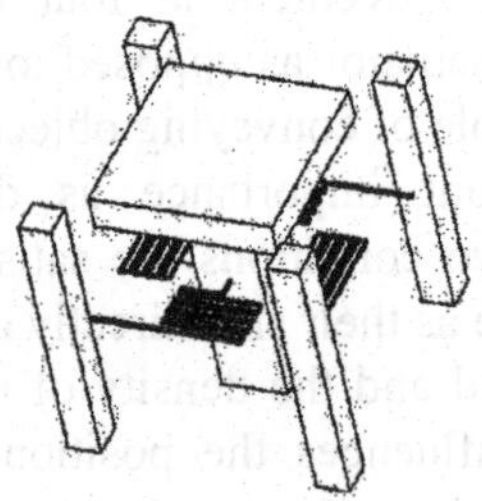

Figure 7. View of the microactuator without the side walls

3.1 Microfeeding functionalities

The air jets coming out of the microactuators form a force field that can be globally controlled through the local control of the nozzles. As a result of this coordinated motion several functionalities, which are specific tasks of a feeding system, can be generated.

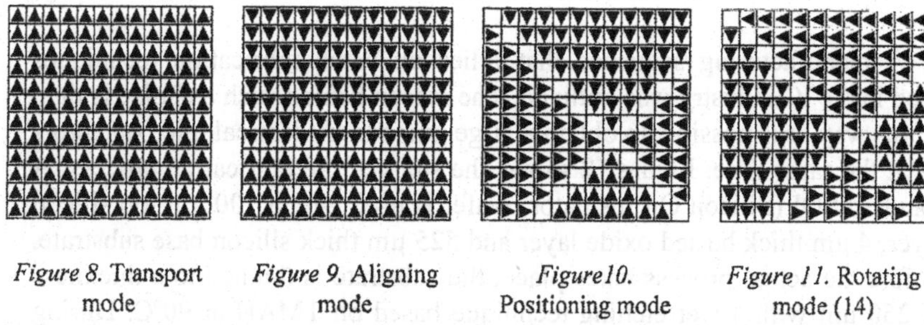

Figure 8. Transport mode

Figure 9. Aligning mode

Figure10. Positioning mode

Figure 11. Rotating mode (14)

For transporting a part (figure 8), all the air jets point in the same direction. If there's the need to align microparts (figure 9), the feeder's surface can be divided into two regions in which the relative force fields move the parts into two opposite directions. The parts align along the border of the two regions. Airflows can be arranged in a way such that the microparts are moved to any specific position (figure 10). The borders between different regions with different force fields act as spatial filters. It is also possible to maintain the position and just change the orientation (figure 11) with four orthogonal force fields. As all the nozzles can be independently activated, the force field due to the air jets is dynamically controllable. Hence, several feeding functionalities can be obtained in cascade: a micropart placed on the feeder is moved to the desired position, its orientation is changed according to the particular needs and then it is moved to a different position which acts as a dead nest. The combination of feeding functionalities also offers the possibility of realising some elementary assembly operations as can be seen in figure 12: once the two microparts are in the positions A and B, the force field is modified as represented. The combined microparts can be fed to a microassembly system as a subassembly thus simplifying the whole process.

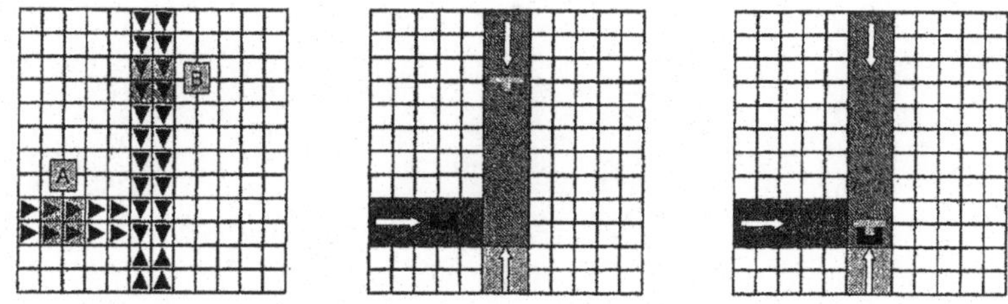

Figure 12. Combination of parts into a subassembly

4. MICROFEEDER FABRICATION PROCESS

A manufacturing sequence that relies on batch fabrication techniques used in the IC industry was outlined. The reason behind such a choice is that, in this way, it is possible to obtain a large number of identical microactuators all at the same time. Figure 12 shows the steps of the fabrication process. A standard SOI (Silicon On Insulator) wafer is used with a 100 μm top silicon layer, 4 μm thick buried oxide layer and 525 μm thick silicon base substrate. Before the actual process takes place, the substrate is thinned to a thickness of 250 μm with a wet etching technique based on TMAH at 90°C. During this step, the front side of the wafer is protected by a simple oxide layer of 50 nm realized by oxidation furnace deposition. In the first step of the fabrication process the lower silicon substrate is machined by deep RIE (Reactive Ion Etching). The material that is not going to be removed is protected with photo resist mask. RIE is chosen because, as an effect of its strong anisotropy, only the material perpendicular to the substrate surface is removed. As a result, high aspect ratio walls can be obtained. The next step is the fabrication of the springs that support the central cursor, again by RIE. In order to have the springs free to contract and expand, the oxide layer is removed by sacrificial layer (BOX) etching. HF (hydrofluoric acid) is used in this step as it helps to avoid sticking problems. After the releasing procedure, the top silicon layer is machined by RIE in order to obtain the clearance for the nozzles.

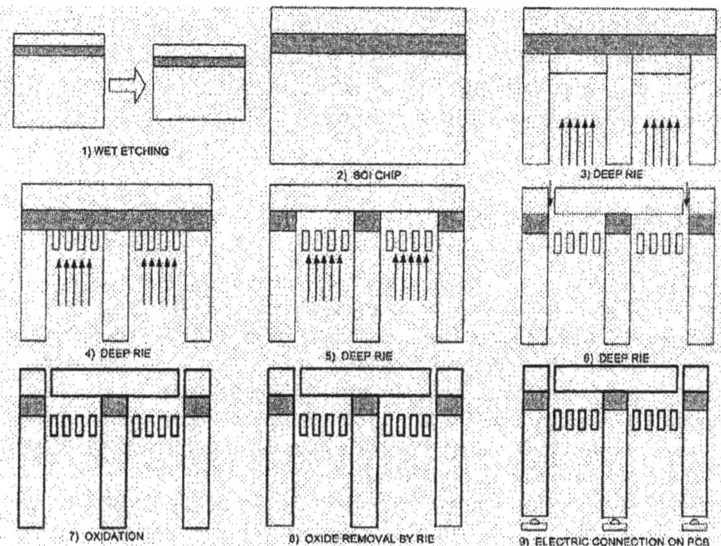

Figure 13. Fabrication process

The whole structure is then oxidised to ensure electrical insulation between the electrodes. In the next step RIE is used to remove the oxide from the bottom of the silicon substrate in order to allow the connectivity to the printed circuit board.

5. CONCLUSION AND FUTURE WORK

A pneumatic microfeeder for high precision assembly based on distributed manipulation was presented. Microparts float over an air cushion and are conveyed to the desired position with the desired orientation through the cooperation of individually controllable airflows. Further studies will be conducted to validate the proposed concept and evaluate its practical applicability using a prototype. For this purpose a manufacturing sequence based on IC-compatible fabrication process was outlined.

In a first stage, the validity of the proposed concept is going to be assessed using visual servoing to implement closed loop control. Ultimately, the IC-compatible fabrication process offers the opportunity to integrate on a single substrate electronic circuits, sensors and processing units. This is an important feature as the number of signals that have to be managed is significant because of the high density of actuators on the array. The integration of sensors and processing units grants each actuator a higher autonomy and reduces the amount of information exchange with the central processing unit; the execution of a motion task is realised through local information exchange among smart actuators. Results of these further investigations will be reported in due course

REFERENCES

1. Geng H., Semiconductor manufacturing handbook, Mc Graw-Hill Handbooks page 22.3
2. Alting L. et al. Micro Engineering, Annals of the CIRP Volume 52/2/2003
3. Koelemeijer Chollet S., Bourgeois F. and Jacot J., Economical justification of flexible microassembly cells, Proceedings of the 5th IEEE International Symposium on Assembly and Task Planning, Besancon, France, July 10-11 2003 pages 48-53
4. Viinikainen H., Uusitalo J. and Tuokko R., New flexible minifeeder for miniature parts, Proceedings of the International Precision Assembly Seminar IPASS 2004, Bad Hofgastein, Austria, Feb 11-13, pages 87-94
5. Huang P. H. et al., Applications of MEMS devices to Delta Wing aircraft: from concept development to transonic flight test, AIAA paper No. 2001-0124, Reno, Nevada, January 8-11 2001
6. Ataka M. et al., Fabrication and operation of polyamide bimorph actuators for ciliary motion system, Journal of Microelectromechanical Systems, Dec 1993 2(4), pages 146-150
7. Goosen J.F.L. and Wiffenbuttel R.F., Object positioning using a surface micromachined distributed system, Proceedings of the 8th International Conference on Solid-State Sensors and Actuators and Eurosensors IX, Stockholm, Sweden, June 25-29 1995, Volume 2, pages 396-399
8. Böhringer K.F., Donald B.R., Mihailovich R. and MacDonald N.C., Sensorless manipulation using massively parallel microfabricated actuator arrays, Proceedings of the IEEE International Conference on Robotics and Automation (ICRA), May 1994, San Diego, California, USA, Volume 1, pages 826-833.
9. Lambert P., Vandaele V., Delchambre A., Non-contact handling in micro-assembly: state of the art, Proceedings of the International Precision Assembly Seminar IPASS 2004, Bad Hofgastein, Austria, Feb 11-13, pages 67-76
10. Moesner F. M. and Higuchi T., Contactless manipulation of microparts by electric field traps, Proceedings of the SPIE's International Symposium on Microrobotics and Microsystems Fabrication, Pittsburgh, October 1997, vol. 3202, pages 168-175
11. Park K. H., Lee S. K., Yi J. H., Kim S.H., Kwak Y.K. and Wang I.A., Contactless magnetically levitated silicon wafer transport system, Journal of Mechatronics 1996 Vol. 6, No. 5, pages 591-610
12. Biegelsen D. Berlin A., Cheung P., Fromherz M., Goldberg D., Jackson W., Preas B., Reich J. and Swartz Lars, Airjet paper mover, presented at SPIE International Symposium on Micromachining and Microfabrication, 4176-11, Sept 2000
13. Fukuta Y., Yanada M., Ino A., Mita Y., Chapuis Y.A., Konishi S. and Fujita H., Conveyor for pneumatic two-dimensional manipulation realized by arrayed MEMS and its control, Journal of Robotics and Mechatronics, 2004, Volume 16, No. 2, pages 163-170
14. Konishi S. and Fujita H., Two dimensional conveyance system using cooperative motions of many microactuators, Proceedings of IROS`96, Volume 2, pages 988-992

PART II

Robotics and Robot Applications for Precision Assembly

"PARVUS" A MICRO-PARALLEL-SCARA ROBOT FOR DESKTOP ASSEMBLY LINES

Arne Burisch[1], Jan Wrege[1], Sven Soetebier[1], Annika Raatz[1], Jürgen Hesselbach[1] and Rolf Slatter[2]

[1]*Technical University Braunschweig, Institute of Machine Tools and Production Technology (IWF), Langer Kamp 19b, 38100 Braunschweig, Germany;* [2]*Micromotion GmbH, Mainz, Germany*

Abstract: The paper describes the development of a micro-parallel-SCARA robot adapted in size to MEMS products. The degree of miniaturization is optimized concerning a smaller structure but high accuracy in a workspace dimensioned to chip card size. The robot supports the mostly used four degrees of freedom with a base area of less than 150 x 150 mm^2. It is the result of a cooperative project between the Institute of Machine Tools and Production Technology at the Technical University of Braunschweig and Micromotion GmbH. This company is an innovative manufacturer of miniaturized zero-backlash gears and actors, which are used as main drives of the robot.

Key words: desktop factory, micro robot, parallel-SCARA, micro-gears

1. INTRODUCTION

Nowadays the trend of miniaturization in product development leads to an increasing gap between the dimensions of the product and the used production systems. Assembly lines for millimeter-sized products often measure some tens of meters. In recent years, the miniaturization of the production systems is discussed in research. Potential for development is seen in economic and technologic benefits of this strategy. On the one hand small dimensions offer the possibility for high modular system designs, improved scalability and flexibility in the manufacturing base of assembly lines. On the other hand small footprints, low power consumption, minor amounts of maintenance and initial costs promise better cost effectiveness. Already in 1990, a research group of the Mechanical Engineering Laboratory

(MEL) in Japan (Tsukuba) estimated that a 1/10 size-reduction of production machines leads to a decrease of energy consumption about 1/100 compared to a conventional factory[1].

Nowadays the MEL with its Desktop Machining Factory is deemed to be a pioneer in the field of micro factory[1, 2]. Basic ideas and questions related to desktop factories are furthermore discussed by Breguet in his paper "Toward the Personal Factory"[3]. Deliberating about advantages and disadvantages, he highlights the vision of conventional factories and desktop factories coexisting in future times. The aim should not be the all-purpose desktop factory but a high modular system adapted in size to MEMS as presented in the concepts of Gaugel[4] et al. or Rochdi[5] et al.. These concepts combine individual assembly modules on a product-neutral platform used for the feeding of parts to individual process stations.

The challenge of the "Parvus Project" was to develop a miniaturized precision industrial robot with the full functional range of bigger models. Independent from its future application area, as a component for a visionary desktop factory or as a miniaturized robot for future assembly lines, some basic requirements had to be determined.

Main questions were the required size of workspace and the striven performance specifications regarding accuracy and velocity. The latter were determined by the presented design studies and the design constraints limited by the available size of system components like servo drives, gearboxes or encoders and switches. An idea of an adequate workspace was given by the size of innovative microproducts by comparing. Product examples, like micro-pumps or labs-on-a-chip as presented by "thinXXS" (see Fig. 1) or the modern RFID transponder led to the size of a chip card.

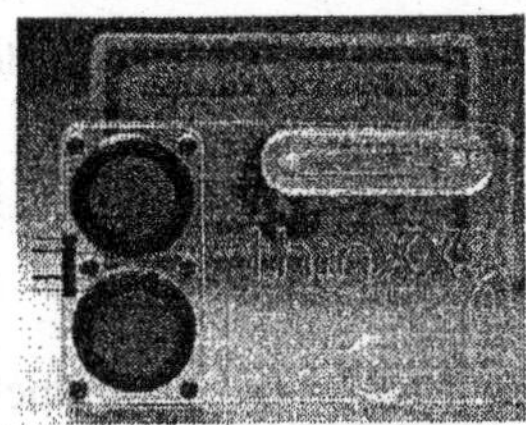

Figure 1. "thinXXS" lab-on-a-chip

2. STATE OF THE ART FOR PRECISION ROBOTS

A survey that categorizes precision robots is shown in Fig. 2. The simplest classification is into serial, parallel and hybrid structures, which in turn can be subdivided into further categories. The first category covers cartesian robots. These are typically very large in comparison to the components to be handled and are often, as a result, very expensive. However, they do provide repeatability between 1 and 3 µm, as demonstrated by, for example, the "Sysmelec Autoplace 411". The second category covers SCARA robots, which have a large workspace in relation to their physical size, but only achieve a repeatability of 10 µm, even in the case of the most accurate designs. In the field of parallel robots, there are few examples in industrial use. The Mitsubishi RP-X is an exception, achieving a repeatability of 5 µm. Most other developments in this area are limited to university research projects, in particular at the Technical University of Braunschweig in Germany, where extensive experience with parallel structures has been gathered, for example, with the Triglide robot, which has achieved repeatability better than 1 µm[6, 7, 8].

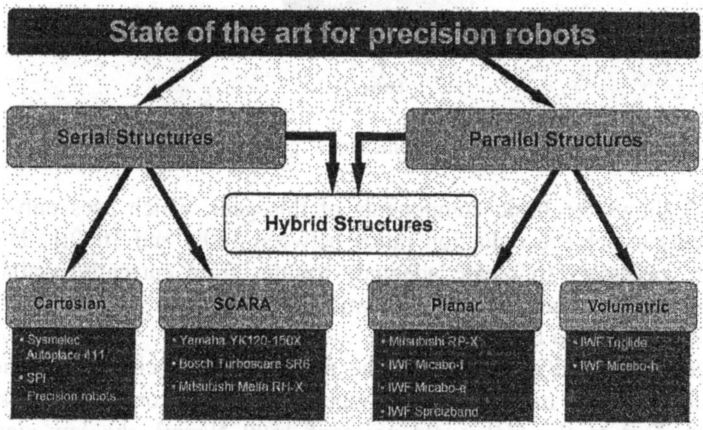

Figure 2. Survey of high precision assembly robots

Nowadays the existing high precision robots are relatively large and expensive. However, a growing market demands for smaller and cheaper robotic devices for precision positioning and assembly is coming up. The minimization of conventional industrial robots is even in progress in commercial products of Yamaha or Mitsubishi. These robots reach a repeatability of 5 µm.

A further miniaturization of such industrial robots is possible because of new enabling technologies, in particular zero-backlash micro-gears and highly dynamic micro-motors with integrated incremental encoders, which allow proven robot arm structures to be miniaturized. Furthermore they allow the use of proven control technology and avoid the complexity of alternative actuator technologies such as piezo actuators[6, 8].

3. INNOVATIVE DRIVE SYSTEMS / COMPONENTS

In recent years commercially miniaturized DC or EC motors and high-resolution encoders came up. In combination with micro-gears, as developed by Micromotion GmbH, these components can serve as adequate drives for micro assembly systems.

The Micro Harmonic Drive® is the only micro-gear currently available that offers the same positioning accuracy as the large-scale Harmonic Drive gears used in industrial robots.

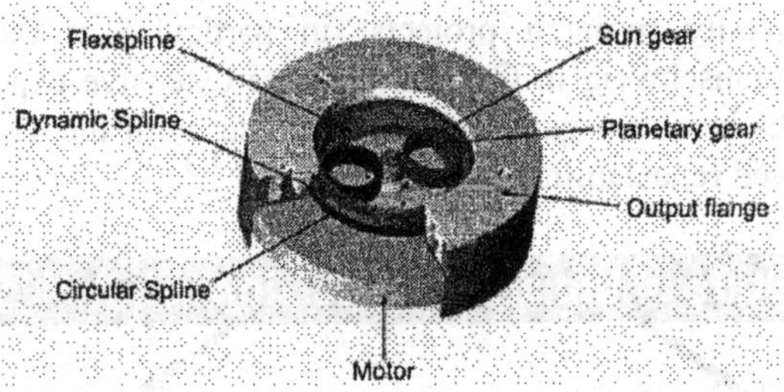

Figure 3. Micro Harmonic Drive® gear components

Fig. 3 shows the basic components of this gear, which uses only six components to achieve reduction ratios between 160:1 and 1000:1. These ratios are necessary to create adequate torques from currently available micro-motors, which are capable of rotational speeds up to 100000 rpm, but only offer torques of a few μNm.

Related to the LIGA-technique, the single gear wheels of the Micro Harmonic Drive® are manufactured by electroplating and consist of a nickel-iron-alloy. Due to the high yield point of 1500 N/mm², the low elastic modulus of 165000 N/mm² and its good fatigue endurance, this electroplated alloy possesses the necessary properties for perfect functioning of the flexible gear wheels of this micro gear system.

By providing an angular repeatability of 10 arc seconds, the Micro Harmonic Drive® gear is the only micro-gear currently available that provides sufficient accuracy for a micro-robot meeting our requirements.

4. ROBOT DESIGN APPROACHES

At the beginning of the design phase, some basic specifications were established and quantitative specifications were fixed (see Table 1). The aim was to find a simple modular structure with a small envelope and easy access to the working area of the robot. The robot should provide four degrees of freedom (DOF) with three orthogonal translational axes and one additional rotational axis.

Table 1. Requirements

Criterion	Value
Workspace	chipcard-size
Footprint (area of robot base)	< DIN-A4
Theoretic resolution	sub-µm
Linear speed	> 100 mm/s

A preselection for a possible kinematic structure of the robot was identified taking a closer look on conventional industrial robots. In this principle design phase, two different plane kinematic structures were compared: the serial "SCARA structure", e.g. used in the Yamaha YK 120, and the "parallel SCARA structure", known from the Mitsubishi RP-1AH.

The decision, whether using the serial or parallel structure, was supported by a detailed positioning sensitivity analysis of both structures. Therefore, the achievable positioning resolution at all points in the chosen workspace of the conceptual designs had to be analyzed. The kinematic transfer functions according the robot drives q and the world coordinates ρ are basically given by the direct kinematic problem (DKP) and the inverse kinematic problem (IKP) of both structures.

$$\underline{q} = (q_1, \ldots, q_F)^T, {}_{F=DOF} \qquad \underline{\rho} = (X_{TCP}, Y_{TCP}, Z_{TCP}, \Psi_{TCP})^T$$

Depending on geometric parameters, such as number of joints, length of arms and distance between supporting points, the analysis first leads to the workspace of the structure. By calculating from estimated backlash of ball bearings and resolution of the robot drives systems, the matlab-analysis shows the forecasted sensitivity of the structure. Sensitivity E as defined by

$$E = \frac{\Delta X_O}{\Delta X_I} \qquad \Delta X_O = output\,; \Delta X_I = input$$

is subdivided into sensitivity of the structure $\underline{\underline{J}}$ and sensitivity of dimension tolerance $\underline{\underline{J}}_{ME}$.

$$\underline{\underline{J}} = \frac{\partial \underline{\rho}}{\partial \underline{q}} = \begin{bmatrix} \frac{\partial x}{\partial q_1} & \cdots & \frac{\partial x}{\partial q_6} \\ \vdots & \ddots & \vdots \\ \frac{\partial x}{\partial \varphi} & \cdots & \frac{\partial x}{\partial \varphi} \end{bmatrix} \qquad \underline{\underline{J}}_{ME} = \frac{\partial \underline{\rho}}{\partial \underline{l}} = \begin{bmatrix} \frac{\partial x}{\partial l_1} & \cdots & \frac{\partial x}{\partial l_i} \\ \vdots & \ddots & \vdots \\ \frac{\partial x}{\partial l_1} & \cdots & \frac{\partial x}{\partial l_j} \end{bmatrix}$$

Based on the results of sensitivity the theoretical repeatability of the robot's structure is forecasted. The iterative analysis of different geometrical parameters aimed to reach a minimized arm length by keeping a chip-card-sized workspace and high accuracy. The theoretical repeatability, based on accuracy data of drive components, is in some areas better than 1 µm.

The sensitivity plots of the optimized geometrical structures for serial and parallel kinematics respectively are shown in Fig 4. The serial structure has a theoretical sensitivity map which only achieves high accuracy in very limited areas of the available workspace. The parallel structure, in comparison, achieves high accuracy over almost the complete workspace and also has a symmetrical sensitivity map. Furthermore, the parallel structure should offer a significantly better dynamic performance, because only the gear motor for the fourth (rotational) axis is carried by the moving arm. Additionally the passive joints of the parallel structure are easier to miniaturize than active joints. Because of these reasons, the "parallel SCARA structure" was chosen as the basic structure for further developments. An argument against the parallel structure would be its more complex control due to the closed kinematic structure. Nevertheless, the steady decrease in the cost of computing power means that in this case the mechanical advantages of the parallel structures are more important.

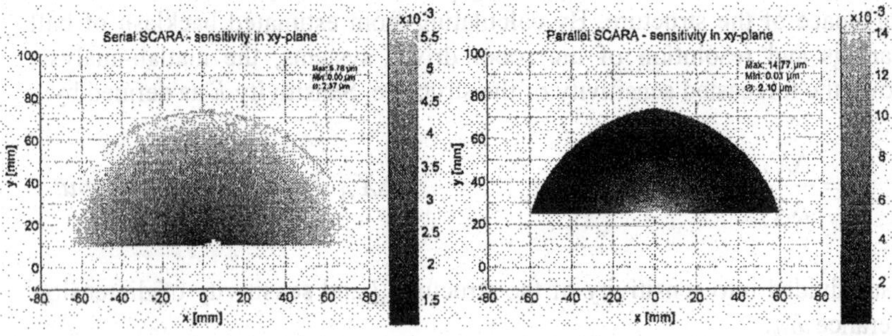

Figure 4. Positioning sensitivity

5. DESCRIPTION OF FINAL ROBOT DESIGN

The results of the design process concerning the basic structure lead to a typical parallel structure shown in Fig. 5. Both active joints A_1 and A_2 are equipped with Micro Harmonic Drive® gears combined with Maxon DC motors (q_1, q_2). They drive the plane structure in x-y-direction, whereas joints B_1, B_2 and C are passive in this case.

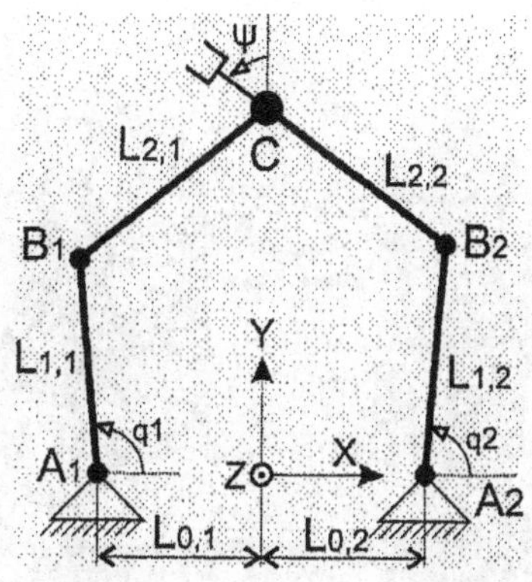

Figure 5. Parallel structure of the robot

Deciding on a plane parallel structure, offering two translational DOF in the x-y-plane. The z-axis is integrated as a serial axis in the base frame of the robot. The easy handling of the whole plane parallel structure driven in z-direction is possible because of its minimized drive components and light aluminum alloy structure. The z-axis is driven by a Harmonic Drive servomotor combined with a conventional ballscrew. Additionally the rotational hand axis Ψ was executed as a hollow rotational axis integrated in the passive joint C as the Tool Center Point (TCP) of the parallel structure. This allows media like e.g. vacuum to be passed along the hand axis. A Micro Harmonic Drive® gear combined with a Maxon EC motor drives the Ψ-axis. All joints of the robot are provided with preloaded angular contact ball bearings that are strained by springs. The resulting joints are nearly free of backlash and have low friction.

All servomotors support a high resolution encoder feedback signal. Additionally the Parvus is equipped with magnetoresistive position sensors for initialization of the robot and an emergency stop function.

The control of the Parvus robot was developed on a real-time system from dSPACE Inc. (Germany). The system features a PowerPC750 digital signal processor (DSP) running at 480 MHz, a digital I/O board, high resolution analog I/O boards, an encoder board and a serial I/O board. The dSPACE system was chosen because of the powerful hardware solution that is available from one supplier and the good operability of the communication software "Control Desk" that was used to create the graphical user interface. For programming the control, dSPACE Inc. gives the possibility to use program codes in "Matlab/ Simulink" and "C".

The first two functional prototypes of Parvus, shown in Fig. 6, are used for further research and presentation at present.

Figure 6. The first two functional models of the Parvus

The first prototype of Parvus fulfills the following specifications shown in Table 2.

Table 2. Technical specifications

Criterion	Value	Unit
Workspace (rectangular)	60x45x20	mm^3
Footprint (area of robot base)	130x170	mm^2
Theoretic resolution	< 3	µm
Linear speed	> 100	mm/s
Rotational speed (Ψ axis)	$187^* / 60^{**}$	rpm
Angular resolution (Ψ axis)	$0.022^* / 0.007^{**}$	°
Payload	50	g

*ratio 160:1 / **ratio 500:1

6. MEASUREMENTS

A matlab-analysis based on the kinematic transfer functions mentioned in section 4 led to a linear speed diagram shown in Fig. 7.

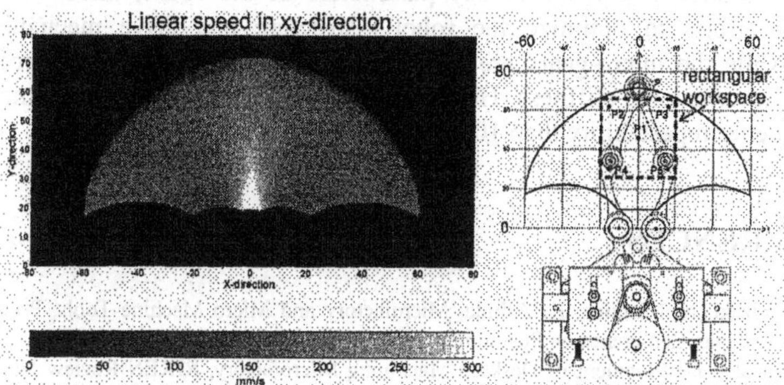

Figure 7. Linear speed diagram and measured workspace of Parvus

Furthermore, performance specifications regarding accuracy were measured at the IWF. The applied test method conforms to the ISO standard EN ISO 9283. The repeatability was measured at five definite points within a rectangular workspace shown in Fig. 7. Not including point P5, the repeatability of this first prototype was between 5 and 10 μm. Fig. 8 shows the plots of the best and the worst result. As expected, the practically measured repeatability is lower than the forecasted high theoretical accuracy. Experiments demonstrate that influences of elasticity in the drive components may be responsible for this effect.

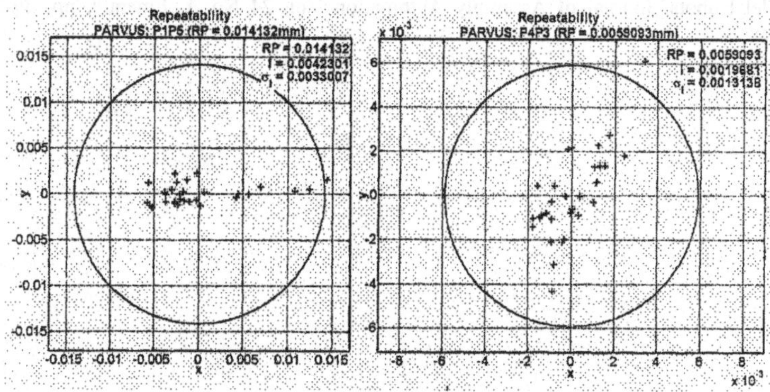

Figure 8. Results of repeatability

7. CONCLUSION AND OUTLOOK

The prototype of the robot (Fig. 6) was demonstrated to the public at the Hanover Fair in April 2005 for the first time. The feedback of industrial partners gives good perspectives using the Parvus in a visionary desktop-factory. Concerning the accuracy parameters of the robot, there will be further efforts to optimize the robot's structure and drive mechanism. Meanwhile the micro-gears are available with an optimized tooth profile, better accuracy and stiffness. Hence, the improved Parvus will guarantee a much better repeatability than 10 μm.

REFERENCES

1. Y. Okazaki, N. Mishima, K. Ashida, Microfactory and Micro Machine Tools, Proc. of Korean-Japan Conference on Positioning Technology, Daejeon, Korea, 2002
2. M. Tanaka: Development of Desktop Machining Microfactory, RIKEN Review No. 34, (April, 2001).
3. J.-M. Breguet, A. Bergander, Toward the Personal Factory?, Proc. of SPIE, Microrobotics and Microassembly III, Vol. 4568, pp. 293-303, (2001).
4. T. Gaugel, M. Bengel, D. Malthan, Building a Mini-Assembly System from a Technology Construction Kit, Proc. of International Precision Assembly Seminar IPAS'2003, (17-19 March, Bad Hofgastein, Austria, 2003).
5. K. Rochdi, Y. Haddab, S. Dembélé, N. Chaillet, A Microassembly Workcell, Proc. of International Precision Assembly Seminar IPAS 2003, (17-19 March, Bad Hofgastein, Austria, 2003).
6. J. Hesselbach, A. Raatz, J. Wrege, S. Soetebier, Design and Analysis of a Macro Parallel Robot with Flexure Hinges for Micro Assembly Tasks, Proc. of 35th International Symposium on Robotics (ISR), No. TU14-041fp, (23-26 March, Paris, France, 2004).
7. J. Hesselbach, G. Pokar, J. Wrege, K. Heuer, Some Aspects on the Assembly of Active Micro Systems, Production Engineering Vol. XI/1, pp. 159-164, (2004).
8. J. Hesselbach, J. Wrege, A. Raatz, O. Becker, Aspects on Design of High Precision Parallel Robots, Journal of Assembly Automation Vol. 24 No. 1, Emerald, pp. 49-57, (2004).

METHODS FOR COMPARING SERVO GRIPPERS FOR MINI AND MICRO ASSEMBLY APPLICATIONS

Ilpo Karjalainen and Reijo Tuokko

Tampere University of Technology, Institute of Production Engineering, Robotics and Automation Laboratory, P.O Box 589 33101 Tampere, Finland. Tel. +358-3-3115 2111, Fax. +358-3-3115 2753, Email: ilpo.karjalainen@tut.fi and reijo.tuokko@tut.fi

Abstract: In the future the assembled parts become smaller and the precision demands increase in many cases. Grippers are one factor of the solution to make a competitive automation assembly system. Grippers have to be flexible and also capable of holding parts without dropping them during accelerations. Other demands for a gripper for mini and micro assembly applications are, among other things, accuracy and good force/size ratio. By using servo grippers also in high-precision applications for miniature-size parts, a noticeably amount of time will be saved. This paper presents two different kinds of developed two-fingered servo grippers with parallel moving mechanics, as well as the developed test systems for servo grippers and some test results.

Key words: Servo gripper, micro assembly

1. INTRODUCTION

In the future the assembled parts become smaller and the precision demands increase in many cases. Grippers are one factor of the solution to make a competitive automation assembly system. Micro gripper research has been widely made in last twenty years [1,2]. So far most commercially available grippers are two-fingered or at most three-fingered [3]. Demands for a gripper for mini and micro assembly applications are, among other things, accuracy and good force/size ratio. By using servo grippers also in high-precision applications for miniature-size parts, a noticeably amount of time will be saved. In most of the servo gripper applications any kind of angular movement of fingers is a problem. In reference [4] weakness of the rotational movements of gripping arms is mentioned and analyzed. This paper presents two different kinds of developed two-fingered servo grippers with parallel moving mechanics, as well as the developed test systems for servo grippers and test results of servo grippers.

2. DEVELOPED SERVO GRIPPERS

In this work the target was to develop a high-precision parallel servo gripper with non-synchronous driven fingers based on piezoelectric linear motors. Target for the gripper was that the structure is simple and the gripper does not drop parts under high accelerations. Mechanically the gripper is based on two standing wave ultrasonic motors (UM), two high precision linear guides and two linear encoders (resolution 0.2 micrometer). The driving stroke each of the individual fingers is 5 mm and the maximum opening is 10 mm. Weight of the servo gripper is less than 300 grams. The gripper is mechanically designed to be compact and volumetrically small. Volumetrical size is 89 mm x 78 mm x 26.5 mm without fingers. The gripper fingers can be freely programmable and servo driven individually to the defined points with exact velocity sequence confirmed via controller's HMI. A camera or a microscope can be easily integrated to the gripper through a hole in the center of the gripper. This solution enables continuous monitoring of the parts during grasping, handling and releasing. Gripper has PC-control including HMI and joystick for driving finger and teaching cycle points. Figure 1 shows a view of the ultrasonic motor-based gripper.

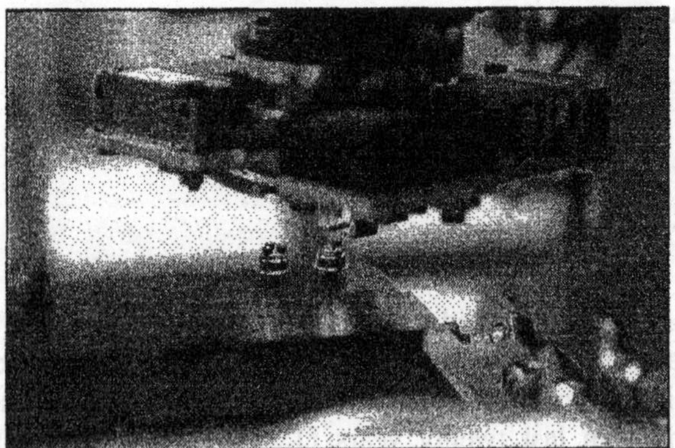

Figure 1. A servo gripper based on PZT motors in assembly work.

A state-of-the-art servo gripper based on ball screws and rotating motors (RM) was built to research this traditional technology also in high-precision applications, see figure 2. The developed gripper is a two-fingered servo gripper with position feedback including the following main components: two small DC-motors with brushes; two high-precision, preloaded ball screws; two high-precision linear guides and two high-resolution encoders (resolution 0.2 micron). This gripper motors has mostly same quantities than the gripper based on ultrasonic motors, like e.g. HMI interface. Fingers are driven without gears changing rotating ratio between motor axes and ball screws. Motors are directly connected to the ball screws with mechanical coupling.

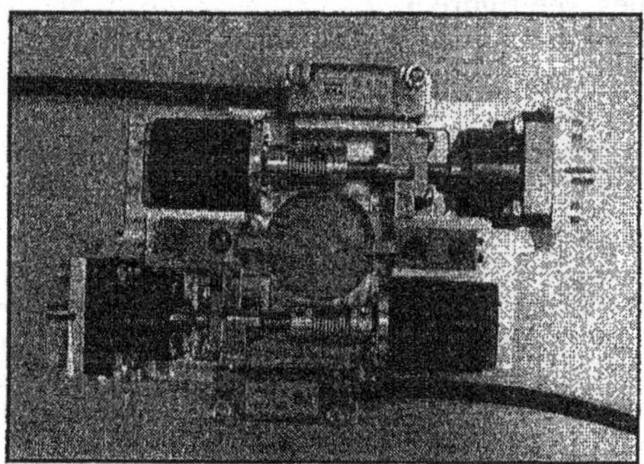

Figure 2. The developed gripper based on rotating DC-motors.

3. TESTING METHODS FOR SERVO GRIPPERS

Basically, servo grippers have requirements similar to those of normal grippers. When comparing servo grippers, it is expedient to concentrate on qualities that are characteristics of servo grippers and that are most important for normal on/off grippers.

Motion-controlled servo grippers are able to carry out movements, which are shorter than the maximum stroke. This measuring problem can be divided to two individual parts: first, measuring linearity, accuracy and repeatability of the movement, and, second measure position and velocity responses of fingers. In this project the accuracy and repeatability of the gripper fingers is measured with a laser interferometer [5] and a plane mirror mounted to a finger axis. A small plane mirror is attached to the gripper fingers one finger at a time. ISO 230-2 standard determines accuracy and repeatability tests of positioning numerically controlled axes [6]. There are several possibilities for measuring positioning and velocity responses of the gripper fingers. In research environments e.g. a laser interferometer with high sampling rate can be used.

A servo gripper's finger-axis can became unstable when undergoing transporting accelerations, due to possible instability of the finger-axis. In industrial applications dropping of the parts during handling is a considerable problem. When parts become smaller and accelerations higher the problem is even bigger. In servo gripper research it is especially important to test the behaviour of the gripper under specific accelerations. Some gripper actuators such as moving coil linear motors or pneumatic servo valves can become unstable even under relatively low accelerations. In Tampere University of Technology a Cartesian robot has been developed [7,8], which is suitable for researching and comparing different grasping and handling techniques, see figure 3.

Figure 3. Servo gripper connected to Cartesian robot [7].

With the robot's microscope also fine positioning of parts can be researched. With this system the grippers can be tested under specified horizontal, vertical and rotational accelerations. With the same system one can study pick and place times in different assembly operations and processes. With the achieved results one can optimize force, size of the gripper and shape of the gripper's fingers.

A servo gripper's finger-axis can became unstable in a grasping situation, because of the grasping force. When force feedback is used in servo gripper, it is also possible to control grasping force. Practical tests have been made with grippers for different materials, control algorithms, and approach velocities. Fingers' grasping force was measured with a 6 DOF force sensor so the direction of each force can also be seen. The other finger at the same time had an opposing flange.

Two-fingered servo grippers can be made in two different ways. Servo gripper's fingers can move parallel or individually. When fingers are moving individually the gripper is more flexible, but problems may arise with the centering of the part to be grasped. Depending on the centering of the part and the deviation in the axis forces the final positioning and the settling time may vary. We have researched centering behavior and settling times of grippers with two individually moving fingers by measuring positioning and velocity responses from both fingers during grasping. Fingers can also be driven synchronously so that one finger is moving to a closed position and the other to an open position. In other words, the gripper is able to move the part during grasping. A finger can also be driven individually, and non-synchronously, and thereby increase flexibility in grasping situations. These qualities were tested with case products.

4. TEST RESULTS FOR THE SERVO GRIPPERS

Using the results of the introduced bidirectional measurement the repeatability (3s) of gripper fingers can be calculated with equation 1, where q is the independent measurement result. The measurement was repeated five times with a stroke of 5 mm.

$$\pm 3S_q = \pm 3\sqrt{\frac{\sum_{k=1}^{N}(q_k - \bar{q})^2}{N-1}} \quad (1)$$

In table 1 the accuracy is presented; repeatability is presented as a worst-case repeatability.

Table 1. Laser measurements results of grippers axes (3σ)

	Accuracy	Repeatability
Finger 1 (rotating DC-motor)	10,5 micron	± 4,1 micron
Finger 2 (rotating DC-motor)	7,1 micron	± 2,4 micron
Finger 1 (ultrasonic motor)	18,6 micron	± 1,3 micron
Finger 2 (ultrasonic motor)	19,4 micron	± 0,8 micron

Accuracy, repeatability, velocity, and settling times were measured for both of the gripper's fingers. The main focus was on accuracy and repeatability tests of the fingers. Another objective was that the axes actual position be able to follow the command position as exactly as possible.

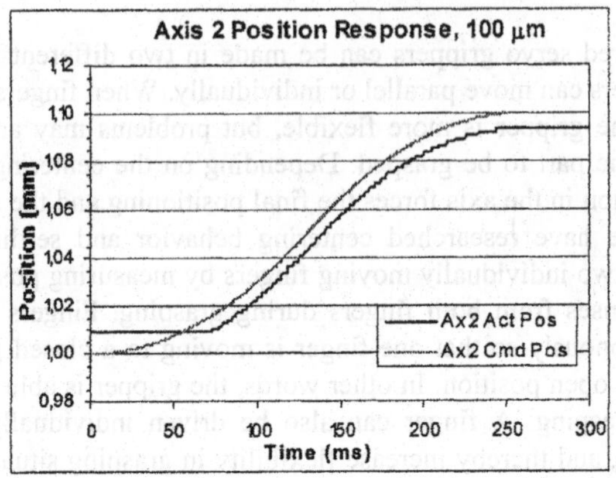

Figure 4. Position response of UM-finger in a 100 micron movement.

The actual position is not able to achieve command positioning with a PID controller, if the driven length is only 100 microns, see figure 4. The problem is similar with both of the axes. The reason is a static friction of the linear guide, and a low force of the UM at low velocities. Increasing the velocity makes movement unstable, in short movements.

Position responses of rotating RM-gripper axis 100 micron movement is introduced in figure 5. Position error in the end of the movement is smaller than with ultrasonic motor axes.

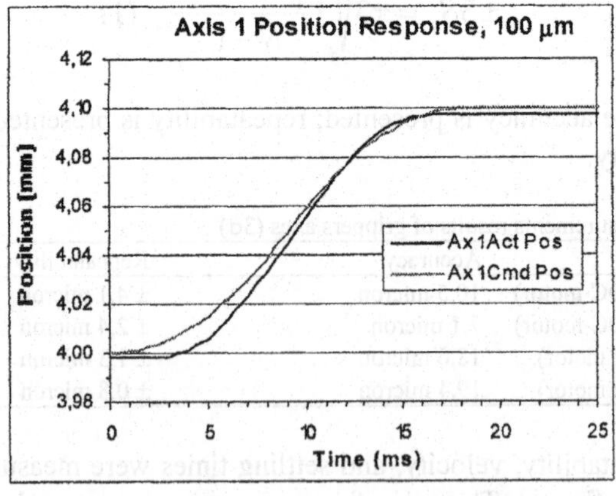

Figure 5. Position response of RM-finger in a 100 micron movement.

Behaviors of the developed grippers under accelerations were tested with the introduced research environment. The grippers did not drop parts under maximum accelerations available: 40 m/s^2 in perpendicular horizontal directions and 30 m/s^2 in vertical direction. The gripper based on ultrasonic motors was able to hold grasped parts under same accelerations also when the power of the ultrasonic motors was turned off. In this situation the grasp hold was maintained with the static friction between the motor's driving tip and the slide.

Grasping force depend on finger approaching velocity, which was 1 to 10 mm/s. Gripper based on ultrasonic motors, grasping force was 8 to 10 Newton. For a gripper based on DC-motors grasping force was little higher, between about 20 to 27 Newton.

Centering capability of the two independently moving fingers in grasping situation was tested. The graph in figure 6 shows the situation when finger 2 stays still and finger 1 starts to grasp the part. Graph shows that the positioning error of finger 2 increases to 18 micron and does not return to the original command position, but new position is stable.

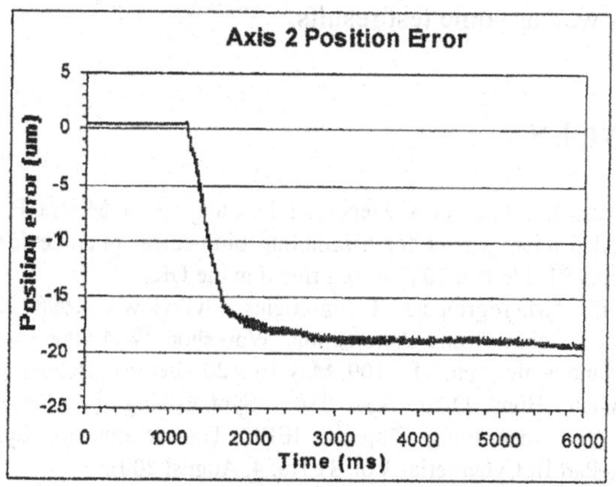

Figure 6. Position error of finger 2 in grasping

Centring behaviour with the gripper based on RM is better than with UM. The reason is lower friction. With RM, axes improve their centring after grasping, but it takes several seconds.

For testing of the gripper's quantities in practice a miniature size gearbox was assembled and disassembled. This gearbox consists of five different parts. This gearbox was used as a case product. Biggest part of the planetary gearbox is the tube, which diameter is 8 millimeter and smallest part is the planetary wheel, the diameter of which is 2,2 millimeter. In assembly process it was difficult to assemble some of the parts. According to our experiences it can be concluded that parallel mechanism is very important feature for a servo gripper when several different size parts are handled with same fingers. When parts are miniature-sized, this is even more important.

Individually moving fingers increase flexibility in grasping situation and together with integrated microscope grasping is easy to do. With a traditional DC-motor, with ball screw technology, it is possible to make a high accuracy gripper within a few microns of repeatability, when all mechanical components are of high precision.

5. CONCLUSION

In this paper two different kinds of developed two-fingered servo grippers with parallel moving mechanism have been presented. Parallel mechanism is very important feature for a servo gripper when several different size parts are handled with same fingers. When parts are miniature size, this is even more important. Moving fingers individually increases flexibility in grasping situation. Proposals for tests and measurements for two-fingered servo gripper with parallel movement have been developed and introduced as well as some test results.

REFERENCES

1. M.A. Carrozza, A. Eisinberg, A. Menciassi, D. Campolo, S. Micera, P. Dario: "Towards a force-controlled microgripper for assembling biomedical microdevices", J. Micromech. Microeng. pp. 271-276 No. 10 (2000). Printed in the UK.
2. Y. Bellouard. "Microgrippers Technologies Overview", 1998 IEEE International Conference on Robotics and Automation, Workshop WS4: Precision Manipulation at Micro and Nano Scales,. pp. 84 – 109, May 16 – 20.. Leuven, Belgium, 1998.
3. Wen-Han Qian, Hong Qiao, S.K. Tso: "Synthesizing Two-Fingered Grippers for Positioning and Identifying Objects, IEEE Transactions on Systems, Man, and Cybernetics–Part B: Cybernetics Vol. 31 No. 4, August 2001.
4. R. Keoschkerjan, H. Wurmus "A Novel Microgripper with Parallel Movement of Gripping Arms", Actuator 2002, 8th International Conference on New Actuators", pp. 321-324, Bremen, Germany, June 10-12 2002.
5. Anonym: "HP Laser Interferometer 5527A/B General Information", 50 p.
6. Anonym: "ISO standard 230-2", 1997, 15 p.
7. Heikkilä, R., Tuokko R. Development of a High Precision and High Performance Mini Robot - Technological Highlights and Experiences, International Precision Assembly Seminar, Bad Hofgastein, Austria, 17 –19 March 2003, pp. 49 –54.
8. I. Karjalainen, J. Uusitalo, T. Sandelin, R. Heikkilä, R. Tuokko: "Intelligent and Flexible End-Effector with Integrated Microscope – A Pilot Platform for Mini and Micro Assembly Research", INES 2001 International Conference on Intelligent Engineering Systems, 16 –18 September 2001, Helsinki, Finland, pp. 307 – 311.

COMPLIANT PARALLEL ROBOTS

Development and Performance

Annika Raatz, Jan Wrege, Arne Burisch and Jürgen Hesselbach

Institute of Machine Tools and Production Technology (IWF), Technical University Braunschweig, Langer Kamp 19b, 38106 Braunschweig

Abstract: In this paper the development of a macro parallel robot is presented in which conventional bearings are replaced by pseudo-elastic flexure hinges. The robot consists of a spatial parallel structure with three translational degrees of freedom and is driven by three linear direct drives. The structure has been optimized with respect to workspace and transmission ratio. Additionally, in simulations with the FEA tool ANSYS different geometrical arrangements and combinations of flexure hinges have been investigated with respect to the dynamic behavior of the compliant mechanism. Due to the symmetrical character of the structure and the optimized design of the combined flexure hinges the structure is very stiff. The experimental measured repeatability of the compliant robot is below 0.3 μm.

Key words: High Precision Robotics, Compliant Mechanism, pseudo-elastic SMA, flexure hinges

1. INTRODUCTION

The suitability of robots, which can be used for micro assembly tasks, depends on the performance specifications referred to accuracy. Usually different terms are used to describe these specifications. Slocum[1] explains these terms in a very demonstrative way: Accuracy is the ability to tell the truth, repeatability is the ability to tell the same story each time and resolution is the detail to which you tell a story. A more formal definition for these terms is given in the ISO standard EN ISO 9283 and specifies appropriate test methods. The accuracy of a robot is generally influenced by systematic and stochastic failures. While systematic failures can be compensated directly by the robot control, if they are known, stochastic failures can not. These can be caused by the travel of bearing cage and clearance in the joints of a robot. Due to their structure, parallel robots are typically built up with a high number of joints. These joints mostly have more than one degree of freedom which may result in a decreasing precision. For high precision assembly tasks in microsystem technology, these problems have been overcome by using flexure hinges instead of conventional rotational joints.[2,3,4] The integration of flexure hinges in parallel structures is relatively simple because, except for the drives, all joints are passive. Since flexure hinges gain their mobility exclusively from a deformation of matter, they do not posses the above named disadvantages of conventional joints. The allowable strains of the deformed hinge material are the limiting factor for the attainable angle of rotation. Basic parameters accounting for the maximal angle of rotation are the geometry, geometrical dimensions and the elastic and plastic material properties. Designing flexure hinges is always a compromise between accuracy, compact over-all dimensions, high mobility, out of axis stiffness and the aspired life cycle of the product. At present flexure hinges are mainly used in small devices like micro-positioning devices and micro-grippers with small angular deflections and limited workspaces, but high accuracies.[5,6,7]

In this paper an approach for using flexure hinges in a macro parallel robot, which offers a larger workspace than typical compliant mechanism is presented. Conventional joints were replaced with flexure hinges made of shape memory alloy. Due to large reversible strains of the used material, maximal angular deflections of the used hinges of ± 30° are possible. This high mobility can be achieved with high kinematic accuracy and a relative small design space at the same time. This offers the possibility to design robots with high accuracy and resolution with sufficiently large workspaces for typical micro assembly tasks.

2. FLEXURE HINGES

A wide variety of different designs of flexure hinges are proposed in the literature.[8,9,10] They are designed monolithic or hybrid allowing for up to 3 degrees of freedom (DOF). In most cases it is tried to achieve high angular deflections with small occurring elastic strains. Using pseudo-elastic SMA for flexure hinges offers the possibility to design hinges with large angular deflections and small kinematic deviations at the same time. For the flexure hinges in this robot a pseudo-elastic CuAlNiFe single crystal SMA is used. This SMA has a superior machinability than NiTi-SMA and offers reversible strains up to 17%. Fig. 1(a) shows the well known flexure beam and flexure notch hinge. Generally, the notch hinge permits smaller angular deflections than the flexure beam hinge but it has an advanced kinematic behavior and is more insusceptible against unintentional deformations and buckling.

Kinematic deviations are caused by the instantaneous centre of rotation which has no fixed position relative to the rigid links but moves during deflection.[11,12] The deviation Δr_{fh} is calculated by the difference of the coordinates of point E (Fig. 1) lying on the deflection curve of the hinges and the ideal joint. The deviation depends mainly on geometry, geometrical dimensions and deflection of the hinge. The larger the part of the hinge where the deformation takes place the larger the deviation[13]. For the compliant robot, pseudo-elastic flexure notch hinges with R = 15 mm and h = 0.15 mm are used (l = 27mm). These geometrical dimensions are an optimum between small kinematic deviations and small occurring strain rates. The resulting deviations are 67 μm at 30° deflection. This flexure notch hinge is comparable to a flexure beam hinge with l_h = 2 mm, though easier to manufacture.

The movement of the instantaneous centre of rotation is also reflected in the kinematic behavior of the robot with integrated flexure hinges (compliant mechanism). The compliant mechanism was simulated by means of FEA. For certain positions of the tool centre point (TCP) the displacements of the drives have been computed solving the inverse kinematic problem (IKP) of the rigid body model. These displacement values where used as constraints for the nodes representing the drives in the FE-model. The difference Δr_{CM} between the analytically computed TCP and the TCP of the compliant model can be determined within the whole workspace with Eq. (1).

$$\Delta r_{CM} = \sqrt{\left(x_{TCP,RBM} - x_{TCP,CM}\right)^2 + \left(y_{TCP,RBM} - y_{TCP,CM}\right)^2 + \left(z_{TCP,RBM} - z_{TCP,CM}\right)^2} \quad (1)$$

Starting from the undeformed initial position of the compliant mechanism the deviations are increasing with increasing displacement of the TCP (Fig. 2). Reasons are the larger deflections of the hinges and the accompanying increasing deviations compared to ideal joints.

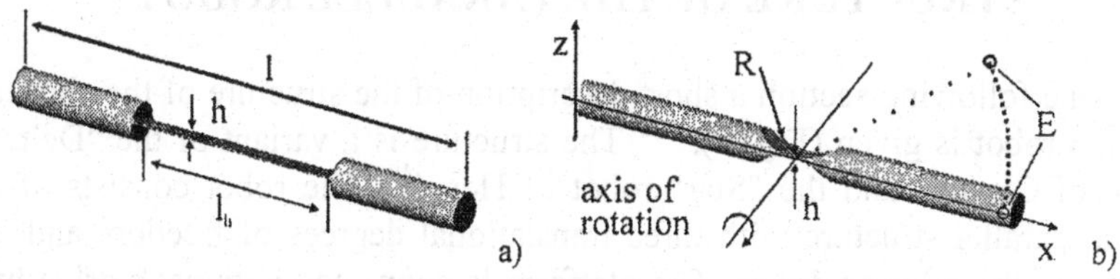

Figure 1. Flexure beam (a) and notch (b) hinge with 1 DOF

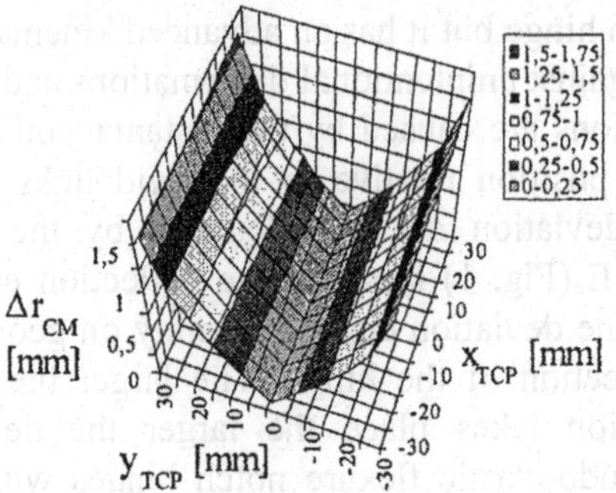

Figure 2. Kinematic deviation of the compliant mechanism (CM) and the rigid body model (RBM) with ideal rotational joints (z_{TCP} = constant)

For example a planar movement in an area of 60 x 60 mm² leads to maximal deviation of about 1.5 mm and if the movement is 60 mm in z-direction the deviation is only about 0.05 mm. The simulations indicate that flexure hinges principally reduce the absolute positioning accuracy. The error is induced when describing the compliant mechanism by means of a rigid body model, neglecting the different kinematic behavior of flexures hinges compared to ideal joints. In consideration of the high number of flexure hinges the TCP deviation is relatively small but not categorically negligible. There exist approaches to increase the absolute accuracy of a compliant robot, e.g. by adding additional actuators[7] or by varying the link lengths in the model depending on the deflection of the hinges[14]. To succeed with these approaches a very accurate calibration of the robot is required. Depending on the application it has to be decided whether the effort to compensate for the deviation is necessary. For many assembly tasks a high resolution and repeatability should be sufficient[15], which are not influenced by these kinematic deviations.

3. STRUCTURE OF THE PARALLEL ROBOT

In the following section a short description of the structure of the chosen parallel robot is given (Fig. 3).[16,17] The structure is a variant of the "Delta" robot of Clavel[18] and the "Star" robot of Hervé[19]. The robot consists of a spatial parallel structure with three translational degrees of freedom and is driven by three linear drives. The platform is connected with each drive by two links forming a parallelogram and allowing only translational movements of the platform and keeping the platform parallel to the base plane. The three drives of the structure are arranged in the base plane at intervals of 120° star-shaped. Thus the structure has a workspace which is nearly round or triangle-shaped. By restricting the angular deflection of the flexure hinges the workspace is reduced and also changes slightly its shape. With a design angle of $\alpha_0 = 34°$ the workspace of the structure is optimized with respect to dimension and shape (Fig. 4). The angle α_0 is the angle between the planes of the parallelograms and the working platform in the initial position of the structure with no deflections of the hinges (Fig. 3).

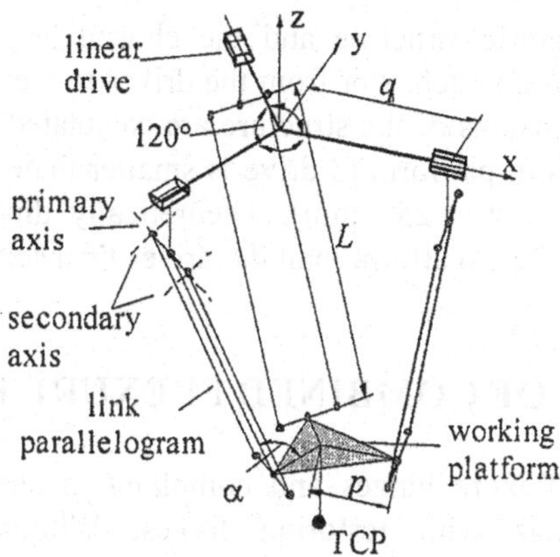

Figure 3. Structure of the parallel robot (Triglide[5])

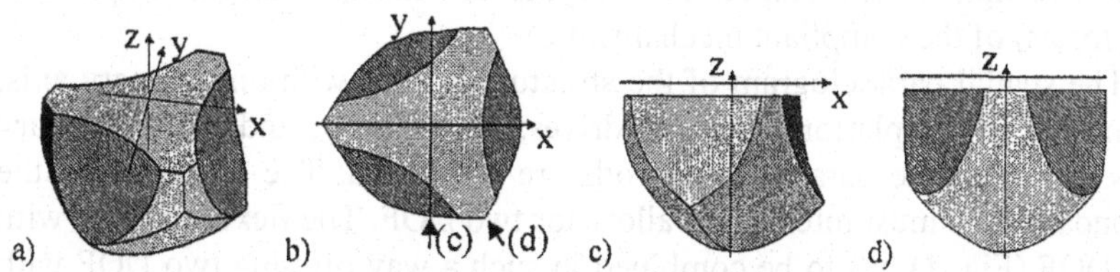

Figure 4. Optimized workspace of the compliant robot: a) spatial view, b) top view

Table 1 gives the results of this workspace optimization. The displacement of the drives is restricted to $\Delta q = 315$ mm and the geometrical dimensions are chosen to L = 350 mm and p = 45 mm. The restriction of the hinges deflections of ± 30° reduces the workspace only minimal compared to the workspace with no angular restriction. The workspace reaches its maximum in the plane with z = -195 mm. The shape is roughly a circle with a diameter of 300 mm. Using flexure hinges with conventional spring steel (maximal elastic strains of 0.5%) the angular deflection is restricted to ± 5°. In this case the resulting workspace of the structure is hundred times smaller compared to the workspace using pseudo-elastic flexure hinges. A restriction of the hinges deflection to ± 20° decreases the workspace by the factor two.

Table 1. Workspace dimensions of the structure

Deflection	x min, max [mm]	y min, max [mm]	z min, max [mm]	workspace
± 180°	-218, 176	-206, 206	-324, 0	1
± 30°	-202, 176	-174, 174	-314, -26	0.88
± 20°	-138, 138	-118, 118	-282, -84	0.45
± 5°	-34, 34	-30, 30	-220, -170	0.008

Due to the symmetric structure and the chosen length of the legs a homogeneous transmission behavior from the drives to the working platform is obtained. The dimensions of the structure are calculated in a way that the transmission ratio t from platform to drive is smaller than 1 within the used workspace (190 mm < z < 250 mm). Theoretically this enables a more accurate positioning of the platform than the drives themselves.[17,20]

4. DESIGN OF COMBINED FLEXURE HINGES

One problem of flexure hinges and compliant mechanism is that the hinges act as springs with restoring forces. Without any damping component, except for the inner material damping, the compliant mechanism normally tends to vibrate due to the compliance. In various FE-simulations, different geometrical arrangements and combinations of flexure hinges have been investigated with respect to the dynamic behavior (natural frequencies and forms) of the compliant mechanism.

The spatial parallelogram of the structure is built with one primary axis, connected to the platform and the drives, respectively, and two secondary axes, at which the parallelogram rods are connected. The primary and the secondary axes must intersect to allow for two DOF. The flexure hinges with one DOF (Fig. 1) has to be combined in such a way offering two DOF with intersecting rotational axes (Fig. 5). This can be done either by arranging the flexure hinges of the primary axis between the flexure hinges of the secondary axes or on the outside of the secondary axes.

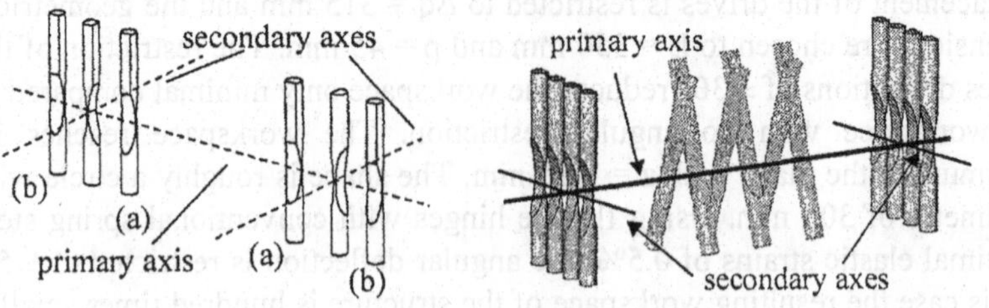

Figure 5. Rotational axes of the flexure hinges and axes of the spatial parallelogram of the structure (left); final design of combined flexure hinges (right)

Fig. 5(left) (a) shows the design variant with the primary axis inside of the secondary axes (i-pa design) and Fig. 5(left) (b) the design variant with the primary axis outside of the secondary axes (o-pa design). With additional spectrum analyses and transient analyses the frequencies of the structures could be determined which have the largest influence. Though the o-pa design seems to be a bit stiffer, the design variant with inner primary axis was chosen for the final robot design due to the possibility of a more compact design. Fig. 5(right) shows the combination of single flexure hinges. Due to parallel and angular arrangement of the hinges, the torsional moments can be better absorbed and transformed into tension and compression forces.

5. COMPLIANT PARALLEL ROBOT

The final design of the compliant robot consists of 84 discrete flexure hinges and is built with CFK rods to minimize the moved mass which is 1120 g (Fig. 6). The robot is driven by linear direct drives with an incremental measuring system with a resolution of 0.125 micrometer. An additional rotational axis can be mounted on the working platform to adjust the orientation of the end-effector. The footprint of the robot is ~1280 x 980 mm^2 and with the actual configuration, a cube with a dimension of 112 x 112 x 112 mm^3 fits into the workspace. Due to symmetrical arrangement of the structure with respect to the support of the working platform and optimized combinations of flexure hinges a quite rigid structure was built and occurring vibrations are damped quite fast. Static FE-simulations of the robot in its initial position leads to a stiffness coefficient of 530 N/mm in z- direction, which is the main joining direction. In the xy-plane the stiffness coefficient is smaller (~200 N/mm), but also more than sufficiently for micro assembly tasks.

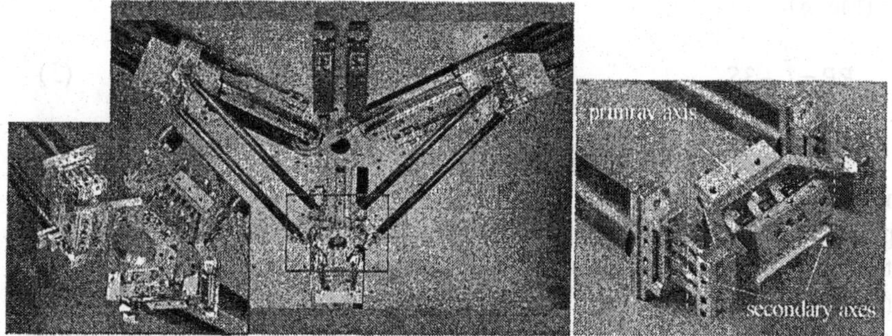

Figure 6. Compliant robot with 3 DOF and optimized design of a flexure hinge with 2 DOF

A modal analysis of the final design leads to natural frequency of $f_1 = 68.7$ Hz, $f_2 = 70.74$ Hz und $f_3 = 120.7$ Hz. The simulations were experimentally validated with deviations less than 8% ($f_{1,ex} = 65$ Hz, $f_{2,ex} = 68$ Hz und $f_{3,ex} = 112$ Hz). The measurement of the second and third natural frequency is shown in Fig. 7. The first frequency could not be measured simultaneously. With measurements against the frame of the robot the first peaks with frequencies below 20 Hz could be identified as vibrations of the frame and basement. These are no vibrations of the structure itself.

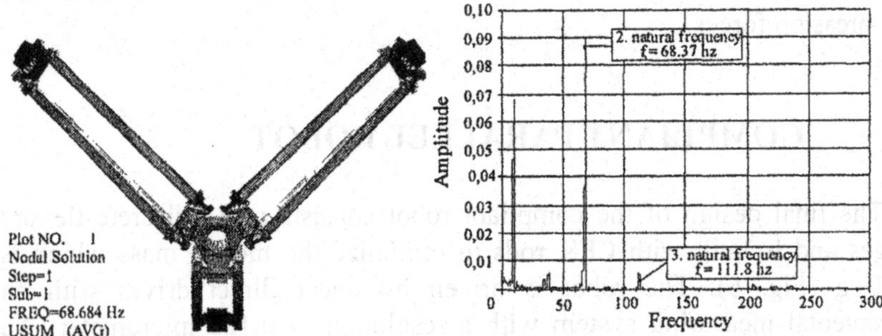

Figure 7. FEM simulation of the compliant mechanism and experimental validation: Plot of the first mode with an natural frequency of 68 Hz (left) and measured natural frequencies

The performance of the robot has been investigated by measuring the repeatability and the variance of multiple direction position accuracy. Capacitive sensors with a resolution of 8 nm have been used and the measures and calculations have been carried out according to EN ISO 9283. The repeatability (RP) is computed according to equation 3 where l denotes the mean deviation from the centroid of subsequent actual poses and S_l denotes the standard deviation. Graphically the repeatability can be described as the radius of a sphere in which 99.7% of all subsequent poses lie (Fig. 8).

$$RP = \bar{l} + 3S_l \tag{3}$$

The variance of the multiple direction position accuracy (vAP) expresses the deviation between the different average actual poses which result from approaching the same desired pose from three directions orthogonal to each other. This value is important for micro assembly tasks in which the end-effector has to move to the same position from different directions. Especially at this kind of robot movements, backlash in conventional joints plays a major role and this value vAP is often much worse than the repeatability RP.[21] The experimental investigations with the compliant robot lead to a repeatability below RP = 0.3 µm (best values were RP_{xyz} = 0.19 µm and in the xy-plane RP_{xy} = 0.13 µm) and a variance of multiple direction position accuracy of vAP = 0.15 µm. These are very good results and could be even improved by increasing the resolution of the linear encoders.

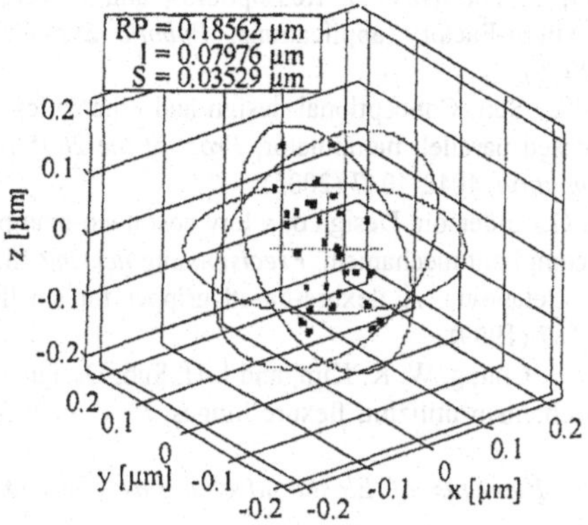

Figure 8. Measured repeatability of the robot

6. SUMMARY

This paper presents the design of a spatial compliant parallel robot with pseudo-elastic flexure hinges. The robot has three degrees of freedom and a workspace larger than 200 x 200 x 60 mm³. The compliant mechanism and the design of the flexure hinges have been optimized with respect to workspace dimension, transmission ratio and stiffness. The large workspace could be achieved due to the pseudo-elastic material of the flexure hinges. The lack of backlash, friction and stick-slip effects within the mechanism lead to a repeatability below 0.3 µm and a resolution better than 0.125 µm, respectively.

REFERENCES

1. A. H. Slocum, Precision Engineering Research Group (07/2004); http://pergatory.mit.edu.
2. S. L. Canfield, J. W. Beard, N. Lobontiu, E. O'Malley, M. Samuelson, and J. Paine, Development of a spatial compliant manipulator, *International Journal of Robotics and Automation* **17**(1), 63-71 (2002).
3. R. Clavel, P. Helmer, T. Niaritsiry, S. Rossopoulos, and I. Verettas, High precision parallel robots for Micro-Factory applications, *Robotic Systems for Handling and Assembly*, 285-296 (2005).
4. W. Dong, Z. Du, and L. Sun, Conceptional design and kinematics modeling of a wide-range flexure hinge-based parallel manipulator, *Proc. of the 2005 IEEE Int. Conference on Robotics and Automation*, 4042-4047 (2005).
5. M. L. Culpepper, and G. Anderson, Design of a low-cost nano-manipulator which utilizes a monolithic, spatial compliant mechanism, *Precision engineering* **28,** 469-482 (2004).
6. M. Goldfarb, and N. Celanovic, A flexure-based gripper for small-scale manipulation, *Robotica,* **17**(2), 181-187 (1999).
7. B.-J. Yi, H.-Y. Na, G. B. Chung, W. K. Kim, and I. H. Suh, Design and experiment of a 3 dof parallel micro-mechanism utilizing flexure hinges, *Proc. of ICRA* **19**(4), 1167-1172 (2003).
8. S.T. Smith, *Flexures: Elements of Elastic Mechanisms* (Gordon & Breach Science Publishers, 2000).
9. N. Lobontiu, *Compliant Mechanisms - Design of Flexure Hinges* (CRC Press, 2003).
10. B. D. Jensen, and L. L. Howell, The modeling of cross-axis flexural pivots, *Mechanism and machine theory* **37,** 461-476 (2001).
11. L. L. Howell, and A. Midha, Parametric deflection approximations for end-loaded, large-deflection beams in compliant mechanisms, *Journal of Mechanical Design* **117**(3), 156-165 (1995).
12. J. Hesselbach, A. Raatz, Pseudo-elastic flexure-hinges in robots for micro assembly, *Proc. of SPIE Microrobotics and Microassembly II*, **4194**, 157-167 (2000).
13 J. Hesselbach, A. Raatz, and H. Kunzmann, Performance of pseudo-elastic flexure hinges in parallel robots for micro-assembly tasks, *Annals of the CIRP* **53**(1), 239-332, (2004).
14. R. Thoben, *Parallelroboter für die automatisierte Mikromontage* (Fortschritt-Berichte VDI Verlag, 1999).
15. P. Schellekens, N. Rosielle, H. Vermeulen, S. Wetzels, and W. Pril, Design for precision: Current status and trends, *Annals of the CIRP* **47**(2), 557-586 (1998).
16. J. Hesselbach, J. Wrege, O. Becker, and S. Dittrich, A micro-assembly-station based on a hybrid 4-dof-robot, *Proc. of the Int. Precision Assembly Seminar*, 55-61 (2003).
17. J. Hesselbach, O. Becker, S. Dittrich, and P. Schlaich, A new hybrid 4-d.o.f.-robot for micro-assembly, *Production Engineering* **IX**(1), 105-108 (2002).
18. R. Clavel, DELTA, a fast robot with parallel geometry, *Proc. of 18th Int. Symposium on Industrial Robot*, 91-100 (1988).
19. J. M. Hervé, and F. Sparacino, Star, a new concept in robotics, *Proc. of the 3rd Int. Workshop on ARK*, 176-183 (1992).
20. A. Raatz, J. Wrege, N. Plitea, and J. Hesselbach, High precision compliant parallel robot, *Production Engineering* **XII**(1), 197-202 (2005).
21. J. Hesselbach, G. Pokar, J. Wrege, and K. Heuer, Some aspects on the assembly of active micro systems, *Production Engineering* **XI**(1), 159-164 (2004).

TEST ENVIRONMENT FOR HIGH-PERFORMANCE PRECISION ASSEMBLY - DEVELOPMENT AND PRELIMINARY TESTS

Timo Prusi, Riku Heikkilä, Jani Uusitalo, Reijo Tuokko

Tampere University of Technology, Institute of Production Engineering, Robotics and Automation Laboratory, P.O. BOX 589, FIN-33101 Tampere, Finland, Tel +358 3 3115 4487, Fax +358 3 3115 2753, timo.prusi@tut.fi, riku.heikkila@tut.fi, jani.uusitalo@tut.fi, and reijo.tuokko@tut.fi

Abstract: This paper presents a test environment enabling the study of factors affecting on the success of a robotic precision assembly work cycle. The developed testing environment measures forces and torques occurring during the assembly, and uses a system based on machine vision to measure the repeatability of work piece positioning. The testing environment is capable of producing exactly known artificial positioning errors in four degrees-of-freedom to simulate errors in work-piece positioning accuracy. The testing environment also measures the total duration of the robot work cycle as well as the durations of all essential phases of the work cycle. The testing environment is best suited for light assembly operations and has measurement ranges of ±36 N and ±0.5 Nm and the vision system has a field-of-view of 6 mm.

The latter part of this paper presents the results of the research done in order to find out how some selected factors affect the assembly forces of robotic assembly. These factors include work piece and process parameters such as work piece material and design (chamfered/straight), positioning tolerances, and robot insertion motion speed.

Key words: Assembly force, assembly process testing, work cycle time, positioning errors

1. INTRODUCTION

Assembly is widely accepted as the most time-consuming part of manufacturing process for industrial goods, and especially in electronics production (Rampersad 1994, Lane & Stranahan 1986, Myrup Andreasen et al 1988).

The productivity of assembly processes and assembly equipment should be maximized in order to provide companies with the best possible return for their investments. The productivity of assembly equipment can be maximized by minimizing assembly work cycle durations and assembly equipment down time. In order to minimize work cycle durations and equipment down time, it is necessary to know what are the causes of faults or errors during assembly – or in other words the factors that affect the success of a precision assembly work cycle. Table 1 summarizes some of these factors that can be divided into four main groups: factors dealing with part, with equipment, with environment, and with the assembly task itself. Factors dealing with the equipment are further divided into three groups of factors dealing with robot, with gripper, and with feeder. Most of them were found in previous researches such as Rampersaad (1994), Linderstram (1995), and Rathmill (1985), but some result from the discussions at our laboratory.

Table 1. Possible factors affecting the success of an assembly work cycle

	Factor
Part	The shape, size, symmetry, geometry and weight (also affects the dynamic behavior of the robot)
	Material; stiffness, vulnerability, elasticity, "slipperiness"
	Design; chamfers and other guiding surfaces, self-alignment, hidden features, catering for grippers, tolerance
	Physical appearance errors such as, dimensional, geometrical, surface, and ruts
	Other; (surface)quality, temperature
Robot	Repeatability, accuracy
	Dynamic behavior; speed, acceleration/deceleration (affected by gripper and part weight)
	Design; lifting capacity, stiffness, compliance
Gripper	Weight (affects the dynamic behavior of the robot)
	Type, accuracy, gripping force, actuation
Feeder	Accuracy, repeatability, reliability
Environment	Contamination, humidity (small particles may jam the peg in the hole)
	(Electrical) interference
	Temperature
Task	Assembly direction, manner of approach, stability of the (base)part
	Type of assembly, cycle time, needed positional accuracy, number of different components

	Factor
	Collisions, impact while extracting part at pick-up, impact when starting insertion
Other	Active control; force and/or vision
	Remote center compliance units (RCC-unit)

2. TEST ENVIRONMENT

In order to study the effects of various factors on the success of a precision assembly work cycle, a novel test environment was developed. The developed test environment is portable, compact in size, and it was designed to be easily applicable to practical cases without major modifications to assembly equipment. It can measure work piece positioning repeatability, forces and torques acting. The forces and moments acting during the insertion phase of the work cycle are measured with a 6 degree-of-freedom force and torque sensor having a measurement range of ±36 N and ±0.5 Nm and resolution of 2.0*10^-3 N and 2.5*10^-5 Nm (Ati 2005). The F/T-sensor is located directly under the assembly location. Both the F/T-sensor and assembly location are assembled on top of precision stages enabling XY movements and rotation and tilt adjustments of the assembly location thus enabling accurately know positioning errors to be made to the assembly location during the insertion of the work piece. Figure 1 shows the test environment and a small scara-type robot (Mitsubishi RP-1H) used in the tests.

In Figure 1 the robot is in the starting position directly above the assembly location. From there, the robot moves above the pick-up location on the left in Figure 1, moves down, grasps the part, moves first up and then back to the starting position. From there starts the actual downward insertion movement during which the forces and torques are measured and recorded. After inserting the part, the robot opens the gripper and moves up back to the starting position. Finally, the work piece positioning repeatability is measured with a machine vision based system.

Figure 2 shows the system used to measure work piece positioning repeatability. It consists of a standard machine vision sensor (Cognex In-Sight 2000), a partially telecentric lens (Edmund Optics), three prisms, and a purpose-design stand. In the system, a right-angle prism (prism 1) divides the camera image vertically into two equally sized parts. The left-hand side of the image turns 90° to left in the image-dividing prism 1, and the right-hand side continues directly. Two more right-angle prisms bend the two

separate optical paths towards the center of the testing environment and the work object.

This set-up enables measuring work piece positioning repeatability in five degrees-of-freedom: XYZ movements and rotations about X- and Y-axes. The horizontal field-of-view of this system is approximately 6 mm in total thus giving field-of-view of 3 mm per half-image. The measurements with this set-up have repeatability of less than 10 μm. (Prusi 2003)

Figure 1. The test environment and a small scara-type robot.

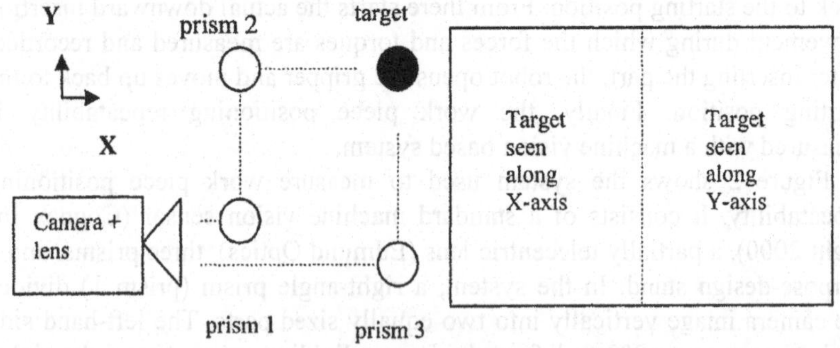

Figure 2. The principle arrangement of the system used to measure work piece positioning repeatability and a resulting image.

Work cycle duration is measured on PC. In addition to the total work cycle duration, the test environment also measures and records the durations of selected phases of the work cycle. These phases are: 1) from the start of the robot work cycle to the moment robot has picked the part, 2) from the start to the moment the actual work piece insertion starts, and 3) the duration of the actual insertion. The moment when the robot has picked the part is detected with optical sensors assembled around the pick-up location. The start and the end of the insertion phase are detected from the F/T-measurements, namely the first moment when any of the six F/T-components exceeds a specified threshold and the moment when all of the six F/T-components are again below the threshold.

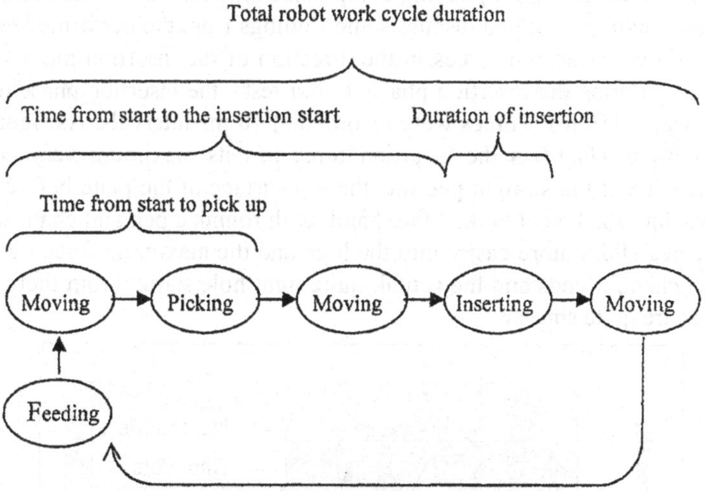

Total robot work cycle duration

Time from start to the insertion start

Duration of insertion

Time from start to pick up

Moving → Picking → Moving → Inserting → Moving

Feeding

Figure 3. Phases of a typical robot assembly work cycle. Figure shows also the phases of the work cycle whose durations are measured.

The operation of the test environment is controlled with a purpose made software running on standard office-PC and Windows 2000. The software not only records the measured data but it also acts as an interface for the operator. With the software, the operator sets up the test parameters, monitors the measurements, and also analyses the measurement data. The measured data is saved in text-format and is therefore easy to import to other applications such as Microsoft Excel for more detailed analysis.

A more detailed description of the testing environment can be found in (Prusi, 2003).

3. TESTS

In order to validate the operation of the testing environment and to study the effects of some work piece and process parameters, a series of tests were run where the robot performed a simple peg-in-a-hole assembly operation. The peg-in-a-hole task is a rather artificial case but it was used because of its simplicity. However, the environment itself does not limit the product to be assembled and studied. From the performed test runs, we can study the effects of the following factors:

- The design of the hole (chamfered / straight).
- The material of the plate where the hole is (aluminum / plastic).
- The design of the steel peg (rounded / straight).
- The effect of the use of an RCC-unit (remote center compliance) from CCMOP (CCMOP 2005) having compliance only in X and Y directions.

Next we will present and discuss some findings from the performed tests. Figure 4 shows measured forces in the direction of the insertion movement (negative Z) during the insertion phase. In our tests, the insertion phase took approximately 150 ms. Forces were recorded at 10 ms intervals. As figure 4 shows, without chamfers the insertion force has its maximum very early when the edge of the straight peg hits the top surface of the plate before the peg slides into the hole. On the other hand, with rounded peg and chamfered hole, the peg slides more easily into the hole and the maximum force occurs when the chamfer ends and the actual, quite tight hole starts. From there on, the forces are quite similar.

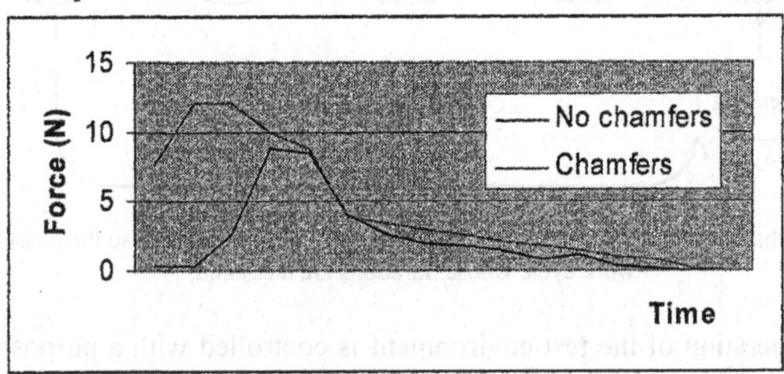

Figure 4. Typical insertion forces with and without hole and peg chamfers.

Figure 5 shows the effect of the rounding in the peg. The graphs show maximum forces measured with varying positioning error of the hole. The upper graph shows the forces when both the hole and the peg were straight whereas in the lower the peg was rounded. With straight peg and large positioning errors, the edge of the peg collides with the edge of the hole and does not slide into the hole. Therefore, only 0.20 mm error could be measured. With the rounded peg, on the other hand, the rounding guides the peg into the hole and the deflection of the robot Z-axis allows even 1 mm positioning error.

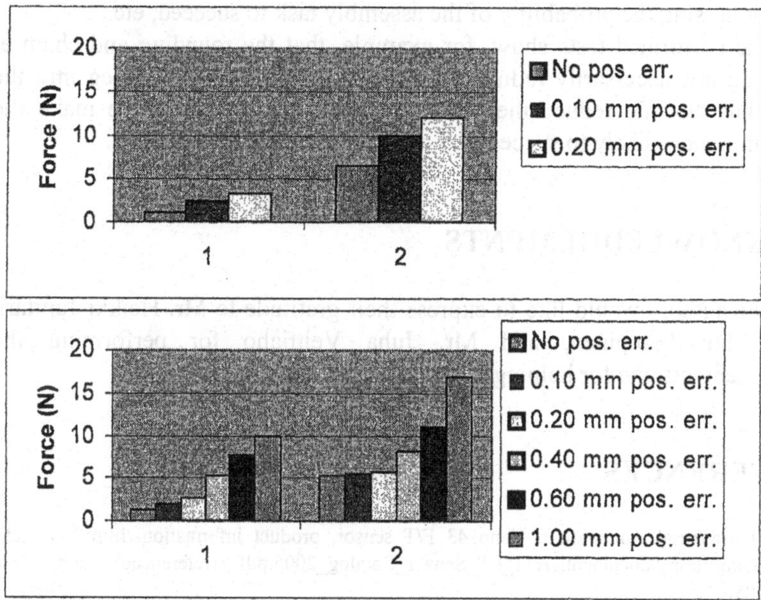

Figure 5. Average maximum forces of tests run with varying positioning error in positive X-direction. In the upper graph, both the peg and the hole were straight whereas in the lover the peg was rounded and the hole straight. The leftmost cluster shows forces in negative X-direction and the rightmost in negative Z-direction (the direction of the insertion movement).

In addition to forces, the test environment also measures the durations of selected phases of the work cycle (fig. 3). These measurements can be used to maximize the performance of the assembly equipment by minimizing the work cycle duration but still confirming that the forces do not exceed safe values. For example, we did some tests with different robot insertion motion speeds and found that in that case, 30% of maximum speed was the highest safe speed and with higher speeds, the initial impact forces were too high.

4. CONCLUSIONS

In this paper, we have presented a test environment for high-performance precision assembly. The test environment has proven to be a suitable tool for measuring and analyzing a typical robotic pick-and-place work cycle. It can be used to study the effects of various work piece and assembly process parameters. It could also be used to confirm the operation of real assembly equipment: are the assembly forces within safe limits, is the equipment as productive as possible or could some movements be faster without compromising the probability of the assembly task to succeed, etc.

The performed tests show, for example, that the rounding and chamfer used do not necessarily reduce the force needed to insert the peg into the hole but they do reduce the initial contact force and therefore make the insertion more likely to succeed.

ACKNOWLEDGEMENTS

The authors would like to express their gratitude to Mr. Heikki Junttari, Mr. Riku Lampinen, and Mr. Juha Veittiaho for performing the measurements and for helping with the analysis of the results.

REFERENCES

1. Ati Industrial Automation, Nano 43 F/T sensor, product information. http://www.ati-ia.com/library/documents/ATI_FT_Sensor_Catalog_2005.pdf (referenced June 29th, 2005)
2. CCMOP, product information of a remote compliance center unit CH50, http://www.ccmop.com/ → Products → Robot Peripheral Tools (referenced June 16th, 2005).
3. Edmund Optics, product information of a partially telecentric video lens, http://www.edmundoptics.com/onlinecatalog/displayproduct.cfm?productID=1963&search=1 (referenced June 27th, 2005)
4. Lane, Jack D.; Stranahan, Judy D. (editors). 1986. Automated Assembly. Dearborn, USA, Society of Manufacturing Engineers. 452 pages.
5. Linderstram, Charlotta. 1995. On Monitoring Interactions for Error Detection in Robotic Assembly, PhD Thesis. Linköping, Linköping University, Department of Mechanical Engineering. 108 pages.
6. Myrup Andreasen, Mogens; Kähler, Steen; Lund, Thomas. 1988. Desing for Assembly, 2nd Edition. Bedford, UK, IFS Publications. 212 pages.
7. Prusi, Timo 2003. Test environment for high-performance precision assembly. Master of Science Thesis. Tampere: Tampere University of Technology, Automation Department. 88 pages.
8. Rampersad, Hubert K. 1994. Integrated and Simultaneous Design for Robotic Assembly. West Sussex, England, John Wiley & Sons. 212 pages.
9. Rathmill, Keith (editor). 1985. Robotic assembly. Bedford, UK, IFS Publications. 365 pages.

SENSOR GUIDED MICRO ASSEMBLY BY USING LASER-SCANNING TECHNOLOGY

Sven Rathmann, Jan Wrege, Kerstin Schöttler, Annika Raatz and Jürgen Hesselbach

Institute of Machine Tools and Production Technology (IWF), Technical University Braunschweig, Langer Kamp 19b, 38106 Braunschweig, Germany

Abstract: To recognize geometric objects of components in assembly processes, in particular of microelectronic or micro system components, nowadays the use of vision systems is preferred. These systems are working very fast but the results depend on ambient conditions, especially of light settings. They deliver 2D object information referring to the image plane of a camera. In most sensor guided assembly systems, additional to vision systems laser displacement sensors are implemented to get information about the third dimension, the components height. In this paper a scanning method is presented which enables object recognition by using a laser positioning sensor and to use the achieved measuring values for a sensor guided assembly process.

Key words: precision assembly; laser scanning; in process.

1. INTRODUCTION

The used laser-scanning method is a technology to scan surfaces or bodies by using a laser beam. In most cases the laser is deflected by a mirror unit. Alternatively the laser or rather the object is moved. Three different methods are known in laser-scanning technology: confocal laser-scanning, 3D-laser-scanning technology as well as laser altimetry. The most interesting technologies for scanning micro components are the confocal laser-scanning and the 3D-laser-scanning. With these technologies it is possible to obtain exact models with sub micrometer accuracy of investigated micro components. A disadvantage of the named methods is that it is not possible to scan the object during production process. However, in high precision assembly processes it is necessary to get exact in-process information about these micro components, for example their position and geometry and the deviation of geometry. For this purpose vision systems can be used. The disadvantage of them is the dependence of light settings. Especially in case of complex geometrical structures with different surface reflection properties, which often occur in micro systems, wrong geometric properties can be detected. For this reason the laser-scanning technology is used to get the desired information about position and deviation of micro components in an assembly process. In high precision assembly systems laser displacement sensors are often implemented. So it is possible to extend the existing assembly systems with a technology in following called In-Process Laser-Scanning (IPLS).

2. CONFIGURATION AND METHODS

IPLS uses a laser, which is mounted on a high accurate micro assembly robot in z-direction of the robot coordinate system (Fig. 1). The sensor is usually used to measure the height of components. In this case it will be used for the scanning process by moving the robot over the component. During IPLS the heights' information of the laser sensor is related to the robot position in x- and y-direction. The accuracy of the scanning process is mainly addicted to the diameter of the laser beam. For experiments, a confocal laser displacement sensor (Keyence, LT-9001) was used. It has a laser beam diameter of 2 µm with a measurement range in z-direction of ± 0.3 mm. Additionally to the measurement of the height via a point, the LT-9001 offers a line scan option. The integrated horizontal oscillating system deflects the laser beam ± 550 µm in x-direction to get height information over the deflected range by an increment size down to 2 µm.

Figure 1. Assembly system with mounted laser displacement sensor (LT-9001)

In-Process Laser-Scanning (IPLS)				
Detection method	edge-point detection		surface-point detection	
Scan method	raster scan	line scan	raster scan	line scan
Data analysis	2 D object recognition	2.5 D object recognition	3 D object recognition	
Application	in process: > recognition of geometric properties in the x-y plane > height calculation of compensation planes for plane surfaces	in process: > recognition of geometric properties in planes with different height ranges	system setup: > calibration of robot / measuring systems process setup: > calibration of process devices in process: > spatial recognition of geometric properties > error diagnosis	

Figure 2. Overview of In-Process Laser-Scanning options

The laser is mounted on a micro assembly robot AUTOPLACE 411 (Sysmelec). The robot offers a repeatability in x-, y- and z-direction of 1 µm[1]. The smallest step size is 0.5 µm. The laser is connected with the robot control via analog interface. The setup of the high accurate robot and the confocal laser displacement sensor offer a range of scanning options. Fig. 2 shows an overview of those different scanning options.

The IPLS is classified in two detection methods. First the simple edge-point detection and second the detection of a surface-point of a component. With the edge-point detection and the two scan methods, raster scan and line scan, it is possible to get information about the components' position and their deviation by using a 2 D or 2.5 D data analysis for the object recognition. The raster scan method works with the point laser of the LT-9001 and the movement of the robot in x- and y-direction. The raster is defined either by a step size in x-direction or y-direction or both (Fig. 3a). The line scan method uses the line scan option of the LT-9001. The oscillating laser gives information in z_l- and x_l-direction while the robot moves only in y-direction (Fig. 3b).

By using the raster scan method an edge-point is detected if the laser height signal is changing significantly during moving over the edge. The data analysis of several edge-points allows the reconstruction of geometric properties of the component in the two dimensions x_r and y_r related to the robot coordinate system.

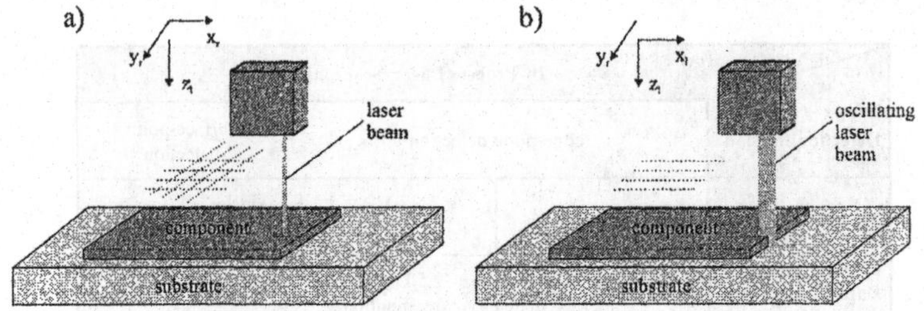

Figure 3. Illustration of the scanning process (a) point scan, (b) line scan

The line scan method allows an exact description of the edge geometry. With this method it is possible to detect edges in spite of disruptions or blurred transition, for instance by evaluating a chosen feature of the edge geometry. This provides an improved stability of the object recognition. In contrast to the raster scan method, the analysis of the line scan information is addicted to a high computation power because of complex mathematical calculations like data projection and coordinate transformation. A reconstruction of the components' geometry can be done using the dimensions x_r and y_r additionally with the filtered z_l-dimension (2.5 D).

An enhancement of the edge-point detection is the surface-point detection. This detection works with the coordinates of the robot, the height information of the sensor system and the analysis of this information to get 3D component information. The surface-point detection works also with the raster scan or the line scan method. Using the raster scan method, the robot moves with defined increments in x_r- and y_r-direction and detects the height information z_l of the component with the laser sensor. The 3D object recognition reconstructs the desired information, for example component position, edge geometry and surface planarity, from the measured scatter plot.

In contrast to the raster scan method, the surface-point detection in combination with the line scan method works much faster. The movement of the robot is restricted to y_r-direction. Because of the small scanning range of the oscillated laser sensor (± 550 μm) this method is only appropriate for surveying of micro components. Using the relation between the x_l and z_l laser information and the x_r and y_r robot information, a reconstruction of the 3D component information related to the robot coordinate system is possible.

3. APPLICATION

In Fig. 2 an overview of the possible applications with the different detection and scan methods is shown.

Edge-point detection with raster scan and line scan is especially used for the detection and recognition of geometric parameters of micro components. With different moving strategies of the robot and analyzing tools recognition, for example of rectangular or circular geometries is possible. In the following chapter an example for the recognition of rectangular geometry is described. Also raster scan can be used for measurements of the height of a component on different positions. The result from this information can be used to adjust the component plane to the working plane. Because of the easy integration into existing micro assembly systems and the simple and fast recognition of object parameters, this detection method is especially qualified for sensor guided assembly.

Surface-point detection can be used for a detection of 3D information of micro components. An advantage of the raster scan method is its possibility to adjust the increment size in x_r- and y_r-direction in a wide range. This allows a scanning over a larger area, for example to measure the waviness of component surfaces. This scan method can be use for error diagnosis in assembly systems as well as for an enhanced detection of geometric properties especially in plane uses.

The small increment size of the laser line option down to 2 µm can be used for a high accurate measurement of micro components. Because of the oscillating principle of the laser optic, the scanning process works without any edge shadowing. For this reason objects with high edge steepness can be detected. An application area can be the calibration of robot axes or implemented devices, like transport devices. Also the setup of new assembly processes or the error diagnosis is possible. Furthermore, the set of 3D geometric information allows an enhanced object and geometric recognition especially in different planes. Chapter 3.2 describes an exemplary 3D measurement of a micro component.

3.1 Edge-Point Detection with Raster Scan

The prior aim of IPLS is to detect the exact position of component edges in the robot coordinate system. Therefore a routine for the AUTOPLACE 411 was programmed. It comprises the following steps:

1. Move close to the edge
2. Move in increments in direction to the edge
3. Read laser signal at the reached position
4. Break movement, if laser signal changes significantly (edge is reached)
5. Store actual robot position

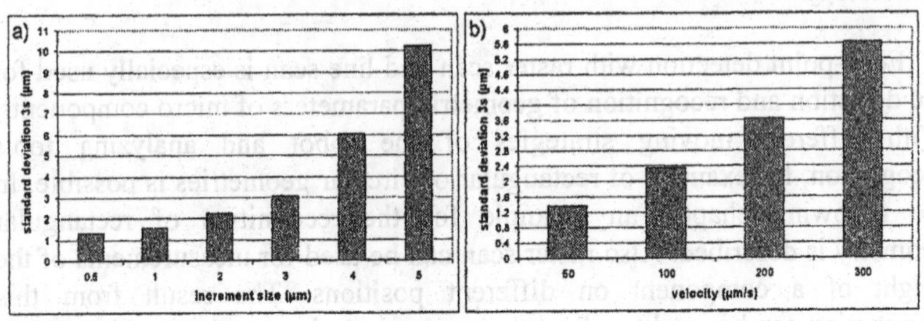

Figure 4. Standard deviation for (a) different increment sizes, (b) different velocities

The reference of the laser to the robot coordinate system was made by a calibration via vision system, which was implemented into the robot's head and was calibrated by manufacture. The vision system has a resolution of 2 µm. After scanning different edge-points at the component it is possible to calculate straight lines, intersections, diameter and angels in the assembly plane.

To validate the scanning method a test component, as shown in Fig. 3, was used. This hybrid component consists of a ceramic substrate and a semiconductor chip with a height of merely 100 µm. The semiconductor chip is smaller than the ceramic substrate. Thus the light contrast between chip and substrate is not adequate to detect edges with a standard vision system.

To get the accuracy of the scanning system, measurements were carried out referring to ISO/TS 14253-2[2]. In Fig. 4a the accuracy with variation of the increment size during the movement to the edge is displayed. Additionally to the laser beam diameter, the accuracy of the scanning process is addicted to the increment size as shown in the figure. This can be traced back to the strict separation between movement and measurement.

The time to detect one point of the edge is addicted on the distance between starting point and edge. It averages 15 seconds per edge-point with an increment size of 2 µm and a distance to the edge of about 300 µm.

A variation to the raster scan method is to move with a constant low speed from the starting point to edge. In Fig. 4b the accuracy of the raster scan method with different velocities is shown. With this variation the accuracy could be increased.

The reason for that is a continuous checking of the sensor signal during movement. So the accuracy is not addicted to the increment size but rather to the addiction of the movement's velocity to the runtime of the robot control. The required time to reach the edge-point using continuous movements is 1.5 seconds with a velocity of 200 µm/s and the same distance to the edge as in the raster scan method.

3.2 Surface-Point Detection with Line Scan

To show the capability of the surface-point detection using the line scan method a micro component was scanned. This micro component is a stator of a horizontal micro electric linear actor developed by the IMT, Braunschweig, within the Collaborative Research Center 516[3]. The actor is working according to the reluctance principle. It mainly consists of a ceramic stator with six coil-systems meandering horizontally around soft magnetic poles. This stator is fixed to a silicon die which has two v-grooves used as ball tracks. The critical assembly task is to insert the passive pole rows of an associated traveler between the active pole rows of the stator. The horizontal gap between the pole-teeth amounts 3 μm over a length of approximately 6 mm. After complete assembly of the actor the gap is adjusted by four slide bars attached to the stator and the traveler. With 40 μm depth of insertion a resulting vertical clearance < 2 μm occurs.

In Fig. 5 the measurement of one and a half coil-system of the stator is shown. The measurement was done with a laser resolution in x_l-direction of 2 μm and a robot step size in y_r-direction of 2.5 μm. An interesting information of those measurements is the edge steepness, which can be resulted with this method up to 92°. Analyzing the scatter plot the edge steepness is in range of the measured values by Seidemann[4].

In Fig. 6 a scatter plot of a layer horizontal through the scatter plot of Fig. 5 in a height of 30 to 45 μm is shown. Analyzing the filtered scatter plot edge recognition is possible.

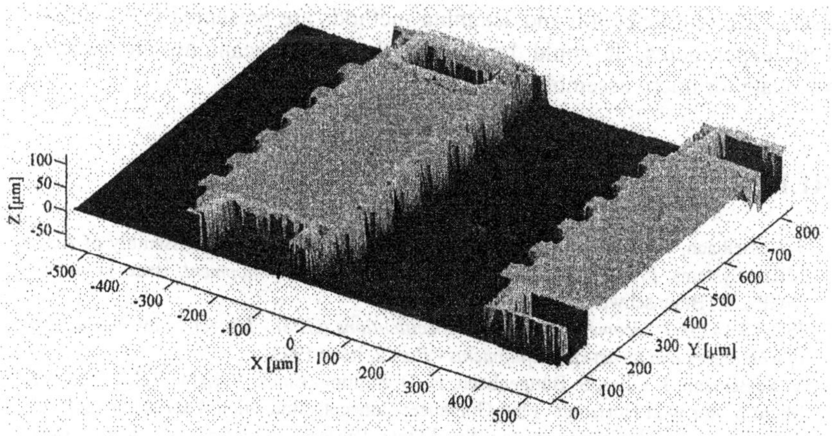

Figure 5. Scatter plot of the micro actuator

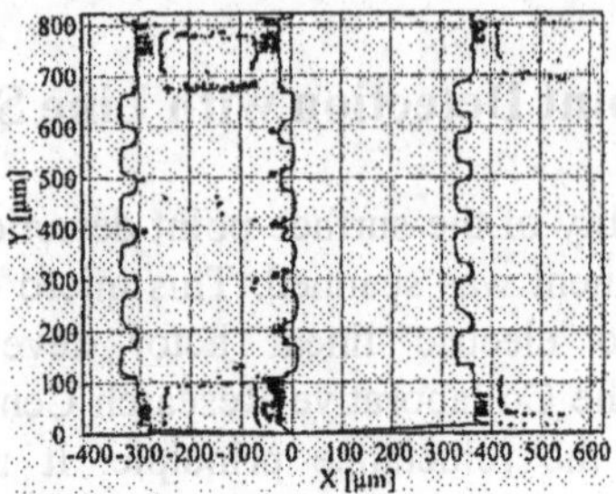

Figure 6. Scatter plot of a horizontal layer of the micro actuator

This information can be used to setup the assembly process, for example to detect starting points for scanning with edge-point detection or to leveling the assembly plane.

4. CONCLUSION

In this paper a laser-scanning technology, called In-Process Laser-Scanning, is presented which can be used in process for measuring micro components. This information can be used for sensor guided assembly. For this method a confocal laser displacement sensor and a high precision robot or tool machine is required. Many micro assembly systems use a laser displacement sensor to get information about the height of components. Therefore the implementation of IPLS is obvious. In this work a laser sensor with a beam diameter of 2 µm and a robot with a repeatability of 1 µm were used. Different detection and scanning methods are presented and validated with experiments. It could be shown that these scanning methods are adequate for micro assembly.

REFERENCES

1. J. Hesselbach, G. Pokar, J. Wrege, K. Heuer, Some Aspects on the Assembly of Active Micro Systems, *Production Engineering. Research and Development*, issue 11, book 1 WGP e.V., Braunschweig, pp. 159-164 (2004).
2. ISO/TS 14253-2, *Geometrical Product Specifications (GPS) - Inspection by measurement of workpieces and measuring equipments - Part 2*, issue 1999.
3. V. Seidemann, J. Edler, S. Büttgenbach, H.-D. Stölting, Linear Variable Reluctance (VR) Micro Motor with Horizontal Flux Guidance: Concept, Simulation, Fabrication and Test, *12th International Conference on Solid-State Sensors, Actuators and Microsystems, Transducers '03*, Boston 2003, pp. 1415-1418.
4. V. Seidemann, *Induktive Mikrosysteme: Technologieentwicklung und Anwendung*, (Shaker Verlag, Aachen, Germany, 2003).

HIGH SPEED AND LOW WEIGHT MICRO ACTUATORS FOR HIGH PRECISION ASSEMBLY APPLICATIONS

Micro Actuators with the Micro Harmonic Drive®

Dr. Reinhard Degen[1] and Dr. Rolf Slatter[2]

[1] *Managing director of Micromotion GmbH, An der Fahrt 13, 55124 Mainz, Germany;*
[2] *Managing director of Micromotion GmbH, An der Fahrt 13, 55124 Mainz, Germany*

Abstract: The trend to miniaturization cannot be overseen. The use of very small electronic and electro-optical components in a variety of consumer and investment goods is leading to an increasing demand for small-scale servo actuators for micro assembly applications in production equipment. The previous generation of micro gears and micro actuators was not suited to this type of application, because of unacceptable accuracy.

The Micro Harmonic Drive® gear was introduced into the market in 2001 as the world's smallest backlash-free micro gear. It is manufactured using a modified LIGA process, called Direct-LIG. This allows the cost-effective production of extremely precise metallic gear components. In the meantime this gear has been implemented in a range of miniaturized servo actuators, which provide zero backlash, excellent repeatability and long operating life.

In addition to the above-mentioned advantages this innovative product also features a central hollow shaft. This allows the design engineer to pass an optical fiber, a laser beam or media such as fluids, compressed air or vacuum along the central axis of the servo actuator. This greatly simplifies the design of machines for micro assembly applications in the semi conductor, consumer goods, medical and optical fields. In this paper we will describe the development history, key features and applications of this innovative drive solution.

Key words: micro actuator; Micro Harmonic Drive®; assembly applications

1. INTRODUCTION

As soon as miniaturised systems and hybrid microsystems need to be manufactured in large series there is a requirement for automated assembly. For small scale products of this type the assembly process is often a major cost-driver, making up to 80% of total production costs [1]. Manual assembly is either too expensive, or does not achieve the required process stability. Automated micro assembly requires, in turn, specialised production equipment for handling miniature components. The assembly process typically requires movements in several degrees of freedom, which are enabled by power transmission components, such as motors, gears, ballscrews etc.

Until recently the physical size of these drive components was much larger than that of both the components to be handled and the necessary workspace, with the result that many machines and robots for micro assembly have dimensions far in excess of the necessary working area [1]. There is now a clear trend to equip physically smaller machines with micro drive systems. These machines have a smaller footprint and often higher assembly accuracy than the previous generation of machine.

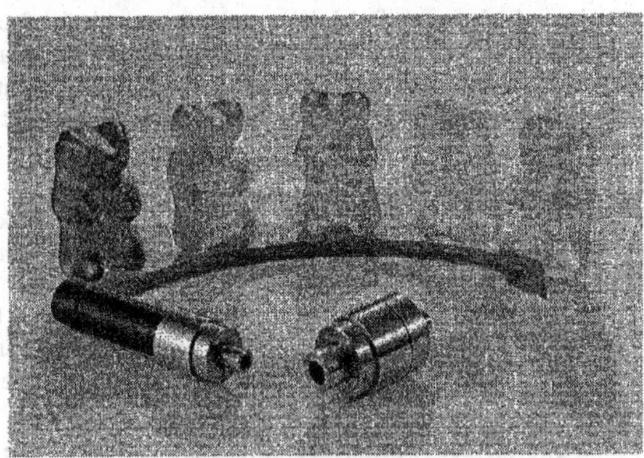

Figure 1. Micro Harmonic Drive gearbox and actuator

Micro gear systems represent a key element in such micro drive systems. Only by using suitable micro gear systems is it possible to apply existing micro motors operating with speeds of up to 100.000 rpm at output torques in the range of some μNm [2] in a wide field of different applications. To access new innovative fields of application in the range of micro drive systems Micromotion GmbH has developed a new generation of high precision and zero backlash micro gear system:
the Micro Harmonic Drive® (see Fig. 1).

2. THE MICRO HARMONIC DRIVE®

Micro-gears are not a particularly recent development and micro-spur gears or micro-planetary gears have been available in the market for a number of years. However, these products suffer from poor positioning accuracy and are therefore rarely used for positioning applications in machines. These previous solutions either have backlash, or only permit very light loads. What is needed are micro-gears that are not only very small in size, but also feature high repeatability, zero backlash, high reduction ratios and a low parts count. These requirements inspired the development of a new micro-gear, the Micro Harmonic Drive® gear [3] (Fig. 1).

This gear was developed by Micromotion GmbH in Mainz, in co-operation with the Institute for Microtechnology, also located in Mainz in Rhineland Palatinate, Germany. The Micro Harmonic Drive® gear is currently the world's smallest zero backlash gear and in combination with a specially developed motor from Maxon Motors, Switzerland, forms part of the world's smallest zero backlash positioning actuator.

The principle of operation is similar to the conventional "macro-technological" Harmonic Drive® gear [3], with the difference that the Wave Generator consists of a planetary gear stage. This enables very large reduction ratios in a small envelope. This is necessary, because most currently available micromotors only produce adequate torque at very high output speeds, typically more than 50.000 rpm, and a high reduction ratio then helps provide sufficient torque at an acceptable speed for practical motion control applications. The planet wheels are hollow and elastically deformable, with the result that backlash can be eliminated by gear pre-loading in the planetary gear stage.

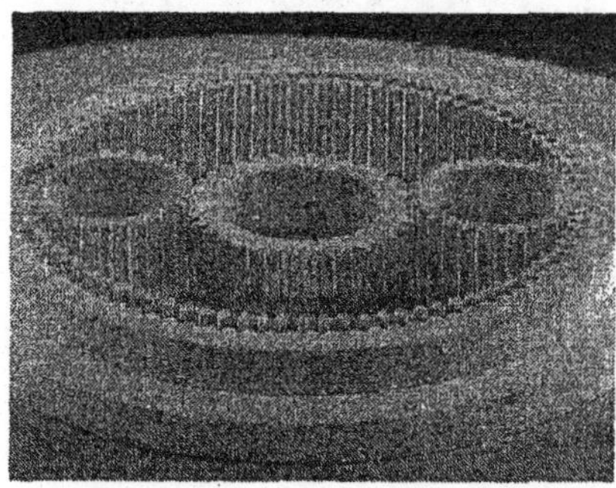

Figure 2. Micro Harmonic Drive® gear component set

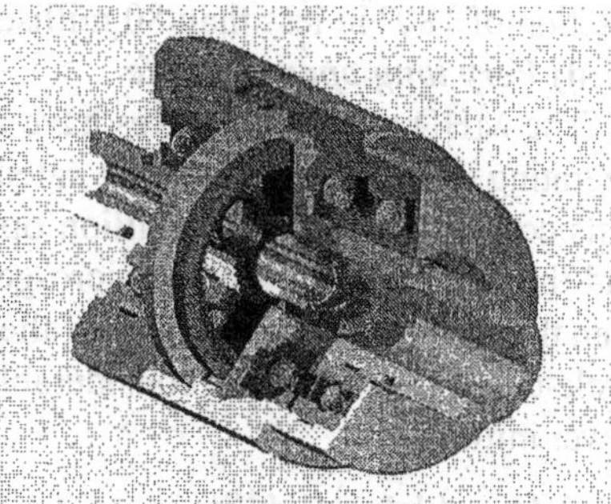

Figure 3. Micro Harmonic Drive® MHD gearbox

The Micro Harmonic Drive® gear component set has an outer diameter of just 6 or 8 mm and an axial length of 1 mm. Fig. 3 shows a REM picture of the component set. It can provide reduction ratios between 160:1 and 1000:1. In order to allow easy integration in a wide range of different applications the component set is mounted inside a micro-gearbox of the MHD series, which is available in two sizes, either with an input shaft or for direct coupling to commonly available micro-motors. [4].

The gear component set is typically mounted inside a gearbox (see Fig. 4) with an output shaft mounted in pre-loaded ball bearings. The gearbox can either be directly coupled to a micro-motor, or can be provided with an input shaft, so that the motor can be mounted off-axis. A hollow shaft with an inner diameter of up to 1 mm passes along the central axis of rotation of the gear box.

This solution provides the machine designer with numerous advantages:

a) ***Miniature dimensions yet zero backlash***

The Harmonic Drive gear stage is backlash-free by nature and the elastically deformable planet wheels eliminate backlash in the planetary stage.

b) ***Excellent repeatability for precise positioning***

The zero backlash of the Micro Harmonic Drive® gear provides a repeatability in the range of a few seconds of arc. This enables positioning tasks to be carried out with sub-µm accuracy.

c) ***High dynamic performance for fast indexing applications***

The high torque capacity and low moment of inertia enable extremely fast accelerations of up to 550 000 rad/s^2 at the input shaft. This corresponds to an acceleration of the motor shaft from 0 to 100 000 rpm in 25 milliseconds. This, in turn, enables extremely fast angular movements e.g. a rotation of 180° in less than 80 milliseconds.

d) ***Very long operating life***

The MHD micro-gearboxes have an operating life of 2500 hours at rated operating conditions, that is, at rated input speed and rated output torque. This corresponds to many million operating cycles in practical applications and the operating life of the micro-gearbox is typically equivalent or longer than the expected operating life of the machine in which it is used. The "life-cycle-costs" are therefore considerably lower than for other solutions with a lower initial cost.

e) ***Very high reliability***

The MHD gearbox has a significantly higher MTBF (Mean Time Between Failure) rating than other microgears. This is mainly the result of the far lower number of parts, compared to other gears. A planetary microgear with a reduction ratio of 1000:1 typically has 25 individual gear wheels, whilst the comparable Micro Harmonic Drive® gear has just 6.

f) High efficiency to avoid power losses

The Micro Harmonic Drive® gear has an efficiency of up to 82% at rated operating conditions. This is also significantly higher than for other micro-gears. The reason lies in the small number of tooth engagement areas. A planetary gear with ratio 1000:1 has 30 regions of tooth engagement, whilst the comparable Micro Harmonic Drive® has just 8.

g) Extremely flat design for compact gearbox dimensions

The axial length of the MHD micro-gearbox is independent of the reduction ratio and is less than half the length of other micro-gearboxes for the same output torque and reduction ratio.

h) Low mass for applications in portable devices or in moving structures

As can be seen from Table 1, the gearboxes weigh just a few grams. In practical applications this means that the moving masses in the machine can be minimised. This, in turn, can contribute to greater thermal stability and lower temperature rise, both of which are essential in high precision machines. Furthermore, this enables higher accelerations and/or smaller feed drives.

i) High reduction ratios for low-loss torque conversion and easy control

The high reduction ratios greatly reduce the load moment of inertia reflected at the motor shaft. The result is that in most practical applications the motor is hardly influenced by the load inertia. In combination with the low input-side moment of inertia of the gear this has the effect that the control of the motor is almost independent of the load inertia over a very large range of load inertias. This makes the control of the motor and setting-up of the control system very easy.

j) Hollow shaft capability

The optional hollow shaft can be used to pass laser beams, air / vacuum supply or optical fibres through the centre of the gear or actuator along the central axis of rotation. This can greatly simplify the design of machines where otherwise the laser beam or fibre would need to be diverted around the actuator.

k) Robust, accurate output bearing arrangement

The high load capacity of the output bearings (preloaded ball bearings in an O-configuration – see Fig. 2) mean that no additional support bearings are needed for the load in most applications. Furthermore, the accurate geometric tolerances (axial and radial run-out less than 5 μm) allow the attachment of load components e.g. mirrors, filters or lenses, directly to the output shaft.

l) ***Applicable under extreme environmental conditions***

The use of high quality materials, such as stainless or high-alloy steels for the gearbox housing, input / output shafts and bearings, provides a high level of corrosion resistance, even for standard MHD micro-gearboxes. The Micro Harmonic Drive® gear, which is manufactured in a high strength Nickel-Iron alloy, can be sterilized and can be used over a very wide temperature range (-70° C - +150° C). It can also be applied in a vacuum [5], using grease, oil or dry lubrication, depending on the specific requirements of the application.

Table 1. Key performance data for MHD gearboxes

Gearbox size		MHD 8		MHD 10		
Reduction ratio		160	500	160	500	1000
Peak torque	[mNm]	14	20	24	36	48
Rated torque	[mNm]	7	10	12	18	24
Repeatability	[arcsec]	10	10	10	10	10
Outer diameter	[mm]	8	8	10	10	10
Weight (with input shaft)	[g]	3.5	3.5	5.7	5.7	5.7

This combination of features makes the Micro Harmonic Drive® gearbox very attractive for precise assembly applications. The high repeatability means that components can be orientated with very high accuracy, while the high dynamic performance means that assembly speed must not be sacrificed.

3. SPECIAL DEVELOPMENTS FOR MICROASSEMBLY APPLICATIONS

In this section two practical examples will be described, where Micro Harmonic Drive® gears are being used successfully in industrial micro assembly applications.

One of the main application areas for micro assembly equipment is in the electronics industry. The production process can be divided into a "front-end" process comprising the lithographic structuring of the silicon wafer and a "back-end" process, starting with the dicing of the wafer into individual chips and ending with the packaging of the electronic components, ready for subsequent final assembly.

So-called "die attach" machines are used in the assembly phase of the "back-end" process. Alphasem AG is one of the world's leading manufacturers of "die attach" machines, which are used to assemble the chips in a protective package and connect the chip to the outside world. To do this the chips, which are today no larger than a piece of dust with dimensions of just 0.25 x 0.25 mm, must be orientated and positioned highly accurately. The new Easyline 8032 machine from Alphasem incorporates a new "Rotary Bond Tool" including a Micro Harmonic Drive® gearbox to realise simultaneously a high accuracy in the range of some milligrad, a short positioning time in the range of some milliseconds and an extreme low weight of about 30 grams. This space and mass optimized unit is used to position the chips with high accuracy at any desired angle of rotation.

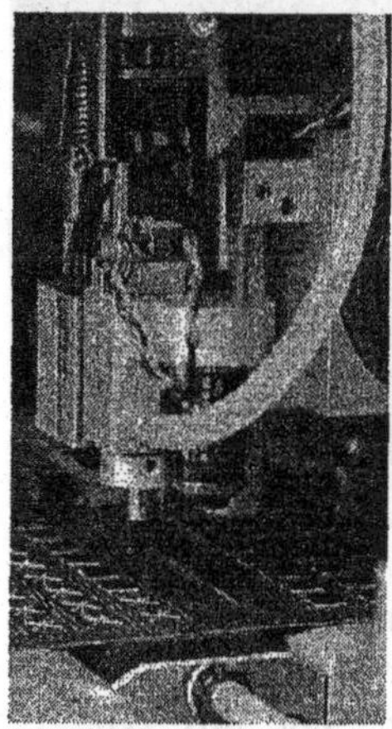

Figure 4. Rotary bond tool with hollow shaft and optical sensor (Alphasem AG)

Figure 5. Rotary bond tool for high speed applications with a total weight of 22 grams and a repeatability of 0.005°

By this customer specific bond tool the motor is mounted off-axis, and so permits a hollow shaft to be passed through the centre of the reduction gear. This hollow shaft is used for a vacuum feedthrough, which is used to hold the chip in place during the positioning and assembly cycle. The hollow shaft also allows the use of an optical sensor, which looks through the centre of the reduction gear and output shaft to check that the chip has been correctly gripped. The figures 5 and 6 illustrate low weight Rotary Bond Tools for high speed applications.

At the heart of the Rotary Bond Tool is a Micro Harmonic Drive® gearbox in a custom-made design. The output shaft is mounted in pre-loaded ball bearings, which ensure that the radial and axial run-out of the output shaft is minimised. The output shaft of the gear box serves simultaneously as fitting for the costumer and process specific tool. Due to the high transmission ratio and the zero backlash of Micro Harmonic Drive® gearbox it is possible to drive these units with simple and robust stepper motors. A section of such a customer specific Rotary Bond Tool from made by Micromotion GmbH is illustrated in Fig. 7. Due to the integration of several functionalities into the gear box it is possible to realise a very low weight and high dynamic system for high accuracy assembly applications.

The complete electromechanical sub-assembly is assembled and tested by Micromotion GmbH and allows chips to be placed with a repeatability of less than 1 µm and high speed. In the field rotary bond tools of this design have achieved more than 30 million cycles without any loss of accuracy.

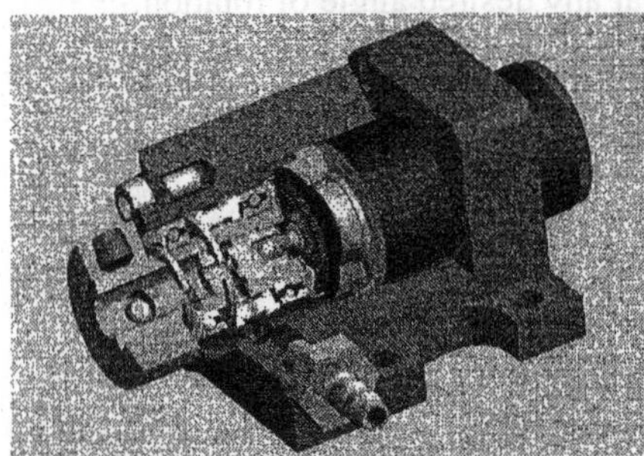

Figure 6. Rotary bond tool section

Figure 7. 3-axis micro-manipulator

The rotary bond tools described above allow highly accurate rotational positioning of a workpiece, but there are many micro assembly tasks requiring movements in three degrees of freedom. For this type of application Micromotion GmbH has developed a 3-axis micro-manipulator (see Fig. 8).

This compact device, with a diameter of only 36.2 mm and an axial length of less than 50 mm, features two linear and one rotational axis. The linear axes are driven using a cam arrangement, which move a small table in X- and Y-directions. The table carries the θ-axis actuator, which drives the tool directly.

This design offers following advantages:

- Sub-µm accuracy
- Easy controllability (stepping motors are used for all axes)
- Low mass (< 50g)
- Highly dynamic performance

Importantly, the long strokes for the linear axes, easy controllability and high stability under production conditions are superior in comparison to solutions based on piezo actuators.

Typically this device is used for fine positioning and is mounted "piggy-back" on high-speed coarse positioning axes. Here the low weight is of particular importance. The trend to shorter assembly cycle times is leading to more dynamic primary positioning axes, typically featuring linear direct drive motors. If the mass of the "piggy-back" micro-manipulator can be minimised then the temperature increase of the linear motors is less for the same duty cycle. This can, in turn, avoid problems due to thermal instability of the machine, which can dramatically affect the positioning accuracy of the machine.

4. OUTLOOK

Micromotion GmbH is continuing to develop the Micro Harmonic Drive® gear in order to further improve its performance. By optimising the tooth profile the peak torque can be increased further, which will enable even more dynamic positioning cycles and so reduce assembly cycle times even more.

Micromotion GmbH is also investigating the application of Micro Harmonic Drive® gears in a project in co-operation with the Institute for Machine Tools and Production Technology (IWF) at the Technical University of Braunschweig. In this project a Micro-SCARA robot with a parallel structure is being developed, which incorporates MHD gearboxes for the primary axes of the parallel arms.

This robot is designed to achieve sub-µm accuracy and is intended to act as a technology demonstrator to open up new applications in the field of micro assembly.

REFERENCES

1. J. Hesselbach, A.Raatz: „*mikroPRO* –Untersuchung zum internationalen Stand der Mikroproduktionstechnik", Vulkan Verlag, Essen, 2002
2. C. Thürigen, W. Ehrfeld, B. Hagemann, H. Lehr, F. Michel: Development, fabrication and testing of a multi-stage micro gear system. Proc. Of *Tribology issues and opportunities in MEMS, pp. 397-402, Columbus (OH)*, November 1997, Kluwer Academic Publishers, 1998
3. R. Degen, R. Slatter: Hollow shaft micro servo actuators realized with the Micro Harmonic Drive®, Proceedings of *Actuator 2002*, Bremen
4. S. Kleen, W. Ehrfeld, F. Michel, M. Nienhaus, H.-D. Stölting: Ultraflache Motoren im Pfennigformat, *F&M*, Jahrg. 108, Heft 4, Carl Hanser Verlag, München, 2000
5. R. Slatter, R. Degen: Micro actuators for precise positionning applications in vacuum, Proceedings of *Actuator 2004*, Bremen

PART III

Design and Planning for Microassembly

AUTOMATED ASSEMBLY PLANNING BASED ON SKELETON MODELLING STRATEGY

H. Bley[1] and M. Bossmann[1]
[1] *Institute of Production Engineering/CAM, University of the Saarland*

Abstract: Because of intensification of the international market situation the industry is forced to extend the product range and to shorten the period of time between the model changes in order to put new products on the market. Standards that allow generating product and resource variants by creating reusable structures (templates) have to be worked out to support the simultaneous engineering process. The challenge is the parallelisation of the product development and the production planning as far as possible to generate a robust production. In process planning as a part of assembly planning, the assembly processes and their sequence have to be defined and the motion path and speed of an assembly process can be visualized and optimized by Digital Mock-Up (DMU). Realising an early connection between product design and assembly planning new strategies have to be developed.

Key words: Design for assembly, Assembly planning, Simultaneous engineering

1. INTRODUCTION

In the Digital Factory [1] assembly planning is supported by collision examinations, ergonomic simulation, process planning, scheduling and material flow simulation. The aim is to support the assembly planning tasks more strongly and to automate routine activities. The assembly planning is based on the bill of materials by which assembly surface matrix, the directed and undirected assembly surface graph as well as the assembly priority graph can be created. These planning tasks are based on the final three-dimensional model. That means, planning tasks have to be integrated earlier in the product development process to evaluate the product model in early design phases with regard to the possibility to assemble the product or to initiate design changes in order to avoid cost-intensive assembly processes as shown in figure 1 [2]. So, standards have to be worked out that allow such analysis without generating additional work for the different departments that are involved in the product development process. Ensuring an improved sequence of operations in assembly planning based on basic design information, new strategies for the data transmission have to be installed to solve the interaction tasks realising a stronger connection and a better transparency along the whole product development process.

Therefore milestones along the development process have to be defined in order to generate a frictionless data exchange between the departments that optimise the sequence and parallelisation of operations.

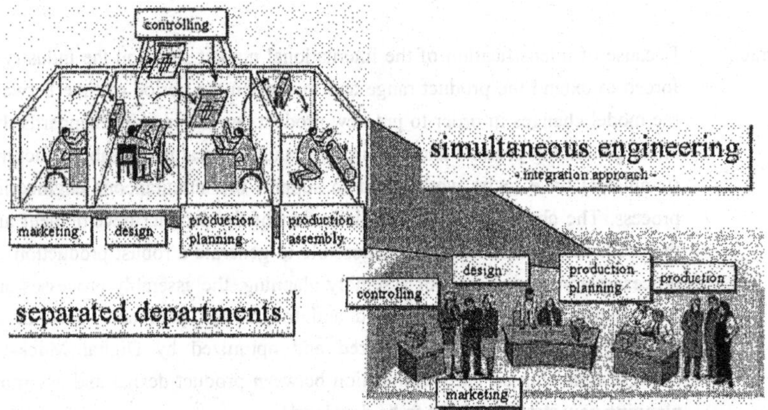

Figure 1. Interdisciplinary Working System to improve the Simultaneous Engineering Process according to [3, 4]

Normally the design department supplies the assembly planning with the finished three-dimensional model and the bill of materials. Then the assembly department has the task to assemble the product and to transform the design-oriented engineering bill of material (E-BOM) into a production-oriented manufacturing bill of material (M-BOM).

Today, there is no real control loop between product design and assembly planning because often there is no time to review the design by the assembly planning department. The design department should take the assembly of the product into account. So, the product is fixed and the assembly planning department has the task to plan the assembly of the product. The data transfer between the design department and the assembly planning department is realised in sequences without real parallelisation, the mentioned departments finish their work completely before the model or other data leave the department. Therefore our aim is to realise an overlapping in time of design and planning work, to install control loops to achieve a better interaction between the different departments. So it is possible to produce cost effective products in a shorter time and with a higher quality. A first important step towards such a development process is the definition of milestones and special data like the function model of the product that have to be transferred to the different departments. Every department has the aim to finish their data for the exchange to climb the next stage in the development process.

In this regard we do not have the aim to produce additional work for the departments, it is our aim to support the departments in routine work such as the data exchange at certain times in a special format for a frictionless data transfer or in other tasks. If there are standardized data formats used for data exchange, it will also be possible to automate the data input. Therefore a stronger and closer contact between the different departments can be realised with the result of an enhanced parallelisation of different work along the development process with the result of a better product transparency as shown in figure 1.

2. FEATURE TECHNOLOGY

The feature technology [5] offers the possibility to implement such standards and to enlarge the product model with the results of an automated support. At the moment, features are primarily used in design to improve routine activities and to save working time. In this area for example drillings are already represented very well with features.

The drilling types as well as the drilling parameters such as length and diameter have to be selected to fix the geometry. Then, the negative volume will be created und subtracted from the three dimensional product model.

Another advantage of using features is a more structured product model that is able to be enhanced with process information or every other needed information. This additional information can be filtered by a macro to separate the information for each department without burden the different departments with unnecessary information.

The newly developed feature elements are called skeleton elements. These elements characterise the function of the area with simple geometry in different colours [6, 7, 8].

Skeleton elements carry their name that represents semantic information. Furthermore the position and the orientation of each element are integrated as a parameter. Specific elements are enhanced with additional parameters to expand the feature with special information about the area to achieve a better product and process description.

These features are integrated into a modern modelling concept called skeleton modelling that allows creating variants by changing parameter configurations [9]. Skeleton modelling can be seen as a template technology for product variants. The skeleton modelling concept has different hierarchy levels: The assembly skeleton level is the topmost level that contains all global parameters and references of the product model. These parameters and references are passed on to the other hierarchy levels such as sub-assembly skeleton and part skeleton. The undermost hierarchy levels do not pass any parameter and references to other elements or other skeleton models. The functionality of the product is visualised by the skeleton elements what offers a fast transparency for the user without an extraction of the whole product because there is no element hidden by another one.

3. AUTOMATED ASSEMBLY PLANNING

Function areas of the product model that are represented by feature elements can be read out by a macro. So, overlapping elements with opposed characters can be found out. These elements represent contact areas of different parts that have to be assembled. Figure 2 illustrates the detection of contact surfaces.

This information can be used to deduce the assembly surface matrix, the undirected and directed assembly surface graph as well as the assembly priority graph. The result is a variety of possibilities to assemble the product model. The aim is an analysis of all assembly sequences to choose the best ones for an automated assembly. Therefore each assembly sequence has to be analysed by a collision check of an automated generated assembly simulation [10]. In the following, it is explained how to generate a kinematical simulation between two parts by using skeleton models with skeleton elements. The feature elements are established in the skeleton model, their information is saved in special areas of the features, so the final position of the movement is fixed for the kinematical simulation by the coordinates of the part in the product situation.

The start position of the simulation is defined in all three axes either by the exploded view of the product or by a coordinate entered manually. The creation of the exploded model that does not represent the assembly sequence but a model without any contact between the different parts is a initial situation of the simulation. The exploded view can be generated automated by positioning the parts in consideration of the dimension of every part. Each component is placed on line with the same orientation of a special part and a connection to the coordinates of the component situation or in a right angle to this line. The decision between these two possibilities is also based on a collision check by the movement of the part in assembly situation in the possible directions.

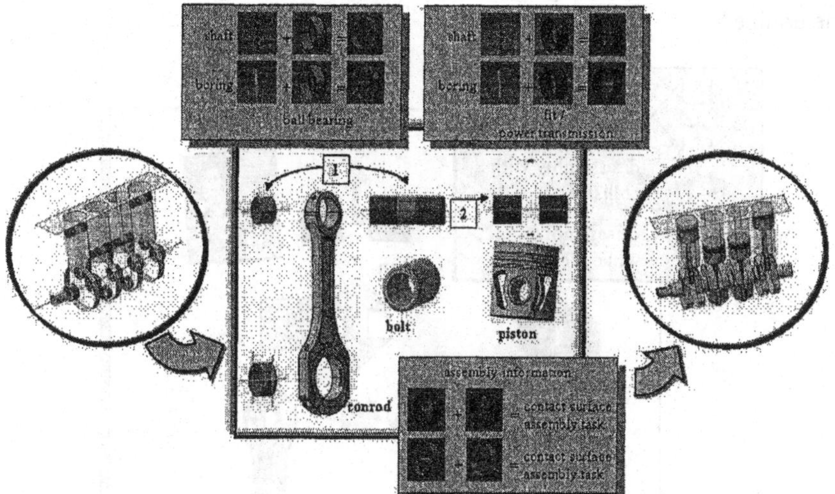

Figure 2. Assembly Operation as semantic Information in Skelton Models

So the path is a straight movement to the component's position of the assembled product. That realises an automated joint process simulation with the lowest effort on the production system because there is no complex movement for the assembly system that increases the cycle time. Figure 3 illustrates the definition of the assembly direction by a collision check while dismantling the product. The movements of the parts can be modified manually to realise another movement to the component's position. But this possibility should be used only, if there is no other idea to assemble the product with straight movements.

Another requirement to reduce the number of possible assembly sequences and to achieve a cheap assembly is to fix moveable components at the assembly simulation. That means, that turnable components are handled as fixed components and rotations are not allowed during the assembly operation of different elements. E.g. the crankshaft does not rotate around its axis during the assembly of the different conrods and the pistons to realise an improved accessibility.

The collision check is based on the automated design realisation of the rough geometry of the components which are given by the skeleton and function elements. The contour is established in the features, so e.g. a shaft is a rotation of a contour line around a defined axis to create a three dimensional body. This rough geometry of the parts is analysed along the joining movement whether there is a collision. The number of the assembly sequences by using the assembly graph must be reduced to minimise the simulation runs. For the assembly sequence the directed assembly graph is consulted and the basic part that takes the other components has to be determined.

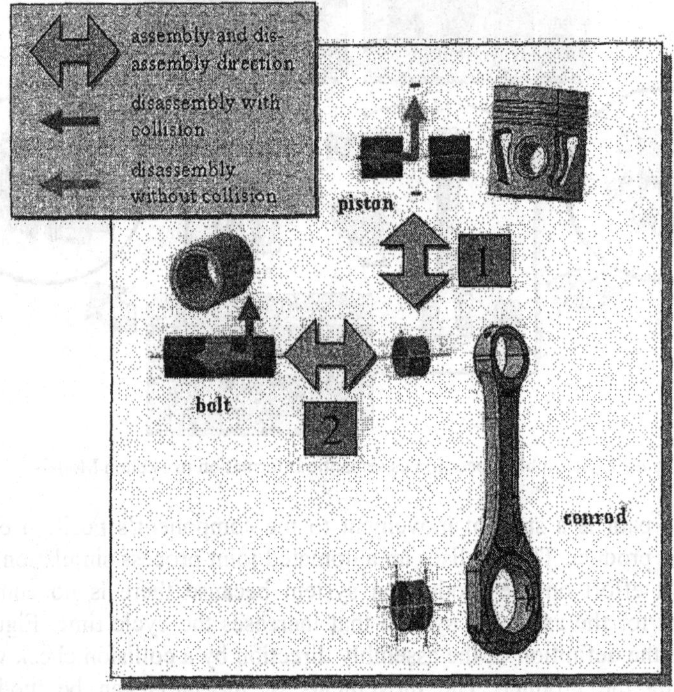

Figure 3. Definition of the Assembly Direction

The directed assembly graph is subdivided into assembly levels [11] in which, for different assembly levels, a defined order is represented. The order within an assembly level but not of the assembly levels themselves can vary. Concerning the assembly sequences it is important to define sub-assemblies as single components that expect only one assembly movement.

Simulations can help to find suitable sub-assemblies in a product model to group all components within the whole product model. The definition of the sequence of the assembly levels is followed by a simulation to check collisions. If there is a collision all assembly orders with the same assembly sub-sequence will be deleted. Therefore every collision minimises the number of assembly possibilities in a large number. The validity of the estimation of the assembly performance must be verified at certain stages to cause a protected model. Now the question can arise, why a planning expert should integrate the assembly simulation at such an early time of the development process. The answer to this question is very simple: Processing steps parallel to product design can be shown and checked by analysing the skeleton model. Such a development process helps to redesign with the consequence of a improved production of the components and a frictionless assembly of the product. It should be only the first step in digital assembly planning to protect the product model against collision, it is also important to have a look at the handling equipment to control its collision. Therefore the possible assembly sequences have to be enhanced with the handling equipment at certain areas of the product that have to be defined by the planning division. [12, 13]

There are two approaches:

1. the use of the real geometry of the grippers
2. the use of a covering geometry which symbolizes the design room of a gripper that is not defined yet.

The expansion of the digital assembly model by the handling equipment is very important to estimate collisions of the equipment with the components or different tools with each other at simultaneous joint events because not only the components can collide during the assembly operation.
Handling equipment, mould cavity geometry or the holding appliances represent "direct" restrictions to the components to assemble the product. The gripper position [13] of each component can be defined as a feature in the product model. So, an automated insert of the real geometry or a place holder that represents the space of the gripper, can be installed in the skeleton model. If the position of the handling equipment is fixed the feature element can choose the best resource for the task without any collision. In this way, the product model is enhanced with the gripper information. The number of assembly orders is reduced by the collision check of the components too, so not all possibilities have to be evaluated.

In the first step of the assembly simulation task only the parts of the product should be simulated. So, it can be avoided to block too much disk space that reduces the calculation performance. Therefore a collision check of the components can be verified faster as an extended simulation model.

The choice algorithm of a possible assembly sequence can run during the creation of the extended simulation model, so parallelisms in the work of one department are possible. After the extended simulation has shown possible assembly orders, this model should be enlarged again by enhancing the handling components and the other components with a direct connection. That way, a creation of a product-resource-model of the assembly system is achieved. An enhancement of the grippers is e.g. the resource "robot" that allows to analyse the working space and the complexity of the movement by joining the parts.

This analyse primarily supports the process planner in positioning the resources. One can imagine that to compare the different positions in different simulation runs to finding an optimised place for all resources to perform their work. The process planner must carry out restrictions to reduce the number of simulation runs, in order to get a better result in less time, e.g. layout restriction.

These three simulation tasks are shown in figure 4. The result is a whole production system simulation in the digital world. With the simulation of the resources the process planner provides the initial information for the control engineering. A control loop should also be integrated between the process planning department and the control department to optimise the production system regarding stability and productivity [14].

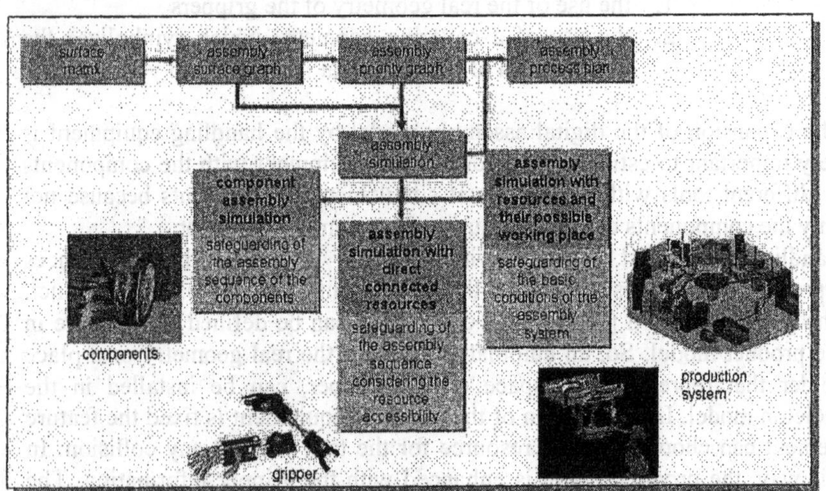

Figure 4. Three Simulation Tasks towards the Digital Model of the Assembly System

Starting with the simulation for an arbitrary assembly order the assembly priority graph can be created and detailed as a next step in the assembly planning process. The assembly priority graph has a dynamic link [15] to the product model which demands an update after product changes in which it is checked again whether the assembly assessment has been lost. This analysis primarily supports the process planner in positioning the resources.

This enhanced parallelisation of product development and production planning can avoid cost-intensive assembly processes by simple improvements in design. Another advantage of the new modelling and planning strategy is an interdisciplinary discussion platform for the departments to improve the transparency in product design and process planning as well as in resource planning. The process and resource information that does not influence the product are integrated in the process model [15]. It is important that there is a dynamical connection between the extended product model and the extended process model. Therefore, an information change that is represented in both models is interlinked with each other to realise an up-to-dateness. The digital copy of the production system can also serve as a test area for the control engineer to evaluate his programs and to make a comparison of the planned value and the calculated cycle time of each single assembly system unit.

After the test runs of the PLC based on the digital model a more robust start-up phase of the production should be ensured within the real factory. At the moment different activities in industries are established in this area of coupling digital and real factory.

Supporting this coupling both models have to be prepared optimally for each other to reach the aim with a minimum of work. Our approach of a gradual expansion of the assembly simulation is a support of the digital planning by an automated creation of kinematics simulation. Supporting routine activities the inhibition threshold for the use of such tools is considerably less. The planning department shall be able to comprehend the approach of the software because the tool shall not be understood as a black box with petitions and editions.

4. SUMMARY AND OUTLOOK

Standardisation of modelling with skeleton elements achieves a model of the product representing the functionality in a transparent way. Using feature technology in combination with the skeleton modelling approach reached the advantage of a well structured product tree and the other advantage of an easy handling by creating product variants. The product model is extended by the formation of feature combinations that enhance the product model with process and resource information through the whole product development process. In doing so, coupling of product design and production planning can be achieved. The department-related relevant data can be selected with different filter technologies.

For the assembly planning task opposite skeleton elements at the same position in the product model symbolise an assembly operation based on a contact surface of two parts. So, it is possible to deduce the assembly contact surface matrix and the assembly contact surface graph of the product model in early development phases.

In the next step the different assembly sequences will be tested automated by a collision check, and the best assembly sequence can be prepared in detail by the assembly planning department. The next step is an enhancement of the product model with direct connected resources to evaluate these sequences again. Then, the assembly model can be extended again with the whole resources to optimise their working position. The result is a digital model of the assembly system representing the initial data for the control department.

REFERENCES

1. VDI, VDI-Richtlinie 4499: Digitale Fabrik, Grundlagen (to be published)
2. Cuiper, R.; Feldmann, C.; Roßgoderer, U.: Rechnerunterstütze Parallesisierung von Konstruktion und Montageplanung. ZWF 91 7-8, pp. 338-341, 1996
3. Ehrlenspiel, K.: Integrierte Produktentwicklung. München, Hanser, 1995
4. Lindemann, U.; Reinhart, G.; Bichlmaier, C.; Grunwald, S.: PMM – Process Methodology for integrated design and assembly planning. Proceedings of the 4th Design for Manufacturing Coference, Las Vegas, Vevade, USA, 1999
5. VDI, VDI-Richtlinie 2218: Feature-Technologie, 1999
6. Weber, C.: What is a feature and What is its Use? – Result of FEMEX Working Group I Proceeding of the 29th International Symposium on Automotive Technology and Automation 1996, pp. 109-116, Florence, ISBN 0-94771-978-4
7. Bley, H.; Bossmann, M.; Zenner, C.: Flexible Process Models in Manufacuring based on Skeleton Product Models. Preprints of IFAC-MIM Conference on Manufacturing, Modelling and Control, Athens, Greece, 2004
8. Bley, H.; Bossmann, M.: Improved Manufacturing Planning based on Localisation of Product Synergy Effects by the Use of Feature Technology. Proceeding of the 38th CIRP International Seminar on Manufacturing Systems, Forianopolis, Bazil, 2005
9. Bär, T.; Haasis, S.; Verkürzung der Entwicklungszeiten durch den Einsatz von Skelett-Modellen und Feature-Technologie, VDI-Bericht Nr. 1614, Düsseldorf, Germany
10. Chaudron, V.; Martin, P.; Godot, X. : Assembly sequences: Planning and simulation assembly operations. Proceeding of the 6th IEEE International Symposium on Assembly and Task Planning, Montreal, Cannada, 2005
11. Bley, H.; Bossmann, M.: Standardisierte Produktmodelle für die automatisierte Montageplanung Featurebasierte Montageplanung unterstützt den Simultaneous Engineering-Prozess, wt Werkstatttechnik - Ausgabe 09-2005, pp. 627-631, 2005
12. Bley, H.; Fox, M.: Entwicklung eines featurebasierten Konzepts zur Montageplanung, VDI Bericht Nr. 1171, Serienfertigung feinwerktechnischer Produkte – von der Produktplanung bis zum Recycling, pp 231-250, VDI-Verlag, Düsseldorf, 1994

13. Dietz, S.: Wissen zur Auswahl von Montagemitteln, seine Aufbereitung und Verarbeitung on CA-Systemen. Dissertationsreihe Universität des Saarlandes, Lehrstuhl für Fertigungstechnik/CAM, Schriftenreihe, Produktionstechnik Band 6, Saarbrücken, 1994
14. Reinhart, G.; Cuiper, R.: Planning and Control of Automated Assembly Systems. MED-Vol.8, Proccedings of the ASME Manufacturing Science and Engineering Division, pp. 325-330, Anaheim, California, USA, 1998
15. Bley, H.; Bossmann, M.; Zenner, C.: Advances towards an Integrated Product and Production Development Process. Proceedings of 2. German - Israeli Symposium for Design and Manufacture - Advances in Methods and Systems for Products and Processes, pp. 129-137, Berlin, Germany, 2005

MORPHOLOGICAL CLASSIFICATION OF HYBRID MICROSYSTEMS ASSEMBLY

Iwan Kurniawan[1], Marcel Tichem[1], and Marian Bartek[2]

[1]*Precision Manufacturing and Assembly, 3mE, TU Delft;*[2]*HiTEC/DIMES, TU Delft*

Abstract: This paper presents a morphological classification of approaches to the assembly of hybrid microsystems or hybrid MEMS. The need for the presented classification comes from the fact that at the moment only limited well structured knowledge is available on how to assemble hybrid MEMS. The classification is based and evaluated on the basis of cases found in literature. The scheme is used in a case study to analyze the assembly process of die encapsulation.

Keywords: microassembly, packaging, hybrid microsystems, MEMS

1. MICROSYSTEMS AND THE ASSEMBLY CHALLENGE

Microsystems. which are also known as MEMS - Micro Electro Mechanical Systems, originate in the field of microelectronics. During the last several years, the development of microsystems has led to systems with more complex functionality. Monolithic integration, even if it can deal with such a system elegantly, is usually outperformed by hybrid solution, where different parts from different technological domains are integrated into composed system, particularly in terms of fabrication yield, overall system cost and time to market.

Despite the impressive developments in the research stage, hybrid microsystems are not yet widely available in the market. The industries' hesitation to introduce such products to the market is mainly caused by the immaturity of the microsystems production process, especially the assembly and packaging process.

The assembly of a microsystem highly depends on parameters like material properties, fabrication processes, and device functionalities. The strategy for the assembly process must be carefully considered and chosen early in the design stage to secure the parameters' compatibility. Therefore, knowledge about the nature of the assembly process, the possible solution, the advantage and limitation, is required.

The aim of this paper is to report on a morphological classification of the assembly process for hybrid microsystems, as the first attempt to put basis for well structured assembly approaches. A case study on die encapsulation is used to demonstrate the usefulness of the classification scheme.

2. THE MORPHOLOGICAL CLASSIFICATION

The morphological classification scheme consists of five category levels (Figure 1). The first two levels, level A and B, are the classification of the parts that need to be assembled together. These parts can be loose parts or sub-assemblies. Levels I, II, and III consist of the assembly step classification, which are transfer, alignment, and joining, respectively. Each category level has several class alternatives, which lead to various possibilities of the overall assembly process configuration.

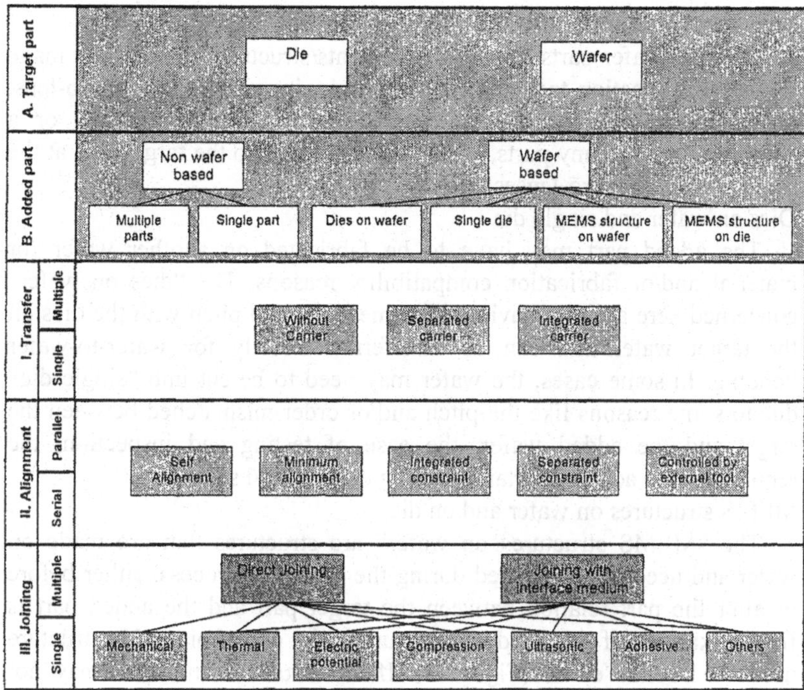

Figure 1. Classification scheme of hybrid microsystems assembly

2.1 Level A: Target part

In the first level, the assembly process is classified according to the type of target part to which other parts will be assembled. Since we consider only wafer based microsystems, the target part is distinguished into the whole wafer and the individual die. The assembly on the individual die is always serial since only a single product will flow through the assembly station, while the assembly on the wafer can be considered as serial or batch wise – depending on whether the assembly operation on the part is done in serial or parallel.

2.2 Level B: Added part

In the second level, the type of part to be added to the target part is considered. Based on the parts origin, two basic cases can be distinguished; i.e. parts that are made by wafer or non-wafer fabrication process. Both wafer and non-wafer parts can be a single part or multiple parts in certain configuration.

- Single and multiple non-wafer parts

  The non-wafer parts are all the elements/structures that are not made by wafer fabrication technology, e.g. bond wire, solder ball, micro-lens, etc. The part can be a single part, e.g. a loose micro-lens, or a configuration of many parts, which will be added to the target part at the same time, e.g. micro-lenses array.
- Dies on wafer and single die

  The added part may have to be fabricated on another wafer for material and/or fabrication compatibility reasons. The "dies on wafer" concerned here are dies having the same order and pitch with the dies on the target wafer and can be transferred directly for wafer-to-wafer bonding. In some cases, the wafer may need to be cut into "single die" due to some reasons like the pitch and/or order mismatched between the target and the added wafer, the ease of testing and inspection, the requirement to achieve certain bonding quality, and so on.
- MEMS structures on wafer and on die

  The "MEMS structures on wafer" are structures that are made on wafer and need to be released during the assembly process, either before or after the part relation between the target part and the added part is fixed. Example of this kind of structure is the out of plane 3D structure used by Dechev et al[1] (Figure 2). If the structure on wafer may not preferable for some technical reasons, e.g. difficulty for alignment, bonding process, etc., usually the wafer is cut into dies before the assembly process.

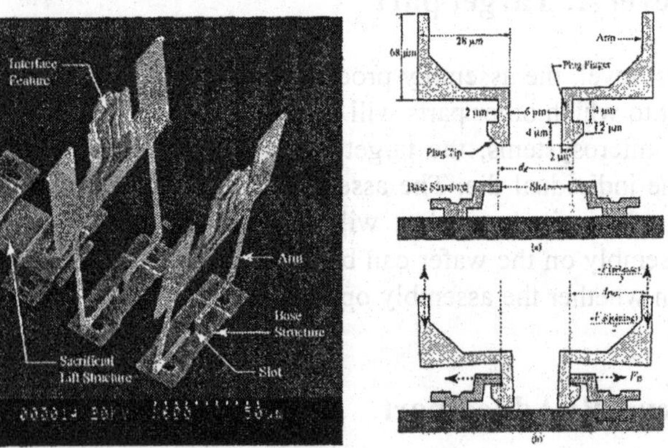

Figure 2. Out of plane 3D structure. Left: microparts joined to perpendicular to base structure. Right: snap-lock plug tip and slot geometry [1]

2.3 Level I: Transfer

The third level of the classification deals with how the added part is transferred to the target part. The number of parts to be transferred, i.e. single or multiple, and the role of transfer carrier, i.e. the interface between the part and the gripper, are considered the most important classification parameters in this level.

- Without carrier

  The part is transferred without the help of any carrier. This kind of transfer requires adequate contact surface(s) on the part to facilitate gripping action. An example of transfer without carrier is the multiple transfer of micro balls demonstrated by Shimokawa et al[2] (Figure 3.a). They used a special design vacuum gripper to pick and place several micro balls at the same time.

- Integrated carrier

  The part is transferred using a carrier that is integrated within the part itself. Usually the carrier and the part are made by the same fabrication process. An example of the integrated carrier is the parent wafer for transferring microsystem structures demonstrated by Boustedt et al[3]. The carrier is removed after the part is transferred and fixed.

- Separated carrier

  The part is transferred in a special carrier such as mould or tray. The main function of a carrier in single transfer is to provide the gripping area, while in multiple transfer is rather to arrange the parts in a desired position. A carrier does not necessarily be made for the specific purpose of the transfer process only. Example shown in Figure 3.b is a HEXIL mould used as a carrier for transferring a cap onto a base part[4]. The mould is retrieved by breaking the tethered structure after the bonding process and then used for making another cap. In this case, the main function of the carrier is as a fabrication mould instead of a transfer carrier.

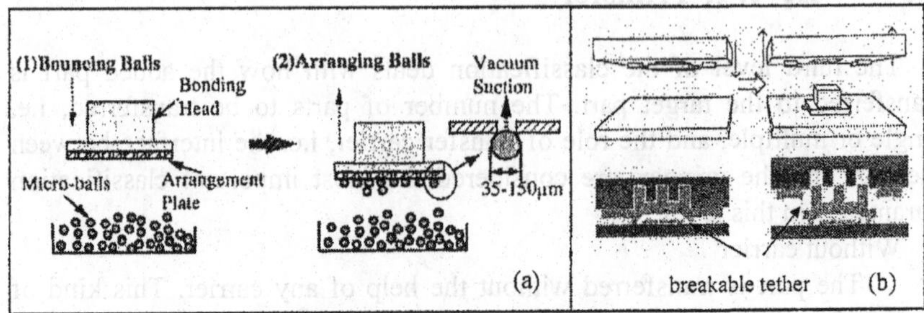

Figure 3. (a) Multiple transfer of micro balls without using carrier[2] (b) Single transfer of a microstructure using a separated carrier[4]

2.4 Level II: Alignment

In this level, the way to achieve the desired parts position, relative to each other, is considered. The classification in this level is done by evaluating how the coarse and the fine alignment between the parts are performed.

- Self alignment

  The parts are put together with coarse alignment (or no alignment at all) and let them adjust the final position by themselves. Some examples of the self alignment are those using the surface tension characteristic of molten solder (Figure 4.a) and the electrostatic attraction between two charged surfaces (Figure 4.b) demonstrated by Tuantranont et al[5] and Zheng and Jacobs[6], respectively.

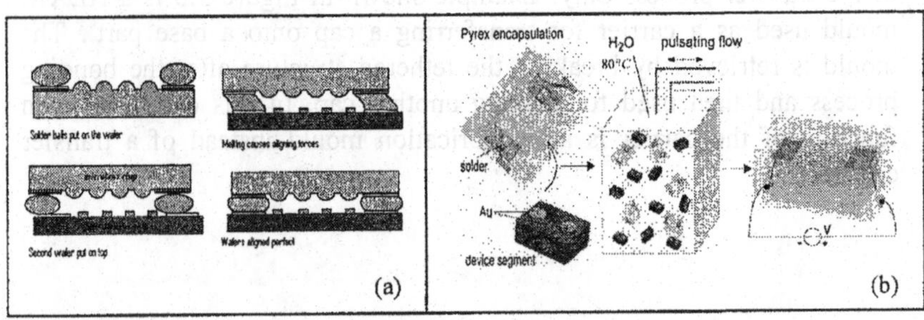

Figure 4. (a) Self alignment by surface tension characteristic of melted solder[5] (b) Self alignment by electrostatic attraction[6]

- Minimum alignment
- The final position can be achieved with only coarse alignment. This kind of alignment is usually applicable when the added part is featureless, thus its orientation or position tolerance is quite loose (Figure 5.a).

- Integrated constraint

  The parts are put together and their final position is determined by a constraining structure integrated on the parts. One example of the integrated constraint alignment is that demonstrated by Syms[7] (Figure 5.b). The stopping mechanism on this part makes the part exactly aligned 45° with respect to the fixed land.
- Separated constraint

  The alignment process is the same with the integrated constraint, but the constraint for fine alignment is provided by a separate structure instead of the structure integrated in part itself.
- Controlled with external tool

  The coarse and fine alignment of the parts are fully controlled by external means, e.g. the gripper controller, the controller of the aligning machine, etc.

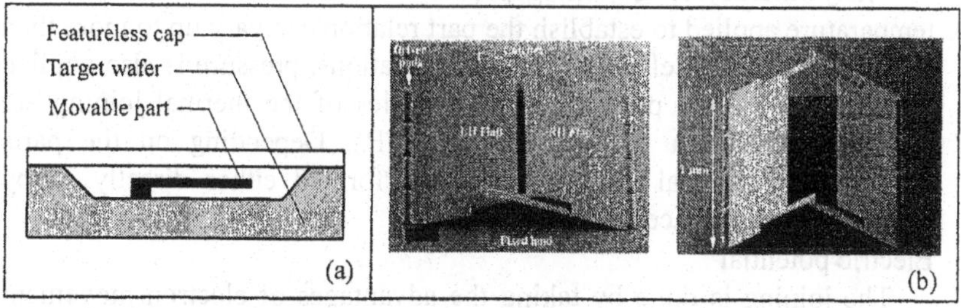

Figure 5. (a) Minimum alignment (b) Stopping mechanism on a micromirror[7]

The categories "serial" on the left hand side of the scheme in Figure 1 means that the alignment of the parts is done sequentially - one by one; while "parallel" means that many parts are aligned at the same time. It should be noted that parallel alignment is not always applicable for microstructures, especially 3D structures.

2.5 Level III: Joining

The classification of joining of the parts is done based on the basic principle used to establish the part relation. As shown in Figure 1, the joining is first classified into two; direct joining and joining with interface medium. Most microsystem parts joining is dependant of the parts material. Certain material couples are compatible to be bonded directly to each other, while others may need some interface medium, such as polymer, metal film, and so on, to facilitate the bonding. This additional layer complicates the process and increases the overall assembly cost.

The joining is classified further according to the physical parameters used to establish the final join. These physical parameters are considered to have potential influence to the joined parts. In this way, the consequences of their application to the part performance can be predicted immediately in the early selection. Since the inventions of new joining methods are still expected to come in the future, category "others" is added in this level. The explanation of each joining class is as follows:

- Mechanical
  Joining by means of mechanical principle is adapted from the principles used in macro domain. However, due to the difficulties imposed by the small size, there is only limited number of principles can be applied for microsystems. One example of the mechanical joining is the snap lock join demonstrated by Dechev et al[1] (Figure 2). There is no interface medium needed between the two parts.
- Thermal
  Many microsystems joining processes involve thermal energy. The temperature applied to establish the part relation can vary up to more than a thousand degree Celsius. In some applications, pressure is also applied to assist the joining process. Two examples of the thermal joining are eutectic and silicon fusion bonding (SFB). Depending on the parts material, the thermal joining can be performed either directly or by employing an interface medium.
- Electric potential
  The joining is done by taking the advantages of electron movement caused by potential different. There are two well-known electric potential joining: anodic bonding and electroplating. Glass-to-silicon is the most popular application of anodic bonding since they can be bonded without any intermediate layer. An example of joining by electroplating has been demonstrated by Pan and Lin[9] (Figure 6.a). They used selective electroplating of Ni to join LIGA structures with microstructure on a glass plate, where Ni served as the interface medium between the two bonding areas.
- Compression
  Joining by means of compression is not so often found in microsystems applications, since only a limited number of materials can be bonded by pressure exertion alone. Gold is usually used as the interface medium for this joining method. One example of compression bonding is that reported by Maharbiz et al. in 1999 [10]. They used compression bonding of Au tether – Au bumps to join a polysilicon lid with a target substrate (Figure 6.b).

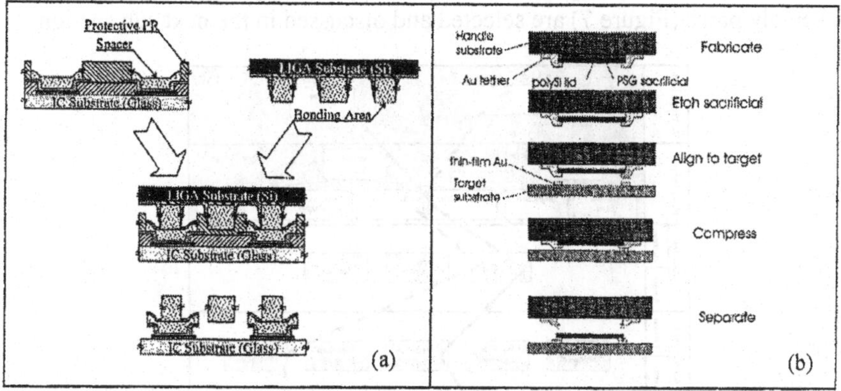

Figure 6. (a) Bonding of LIGA structure by Ni selective electroplating[9] (b) Compression bonding of Au-Au[10]

- Ultrasonic

  This joining method is based on the phenomena of metals softening when exposed to ultrasonic wave[11]. The most common applications of ultrasonic bonding are Al or gold wire bonding for electrical connection. The use of interface medium is dependent on the part material in use.
- Adhesive

  Quite a large variety of adhesives is available in the market for bonding applications. In general, a strong adhesive bond can be achieved by applying enough heat and pressure, using hardener, or by photo curing. The physical properties of adhesives vary and can be chosen according to the application needs, e.g. isolation, electrical conductor, or even optical join.

The terms "single" and "multiple" are added in the classification scheme to show that the process throughput is influenced by the selected joining method.

3. CASE STUDY: DIE ENCAPSULATION

Microsystems are very sensitive and need to be protected from any disturbance imposed by the environment. This is usually done by covering the microsystem die with a suitable cap either during or after the fabrication process.

The classification scheme described in section 2 can be used as a tool to find alternatives for such an assembly case. By combining different alternative in each classification level, many assembly paths can be chosen for die encapsulation. Each path will have its own advantages as well as disadvantages, which may set a limit to its applicability. As an example, two assembly paths (Figure 7) are selected and discussed in the next sub section.

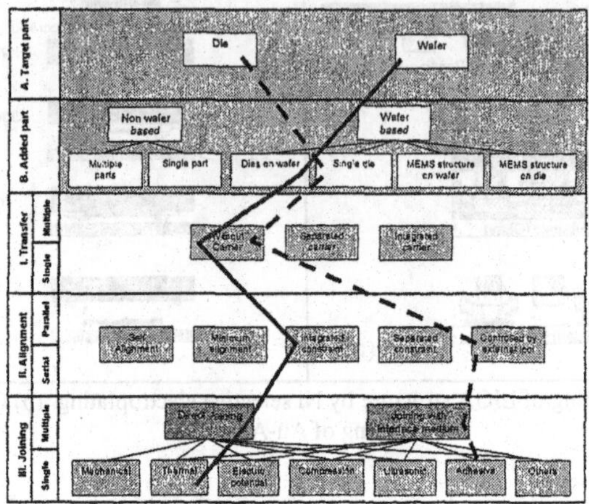

Figure 7. Alternatives of die encapsulation path

3.1 Die to die assembly

The first chosen path is die-to-die assembly shown by the dashed line in Figure 7. Both the target and the added part are cut into singular dies (Figure 8). The assembly process is performed by transferring the dies without using a carrier. The final alignment is done on the alignment machine by matching the images on the cap to those on the microstructure die. The final position is then fixed by using adhesive.

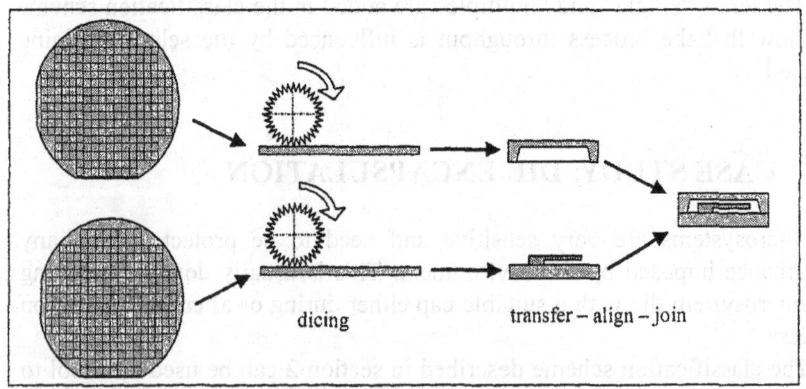

Figure 8. Die-to-die assembly

The main disadvantage of die-to-die assembly is that it is very time consuming and rather expensive. The advantage, on the other hand, is that each microsystem can be inspected and tested prior to the encapsulation process. Thus, only the "known good die" will be processed further and the defected die can be rejected immediately or sent away for reparation. Furthermore, the bonding quality of each dies assembly can be controlled individually.

The selection of alternatives in each classification level may give certain consequences to the device functionality, which can restrict their applicability. Microsystems like absolute pressure sensor or resonator, for example, may need special environment condition, e.g. vacuum, dry, inert gas filled environment, and so on, to be able to perform optimally. For such microsystems, hermetic sealing is an important requirement. Thus, adhesive or other joining method using polymer as the intermediate layer can be discarded immediately, since they do not capable of hermetic sealing. When the hermetic sealing is not an important issue, however, these alternatives are very interesting due to their cheap cost.

3.2 Wafer to wafer assembly

The second path is shown by the full line in Figure 7. The target and the added parts are wafer and dies (caps) on wafer, respectively. The caps wafer has pitch and order that match to those of the corresponding microsystems on the target wafer. The wafer-to-wafer assembly process is illustrated in Figure 9.

The wafers are transferred without using a carrier. The wafers dimension is usually quite big that they can provide enough gripping space. Unless it is required for fabrication purpose, like in the case of nickel caps made by electroplating shown by CT Pan12, normally no part carrier is needed for wafer transfer.

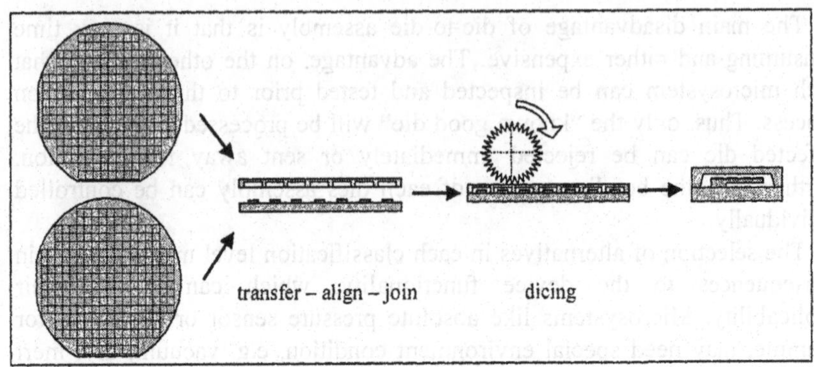

Figure 9. Wafer-to-wafer assembly

The alignment of the two wafers is done by integrated constraint. The shape and mechanism of the constraining structure can vary according to available space on the wafer. V-grove – mesa combinations are commonly used for this kind of alignment.

The aligned wafers are joined together using thermal bonding. There are two popular methods to thermally join silicon wafers, i.e. silicon fusion bonding (SFB) and eutectic bonding. SFB is usually used to bond silicon to silicon, to silicon oxide, or to glass. Eutectic bonding involves atom diffusion at eutectic temperature of an interface medium alloy to form the bond. The most common interface medium used for eutectic bonding is gold alloy.

Both SFB and eutectic bonding methods involve high bonding temperature, which may cause thermal stress, doping contamination, or other damaging effect to the microstructure. These negative effects can be reduced by localizing the heat on the bonding region. Lin and his group have demonstrated this principle by using resistive heater deposited on the bonding area (Figure 10) for fusion and eutectic bonding13, 14, 15, 16. Other researchers have demonstrated similar approach by using inductive heater17, microwave18, and laser19 as the local heat source for eutectic bonding. All these joining methods, however, are more expensive than the conventional SFB due to the need for additional masks and deposition processes of the isolation and micro heater layer.

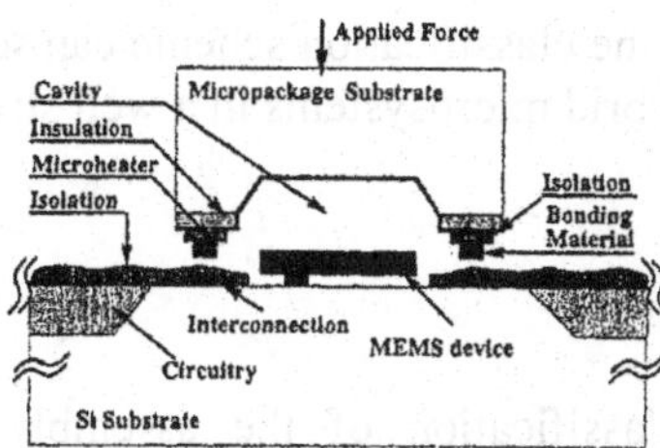

Figure 10. Bonding with localized heating[13]

The disadvantage of wafer-to-wafer assembly is that all the microsystem dies on the wafer, either good or bad, are encapsulated simultaneously. Thus, there is a chance that unnecessary costs are spent to assemble bad products. Moreover, the bonding quality of each die is highly influenced by the wafer planarity. If the wafers are not planar, there will be non-uniform gaps between the dies on the two wafers when they are brought into contact. Some caps might not be properly bonded to the microsystem dies due to wider gaps between them.

The main advantage of wafer to wafer assembly is the high throughput of the batch wise process, which also means shorter production time. If the quality of the wafer fabrication process is high, i.e. small percentage of bad die and good wafer planarity, the total assembly cost can be significantly reduced.

4. DISCUSSION

Using the classification scheme, alternative paths to assemble microsystems can be explored in a systematic way. Two examples have been shown as an illustration in the previous section. Ideally, each path alternative needs to be analyzed and checked whether they can fulfill all the requirements, e.g. technical, economical, safety, environmental requirement, etc, which are usually unique for each microsystem. The best solution can then be found by comparing each alternative based on a set of selection criteria.

The classification scheme can also be used as a starting point to derive design guidelines. Consider the alignment level in the scheme, for example, it is quite obvious that the self alignment and minimum alignment are more preferable than others due to their simplicity. Thus, it is desirable to design the parts so that they can be aligned using the two methods. For example, all the microstructures are put on the target part, while the added part is kept featureless; hence, the two parts can be aligned using minimum alignment.

With design guidelines, the classification scheme can serve as a tool to solve the assembly cases of hybrid microsystems in a well structured way.

5. CONCLUSION

A morphological classification of the assembly process for hybrid microsystems has been presented in this paper. A case study demonstrated that the scheme can be used to find alternative paths for assembly process.

Future research will be performed to further explore the scheme and to derive design guidelines for microsystems assembly with the help of this scheme.

REFERENCES

1. N. Dechev, W.L. Cleghorn, and J.K. Mills, *Microassembly of 3-D Microstructures Using a Compliance, Passive Microgripper*, Journal of MEMS, vol. 13, no. 2, April 2004.
2. K. Shimokawa, E. Hashino, Y. Ohzeki, and K. Tatsumi, *Micro-Ball Bump for Flip Chip Interconnections*, IEEE Electronic Components and Technology Conference, 1998.
3. K. Boustedt, K. Persson, and D. Stranneby, *Flip Chip as an Enabler for MEMS Packaging*, IEEE Electronic Components and Technology Conference, 2002.
4. M.J. Madou, *Fundamentals of Microfabrication*, 2nd edition, CRC Press LCC, pp. 482-486, 2002.
5. A. Tuantranont, V.M. Bright, W. Zhang, J. Zhang, and Y.C. Lee, *Self-Aligned Assembly of Microlens Arrays with Micromirrors*, Proc. SPIE, vol. 3878, pp. 90-100, 1999.
6. W. Zheng and H.O. Jacobs, *Shape-and-Solder-Directed Self-Assembly to Package Semiconductor Device Segments*, Applied Physics Letters, vol. 85, no. 16, October 2004.
7. R.R.A. Syms, *Surface Tension Powered Self-Assembly of 3-D Micro-Optomechanical Structure*, Journal of MEMS, vol. 8, no. 4, December 1999.
8. V. Kaajakari, and A. Lal, *Thermokinetic Actuation for Batch Assembly of Microscale Hinged Structures*, Journal of MEMS, vol. 12, no. 4, August 2003.
9. L.W. Pan and L. Lin, *Batch Transfer of LIGA Microstructures by Selective Electroplating and Bonding*, Journal of MES, vol. 10, no. 1, March 2001.
10. M.M. Maharbiz, M.B. Cohn, R.T. Howe, R. Horowitz, and A.P. Pisano, *Batch Micropackaging by Compression-Bonded Wafer-Wafer Transfer*, IEEE, 1999.
11. G.G. Harman and J. Albers, *The Ultrasonic Welding Mechanism as Applied to Aluminum- and Gold-Wire Bonding in Microelectronics*, IEEE Trans. on Parts, Hybrids, and Packaging, vol. PHP-13, no. 4, December 1997.
12. C.T. Pan, *Selective Low Temperature Microcap Packaging Technique Through Flip Chip and Wafer Level Alignment*, Journal of Micromechanics and Microengineering, vol. 14, pp 522-529, 2004.
13. L. Lin, *MEMS Post-Packaging by Localized Heating and Bonding*, IEEE Trans. on Adv. Packaging, vol. 23, no. 4, November 2000.
14. Y.T. Cheng, L. Lin, and K. Najafi, *Localized Silicon Fusion and Eutectic Bonding for MEMS Fabrication and Packaging*, Journal of MEMS, vol. 9, no.1, March 2000.

15. Y.T. Cheng, L. Lin, and Najafi, K., *Localized Bonding with PSG or Indium Solder as Intermediate Layer*, IEEE, 1999.
16. Y.T. Cheng, W.T. Hsu, K. Najafi, C.T.-C. Nguyen, and L. Lin, *Vacuum Packaging Technology Using Localized Aluminum/Silicon-to-Glass Bonding*, Journal of MEMS, vol. 11, no. 5, October 2002.
17. H.-A. Yang, M. Wu, and W. Fang, *Localized Induction Heating Solder Bonding for Wafer Level MEMS Packaging*, Journal of Micromechanics and Microegineering, 2005.
18. N. Budraa, B. Ng, D. Wang, S. Ahsan, Y. Zhang, E. Cho, B. McQuiston, and J. Mai, *Microwave Processing Techniques for High Density Interconnects and Hybridization*, IEEE, 2004.
19. A. Mohan, C.B.O. O'Neal, A.P. Malshe, and R.B. Foster, *A Wafer Level Packaging Approach for MEMS and Related Microsystems Using Selective Laser-Assisted Bonding (LAB)*, IEEE Elec. Comp. and Tech. Conference, 2005.

FIRST STEPS IN INTEGRATING MICRO-ASSEMBLY FEATURES INTO INDUSTRIALLY USED DFA SOFTWARE

Timo Salmi* and Juhani Lempiäinen**
** VTT Industrial Systems, P.O.Box 1702, FIN - 02044 VTT, Finland** Deltatron Oy, Soidintie 14 , FIN - 00700 Helsinki, Finland*

Abstract: This paper discusses the problems of micro-scale part manipulation and assembly. Parts less than 10x10x10 cm3 in size need special attention when they are designed. There is a need for handling tools, feeding systems, and fastening methods that resemble macro-scale industrial systems but are of a special nature. No effort has been made prior to this presentation to collect the data in any systematic form for designers. The Finnish ***DFA-Tool*** software version 3.0 is the first attempt to try to formulate this problematic in a systematic and suitable presentation for product designers.

Key words: Micro-Assembly, Design for Assembly, DFA, Assemblability analysis

1. THE DESIGN FOR ASSEMBLY APPROACH

Design for Assembly (DFA) rules have over the years been recognized as a powerful and valuable methodology for developing and analyzing producible products. The DFA issue has been approached in different ways. First, it has been given recommendations and rules on how to design products or product components so that they are easy to assemble manually or automatically (e.g. Andreasen et al 1983). Easy automatic assembly means that the product primarily is possible to assemble automatically. The assembly process can be performed by quite simple and commonly accepted known technology, which means that the process is expected to be very reliable and well known. The second approach is methods for analysing the assemblability of products (e.g. Boothroyd et al 1983). Typically the results indicate producibility in the form of an index and an estimated assembly time. Nowadays, most analysis methods have been developed as software tools. The third approach is to study how to develop the design process so that producibility issues are optimally considered. All of these aspects should be implemented if an industrial company is to benefit from DFA. The significance of DFA may be summarized as a precondition for automatic assembly and/or profitable high volume production. With the use of DFA methods during the early design phase, there is a good chance of avoiding the assembly problems that usually emerge during production start-up and even later if DFA methodology is not used.

The DFA design rules and DFA tools are based on the traditional assembly and on the technical solutions that are commonly used in that area. There are no commercialized DFA tools that can give advice in the micro-specific assembly problems.

This paper considers the problems of creating DFA rules for micro-part assembly and introduces the development of a method to handle some problems that are typical especially of micro-assembly. The methods have also been developed in the form of software and included in an industrially used DFA tool for industrial field tests.

2. DIFFERENCES BETWEEN MICRO AND MACRO ASSEMBLY

When entering the micro world there are two main issues that have special importance: The physical behaviour of parts is changing, as are the feeding, manipulation, and insertion techniques used. When scaling down, the gravitational and inertial forces may become insignificant compared to adhesive forces such as electrostatic, surface tension or van der Waal's forces. The micro-forces that exceed gravity can, for example, have the effect that as a gripper approaches a part, electrostatic attraction causes the part to jump off the surface into the gripper with an orientation only dependent on initial charge distributions. And when the part is placed in the assembly position, it may remain in the gripper tool instead of connecting to the base part. These phenomena make accurate placement of the micro-part difficult.

Micro-assembly in many cases meets the technical limitations of traditional technology. Partly because of the limitations and partly because of the new phenomenon, the following new production technologies have been introduced for micro-assembly:

- New grippers, such as (Tichem et al 2002, Lambert et al 2004) contactless grippers, capillary grippers
- New feeding methods, such as different types of reel magazines, adhesive and blister tape cavities, magazines, standard pallets and gel packs, blade feeders, hopper feeders with indexing capabilities
- The use of machine vision technology in different phases of the assembly process. Machine vision is used very often in micro-assembly to compensate various types of positioning errors.
- New joining methods.

3. DESIGN FOR MICRO-ASSEMBLY

The majority of traditional DFA rules are valid in micro-assembly. This question was studied more deeply by Eskilander & Salmi in 2004. However, some new guidelines are needed especially for micro-assembly. Most design rules have been formalized from the analyses of stepwise process chains and commonly used assembly equipment, and from the study of function principles and technical limitations. There are at least two kinds of difficulties when creating new design rules for micro-assembly. First, the technical limitations, e.g. size of the part, are not easy to determine. Although there are some common practices, someone will often cross the line with a slightly different technical solution. Therefore a huge number of tests or published research results would be needed. Second is the divergence of technical solutions. The new phenomenon has raised new technical solutions alongside the traditional ones. Many of the innovations are either in the laboratory phase or not widely used. And micro-assembly technologies are developing rapidly. The different technologies also place different requirements on product design. When there are several different potential strategies for solving an assembly problem, it is not easy to formulate design rules. In this kind of situation it should be considered for which purposes these rules should be formulated. It would seem that integration of the product design, the process and the production equipment is even more essential in the micro world than in traditional assembly. Thus the product should be designed to be manufactured with a particular method and manipulated with a particular handling method, and so on. The requirement that the product should be assembled or manufactured easily with whatever commonly used method is too tight. The interaction between product and production design is such that product features limit possible handling technologies, and on the other hand the selected handling technology places requirements on the product design.

In practical situations, the handling of parts is one of the key issues when designing miniaturized parts. The rules of micro-parts assembly actually resemble those for automated assembly, because manipulating parts of any size smaller than 10x10x10 mm^3 must be done with the help of assembly grasping tools.

When scaling down the products and parts, it is very difficult for product designers and assembly system designers to realize when the limit is reached at which the gravitation force is no longer the most prominent force in the manipulation of a part. The need for clean, aseptic, and antistatic product assembly work conditions is often highlighted in small part assembly. That is reflected in the need for assembly gloves to cover hands, extra work operations due to the spraying of ionised air, and the special materials to ensure ground potential around the workplace. The need to ensure free sight at the point of assembly also has another meaning, as human sight resolution is not enough to guarantee the success of assembly operation.

The ultimate need for aseptic handling of parts raises new challenges even when grasping tools are not utilized. The use of rubber or cotton gloves in any case reduces the natural manual touch in assembly work. The practical speed reduction has been found to be around 10% of the assembly manipulation speed.

Some rules of thumb for micro-part manipulation can easily be generated for the appearance of the part (additional to the numerical entries of size and weight):

- Does it have a stable and indexed position on feeding?
- Does it have a relatively flat surface to grasp with vacuum/Bernoulli?
- Does the flat surface cover a major area of the part to grasp with vacuum/Bernoulli?
- Does the stiffness of the part allow grasping from the sides?
- Are the part sides parallel to each other so that finger-force grasping is possible?
- Does the part surface or shape allow any grasping force?

When interviewing companies on the joining methods used in their micro-assemblies, screwing is the most prominent assembly fastening method used for parts. This is no surprise, since 80% of the fastenings on the macro-scale are also done with screws. On the micro-scale there is thus a huge need to estimate the scaling factor and time constraints. When scaling screwing operations down, it seems obvious that all the design rules apply as they are of macro-scale down to a screw size of 0.8 mm in diameter. That seems to be the ultimate limit for feeding at the moment. Around this limit the torque control is difficult to obtain (around 1 Ncm) and the mechanical stiffness of the parts also limit the speed of screwing (from 1200 rpm down to 300 rpm), thus causing major slowing down of the assembly process.

4. MICRO-ASSEMBLY FEATURES IN A DFA TOOL

The engineering office Deltatron Ltd has developed a software product for assemblability analysis called ***DFA-Tool®***. The most modern version of the software, v3.0, now includes some of the constraints and rules of micro-part assembly described above. It also gives a hint when some other force than gravitation is most prominent in the assembly operation. This helps the designers to understand the character of the assembly operation and paves the way to selection of the most feasible assembly tool. This information is essential both in manual and in robotized assembly.

By asking the part appearance questions, it is possible to evaluate what is the most prominent force in the assembly and which grasping methods are still available for this particular part. The grasping forces for micro-parts are well presented by Fukuda & Arai [1999] and Böhringer et al [1999]. They are roughly modelled in the ***DFA-Tool*** software. Based on the same database, the answers to the above questions on part appearance and nature will generate a list of possible grasping methods.

When selecting the screw type and size in the assembly construction, the software calculates the most feasible assembly time of the screw. In multiple screwing operations with automated feeding, ***DFA-Tool*** manages the extra manipulation times.

The software tool for micro-assemblability is one way of collecting know-how of common practices that the industry currently utilizes for micro-part manipulation. The development of this tool is an iterative process, and the interaction between the industrial user community of this technology and the software developers is the key element. In this case the beta-testers of this study are the world leading electronics and optoelectronics manufacturing companies. The ***DFA-Tool*** beta-tested version 3.0 will be introduced at the IPAS seminar in 2006.

REFERENCES

1. Andreasen, M.,M., Kähler, S., Lund, T., Design For Assembly, Design for Assembly. Berlin, Heidelberg, New York, Tokio; IFS Publications Ltd and Springer-Verlag. 189 s. ISBN 0-903608-35-9, 1983.

2. Boothroyd, G., Dewhurst, P., Design For Assembly, Handbook. Amherst: University of Massachusetts. 1983.

3. Böhringer, Karl F., Fearing Ronald S., Goldberg, Ken Y. Microassembly. In Handbook of Industrial Robotics,, Second Edition, edited by Nof, Shimon Y. John Wiley & Sons, Inc. NEw York, Chicester, Weinheim, Brisbane, Singapore, Toronto.. 1999. pp. 1045 - 1066.

4. Eskilander, Stephan; Salmi, Timo, Are traditional DFA methods valid in micro assembly? Proceedings of the International Precision Assembly Seminar (IPAS2004), Bad Hofgastein, 11-13 Febr. 2004. The Precision Manufacturing Group, School of Mechanic, Materials, Manufacturing Engineering and Management, The University of Nottingham (2004), pp. 19 - 26.

5. Fukuda, Toshio, Arai, Fumihito, Microrobotics, In Handbook of Industrial Robotics,, Second Edition, edited by Nof, Shimon Y. John Wiley & Sons, Inc. NEw York, Chicester, Weinheim, Brisbane, Singapore, Toronto.. 1999. pp. 187 - 198.

6. Lambert, P., Vandaele, V., Delchambre, A., Non-Contact Handling in Micro-assembly: State of the Art. Proceedings of the International Precision Assembly Seminar (IPAS2004), Bad Hofgastein, Austria, 11-13 Febr. 2004. The Precision Manufacturing Group, School of Mechanic, Materials, Manufacturing Engineering and Management, The University of Nottingham (2004) pp. 67 - 76.

7. Tichem, Marcel, Lang, Defeng, Karpuschewski, A Classification Scheme for Quantative Analysis of Micro-Grip Principles; Proceedings of the International Precision Assembly Seminar IPAS'2003, Bad Hofgastein, Austria, 17-19 March 2003, pp. 71 - 78.

TOLERANCE BUDGETING IN A NOVEL COARSE-FINE STRATEGY FOR MICRO-ASSEMBLY

Vincent Henneken and Marcel Tichem
Precision Manufacturing and Assembly, 3mE, TU Delft, Mekelweg 2, 2628 CD Delft, The Netherlands

Abstract: This paper presents the tolerance budgeting for the coarse assembly step in a novel coarse-fine strategy for micro-assembly. The final assembly step is to be performed using MEMS-based functionality that remains part of the product. Total tolerance build-up due to the coarse assembly and dimensional inaccuracies has been determined for the alignment of a single mode optical fibre to a laser diode, a challenging problem in optical telecommunication.

Key words: tolerance budgeting; product-internal assembly functions; MEMS technology; fibre-chip coupling; silicon optical bench (SiOB); mechanical stop; fluid self-alignment.

1. INTRODUCTION

The purpose of this paper is to discuss key issues concerning the coarse assembly step in a novel coarse-fine strategy for micro-assembly, of which the feasibility is being investigated in a structured approach. The need for reducing effort related to the handling of small parts explains the idea for focusing on quite different methods for assembly[1]. This research focuses on the investigation of the method of micro-assembly by means of *product-internal assembly functions*. In this method assembly functionality is created, which is integrated with the product to be assembled, and which remains as part of the product. This functionality includes part actuation, position sensing and part freezing. The method is applied in a two-stage process. In the first stage coarse positioning of components is achieved using product-external assembly functions, typically by a (semi) automatic production machine or a human operator. The final, accurate positioning is performed on basis of product-internal assembly functions. Overall, this aims to result in lower total production cost and/or improved product quality. Micro Electro Mechanical System (MEMS) technology has been selected for use as supporting fabrication technology, because of the small attainable feature sizes and very high accuracy, and it is potentially low-cost due to its possibility of batch-wise processing. The potential use of MEMS-based structures for the purpose of micro-assembly has been reported on before (*e.g.* by Syms[2]), but its feasibility has never before been investigated in a structured approach. In this investigation it is the aim to research all aspects that influence under which circumstances product-internal functionality can be used for the assembly of that particular product. This paper mainly deals with issues related to the coarse assembly and with the interactions between the coarse and fine assembly functionality/steps, which play an important role within the method of micro-assembly using product-internal assembly functions, and which are closely connected with a high level of complexity. Describing this complexity and the possible choices with their accompanying consequences is the aim of this paper. Investigating these issues is considered very relevant for the attainability of success for the concept, and at the same time it can provide a broader insight in the assembly process in general.

2. APPROACH

The approach that is taken for the described exploratory research is that of a *case investigation.* The industrial case considered for investigation is the alignment of a single mode optical fibre to a laser diode, which is a challenging problem in optical telecommunication and sensing applications. This is embedded in the small dimensions of the laser output waveguide (typically 2 μm x 0.4 μm) and the optical fibre (8 μm core diameter; overall fibre diameter: 125 μm). Alignment accuracy in the order of 0.1 μm in the lateral dimension has to be maintained to achieve sufficient coupling efficiency over the economic lifetime of the system. Highly expensive machines or delicate manual handling are normally employed for the needed alignment and fixation steps, which last typically in the order of minutes. Normally, this is performed using *active* alignment, *i.e.* alignment while measuring the optical performance efficiency of the assembly.

In the project a number of concepts for thermal actuation of the fibre tip in one or two degrees of freedom perpendicular to the fibre optical axis have been developed. One of these developed concepts is used as starting point for the investigation of the coarse assembly process. This concept, which is depicted in Figure 1, concerns a horizontal planar layout in which the fibre is placed in an in-plane groove allowing two integrated thermal actuators to move the fibre tip in- and out-of-plane independently. The entire device is built up from one monolithic part, with silicon as base material. For its realisation highly accurate IC-compatible lithographical processing is used.

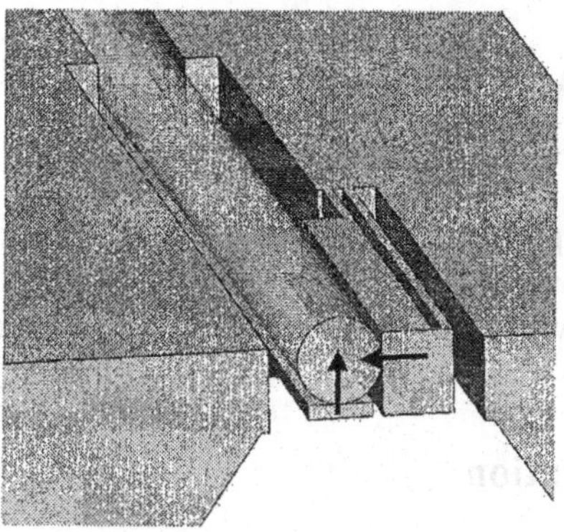

Figure 1. MEMS-based actuator concept for fibre tip positioning used as starting point in the coarse assembly investigation.

The first MEMS-based demonstrators for positioning the fibre in two degrees of freedom perpendicular to the fibre axis have readily been fabricated and are employed for investigating of (part of) the aspects listed here above. A more in-depth treatment of the design process has been covered in a previous publication[3]. A close-up SEM image of a fabricated fibre actuation structure is shown in Figure 2. The typical length of the actuators is around 1800 µm, with a total travel distance of 15-20 µm in the directions perpendicular to the fibre axis. The final fine fixation is also under investigation within the project, but this aspect is not topic of the present paper.

Figure 2. SEM image of fabricated in-plane fibre actuation structures.

The tentative choices that have been made for the specific case under consideration will be presented in detail. Aim of the investigation is to develop micro-product and micro-assembly guidelines that have broader application than the specific optoelectronic case considered.

3. TOLERANCE ANALYSIS FIBRE-CHIP COUPLING

Tolerance build-up is the overall result of all decisions concerning the *configuration*, *fabrication* and *assembly* of the device. This section will successively deal with all three items for the specific case under consideration.

3.1 Configuration

In practice, positioning a part cannot be seen separately from its immediate environment. A position relation always must be established with respect to another part or to a base structure. In the case considered the important position relation is the optical path between the single mode fibre tip and the laser diode exit facet. For functional reasons it is not possible to connect both parts directly to each other. Instead this position relation needs to be established indirectly, thereby creating a part stack of which the tolerance build-up must be controlled such that the accuracy of the optical path remains within the desired range for each individual degree of freedom.

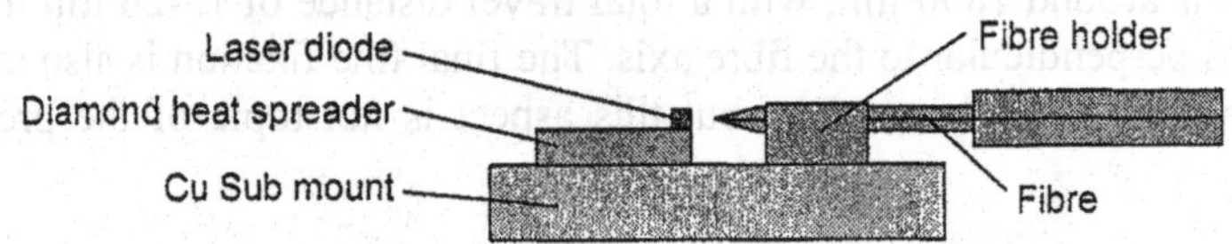

Figure 3. Conventional layout fibre coupled laser diode package.

A conventional layout is presented in Figure 3. A commonly used base structure (or *sub mount*) is copper due to its good heat conductivity. Optionally a diamond heat spreader is positioned in between the copper sub mount and the laser diode for even better heat regulation capabilities. It was found that the copper-diamond stack could be replaced with a single silicon layer without negatively influencing the thermal properties too much. Thereby opening up the possibility of creating lithography-based alignment features on the sub mount or potentially even integrating the sub mount and the actuator part into one single silicon part, which is from assembly point of view the most attractive alternative. However, integrating all assembly functionality into a single silicon component with the required accuracy using MEMS-based technology is very challenging. For this reason it was decided to focus on an assembly solution with separate silicon base part and actuator part (see Figure 4 for a schematic representation of this layout).

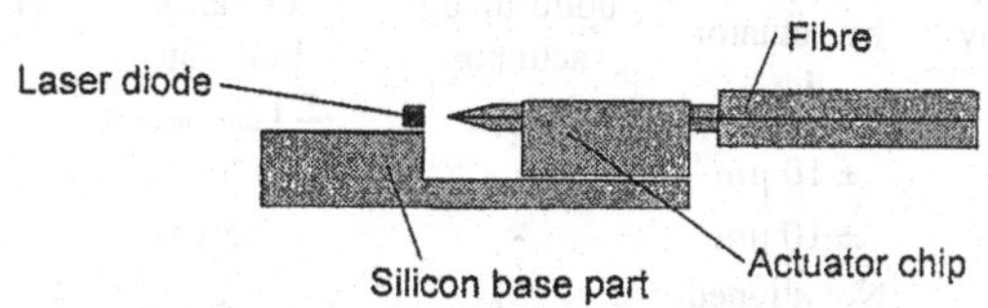

Figure 4. Proposed alternative layout fibre coupled laser diode package.

3.2 Fabrication and assembly related issues

Beside the configuration other tolerance build-up influencing components are fabrication and assembly related: dimensional inaccuracies, ranges of product-internal assembly process, parasitic tolerance build-up induced by internal actuator (only in the directions that are not fine positioned by the product-internal assembly functions), positioning inaccuracies, and shifts due to the fixation process. In the remaining part examples will be provided of each.

3.2.1 Functional accuracy requirements

As mentioned previously, the alignment tolerance between the laser diode and the fibre tip is very critical in the directions perpendicular to the fibre axis (± 0.1 μm). According to the sign convention generally used in fibre to laser diode alignment these are the x and y-directions, see Figure 5.

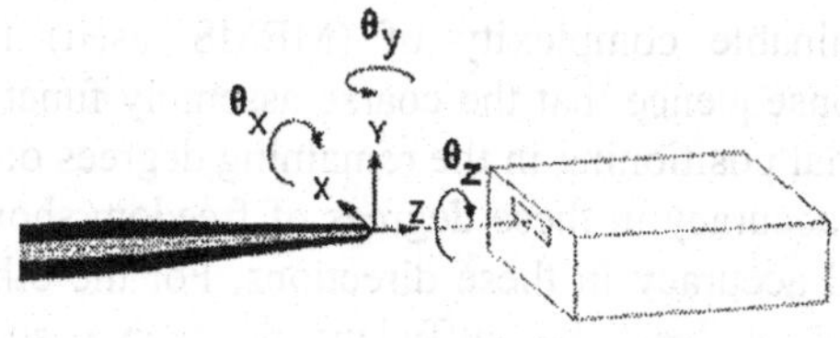

Figure 5. Sign convention fibre to laser alignment.

Naturally, the required accuracies are not the same in all directions. An overview of the required tolerances in all degrees of freedom is presented in the first column of Table 1. All values are displayed in upper (positive number) and lower (negative number) specification limits.

Table 1. Tolerance build-up for all individual degrees of freedom

DOF	Required tolerance by functionality i_{req}	Maximal correction by actuator i_{act}	Parasitic tolerance build-up by actuator $i_{act, par}$	Cumulated dimensional tolerance build-up $\Sigma i_{dim, inacc}$	Maximal allowed tolerance build-up of coarse assembly process $\Sigma i_{CA, allow}$
x	± 0.1 µm	± 10 µm	-	± 4.5 µm	± 5.5 µm
y	± 0.1 µm	± 10 µm	-	± 7.5 µm	± 2.5 µm
z	± 0.5 µm	Not aligned by PIAF	Insignificant	Insignificant	± 0.5 µm
θ_x	± 1.5°	Not aligned by PIAF	- 0.5° - 0°	Insignificant	- 1° + 1.5°
θ_y	± 1.5°	Not aligned by PIAF	- 0.5° - 0°	Insignificant	- 1° + 1.5°
θ_z	∞ rotational symmetric fibre end	Not aligned by PIAF	-	Insignificant	∞
	± 10° non-rotational symmetric fibre-end	Not aligned by PIAF	-	Insignificant	± 10°

3.2.2 Consequences product-internal assembly functionality

As starting point for the coarse assembly investigation an actuator chip concept has been taken that is capable of positioning the fibre tip in the x and y-directions perpendicular to its optical axis. Although the required accuracy in z-direction is also quite strict, this limitation is made to control the complexity in fabrication of the MEMS-based fine assembly functionality to a manageable level.

The limited attainable complexity of (MEMS-based) fine assembly solutions has as a consequence that the coarse assembly functionality needs to take care of the final positioning in the remaining degrees of freedom. The attained positioning accuracy in these degrees of freedom should be at least equal to the required accuracy in these directions. For the other degrees of freedom this should at least be sufficient to enable successful final positioning by the product-internal micro actuation functionality.

The maximal correction by the product-internal actuators is 15-20 µm, but due to the fact that fine alignment of the fibre tip takes place by bending of the fibre end, unwanted parasitic tolerance build-up is induced in some of the unactuated directions. Of these, the rotations θ_x and θ_y of the fibre around the x and y-axis are considered not negligible and are in the order of 0.5° in the negative direction, see columns 2 and 3 in Table 1.

3.2.3 Dimensional inaccuracies

In order to be able to design the product-internal and the coarse assembly process, knowledge needs to be gathered on dimensional accuracies of all parts within the optical assembly stack. The optical assembly stack consists of a silicon sub mount, a GaAs distributed feedback (DFB) laser, a product-internal actuator chip and a single mode optical fibre. The intermediate bonds between these parts (where applicable) also account for part of the tolerance build-up, but these will be treated separately. The contributions of the individual parts to the total tolerance build-up due to dimensional accuracies are in a comparable range. For example, the tolerances on the outer dimensions of a typical laser diode are quite large, ± 10 μm, due to the fabrication processes used, but these are normally not employed in the assembly process. Instead, fiducial marks are used for the positioning of the laser diode in the horizontal plane, which are made using lithography with a typical accuracy of ± 1 μm relative to the laser facet. When the laser diode is placed with the laser facet close to the bottom side ('junction-down'), then the position of the laser facet can vary only as much as ± 2 μm in y-direction.

For the actuator part and the silicon base part, the tolerances in x and y direction are also IC processing-based, and are therefore both in the same range as for the laser diode facet, ± 1 μm and ± 2 μm respectively.

Most tolerance build-up in the optical fibre is in the x and y direction due to core-cladding eccentricity. Maximal tolerance build-up between core and cladding is calculated to be ± 1.5 μm.

In the dimensional inaccuracy analysis the z-direction is not taken into account since the final accuracy in this direction will be achieved by positioning the fibre tip directly relative to the laser diode facet. Therefore any dimensional inaccuracies of the components are insignificant for reaching the accuracy in this direction. The cumulated tolerance build-up due to dimensional accuracies is displayed in column 4 of Table 1. The rightmost column represents the maximum allowed tolerance build-up of the coarse assembly process. This can be read as a requirements list for the coarse assembly process; the coarse assembly process must comply with these limits in order to fulfil the overall assembly demands. It can be clearly seen that the demanded accuracy in the z-direction is now by far the most critical and is likely to deliver the largest challenge for the coarse assembly process.

3.2.4 Passive alignment structure fabrication

Besides meeting the specific accuracy requirements *cost* is the main driving factor for the decision which positioning solutions will be employed for the individual degrees of freedom. The addition of product-internal assembly functionality is expected to increase the assembly cost, therefore it is argued that a more low cost coarse assembly process must be developed to achieve an overall cost-effective assembly alternative compared to the conventional assembly methods. In general, assembly cost is lower if the required accuracy to be reached is also lower. Unfortunately, only in certain degrees of freedom this accuracy demand is lowered by the application of product-internal assembly functions. This does not permit the use of cheaper, less accurate assembly methods for the overall coarse assembly process. A possible solution is the use of features in the product that assist in achieving the required positioning accuracies, thus by obtaining the accuracy from the fabrication process instead of the assembly process. This approach does enable a further decrease of the assembly cost. Since MEMS-based technology is employed for realising the product-internal assembly functions, it may not involve much additional cost and effort on the fabrication side to include also such features to assist in achieving the desired accuracies in the coarse assembly process.

For the accurate fabrication of passive alignment structures, two technologies are available, both based on lithography for reaching the necessary (in-plane) dimensional accuracy:

- Alignment with the use of mechanical stops (compliant positioning)
- Alignment with the use of liquid surface tension (fluid self-alignment)

A *mechanical stop* can be defined as an embedded functional surface in a part to be assembled which constrains the fine motion between the part to be assembled and the sub-assembly during alignment. For each degree of freedom two contact areas and one alignment motion need to be defined. The contact areas have to be mated in the fine motion step by compliant actuation forces.

Alignment accuracy can also be increased using *surface tension*. In order to join two parts together, bonding materials like solder or adhesives need to be in a liquid phase. Surface tension forces occur, which can cause relative movement between parts to be assembled. By smart positioning of wettable surfaces during the fabrication process and by accurate control on the volume of the bonding material, surface tension within the liquid can be used for accurate positioning of small parts during the coarse assembly process (see Figure 6). Earlier-performed research[4-10] has shown that surface tension forces within solder are able to align parts with high accuracies. Parts with an initial alignment of ± 20 μm can be aligned up to ± 1 μm and 0.1°

accuracies in the in plane (x, z) directions and ± 0.5 μm in y-direction (height).

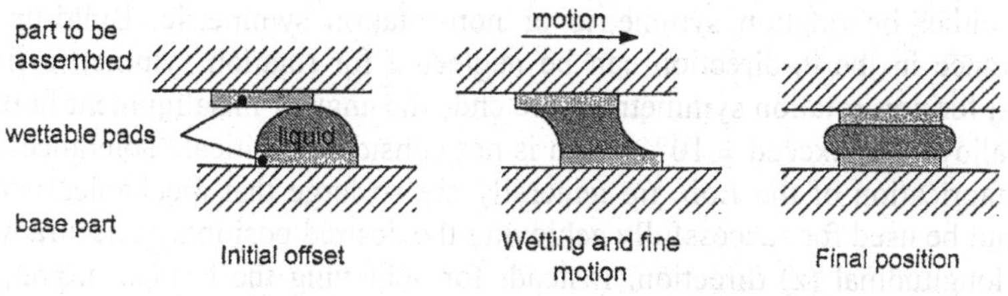

Figure 6. Alignment using liquid surface tension.

In various researches the above two technologies have been applied in silicon to attain high alignment accuracies for optical applications. The silicon base parts in these researches are generally referred to as *silicon optical bench* (SiOB) or *silicon waferboard.*[6,8,11-14] The present research could be considered as an extension to this concept, in which also active assembly functionality is integrated with the product. In the specific case considered as well both mechanical stops and fluid self-alignment have been utilised.

3.2.5 Part relation tolerances

Three main part relations need to be established:

- Assembly of the actuator part to the base part;
- Laser diode to base part assembly;
- Fibre insertion into the actuator part.

Assembly of the actuator and of the laser diode are proposed to be performed using solder self-alignment techniques. Wettable solder pads are to be integrated into the actuator part and the laser diode as well as in the silicon base part. Integrating fluid self-alignment features using MEMS technology is simpler than creating mechanical stops with comparable out-of-plane accuracy. This is the reason that for the two planar part relations (laser diode to base part and actuator part-base part) this option has been selected. For the fibre-actuator part relation mechanical stops are more appropriate due to the more complex contact interface and the fact that it is very difficult to create high accuracy alignment features in the fibre part. Due to the specific characteristics of the fibre part it is only possible to constrain the position in four degrees of freedom (x, y, θ_x and θ_y). Tolerance build-up in these directions between chip and fibre are expected to be around – 0.5 μm to 0 μm and – 0.1° to 0° respectively due to a deliberate alignment of the fibre relative to the actuator end effector with a negative offset (so that the parts in all cases make contact with each other).

In the specific case under consideration a lensed fibre tip is used. This tip can either be rotation symmetric or non-rotation symmetric. Build-up of tolerance in the θ_z-direction can be neglected for rotation symmetric fibre ends; for non-rotation symmetric fibre ends the angular misalignment in θ_z is not allowed to exceed ± 10°, which is not considered critical. Tolerances on the fabrication of the fibre tip generally are so large that mechanical stops cannot be used for successfully achieving the desired positioning accuracy in the longitudinal (z) direction. Instead, for achieving the critical alignment tolerance in z-direction of ± 0.5 μm of the fibre tip relative to the laser diode facet (optimal offset 4-5 μm) still a highly accurate insertion is needed, which is an undesired situation. The fibre could be maintained in place for example by mechanical clamps integrated in the actuator chip. After this final coarse assembly step the alignment of the fibre tip in x and y-direction can be performed actively using the product-internal assembly functions.

An overview of the attainable coarse assembly tolerance build-up for the three main part relations is presented in Table 2. A schematic side view of the individual parts to be assembled is given in Figure 7. The attainable tolerances for the laser diode and the actuator part relative to the silicon base part depend entirely on the fluid self-alignment process and are taken from literature.

Table 2. Overview of attainable tolerances coarse assembly process

DOF	$\Sigma i_{CA,allow}$ Maximal allowed tolerance build-up of coarse assembly process	i_{CA} Actuator - base part (solder self-alignment)	i_{CA} Laser diode - base part (solder self-alignment)	i_{CA} Fibre - actuator (enforced by actuator design)	$\Sigma i_{CA,attain}$ Attainable tolerance build-up coarse assembly process
x	± 5.5 μm	± 1 μm	± 1 μm	- 0.5 μm 0	- 2.5° + 2°
y	± 2.5 μm	± 0.5 μm	± 0.5 μm	- 0.5 μm 0	- 1.5° + 1°
z	± 0.5 μm	± 1 μm	± 1 μm		± 2 μm
θ_x	- 1° + 1.5°	Insignificant	± 0.1°	- 0.1° 0	- 0.2° + 0.1°
θ_y	- 1° + 1.5°	Insignificant	± 0.3°	- 0.1° 0	- 0.4° + 0.3°
θ_z	± 10°	Insignificant	± 0.1°		± 0.1°

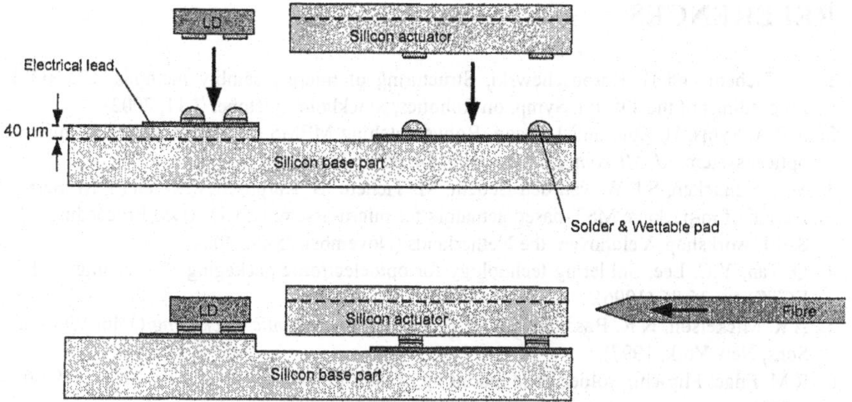

Figure 7. Schematic order coarse assembly steps.

4. CONCLUSIONS

For specific applications the method of micro-assembly by means of product-internal assembly functions could be a promising alternative to conventionally used methods for micro-assembly tasks. A significant degree of design complexity is involved, especially in realising the desired positioning accuracies, due to the existence of a high level of interrelations between a number of closely connected issues that play a role within this assembly method.

Two basic approaches can be distinguished for obtaining the desired positioning accuracies during the coarse assembly process: (1) obtaining the desired coarse positioning accuracy from the assembly process, or (2) from the fabrication process. The last is considered essential for achieving high positioning accuracies at relatively low cost.

At this stage no optimal overall assembly solution has been found for this specific case under investigation. In this research a sequential approach has been taken from fine to coarse assembly; a more optimal result should be feasible in a concurrent development approach.

ACKNOWLEDGEMENT

This research belongs to the Delft Centre for Mechatronics and Microsystems of the TU Delft. It is funded by the Dutch government programme IOP Precision Engineering as part of the project IPT02310 Technologies for in-package optical fibre-chip coupling.

REFERENCES

1. M. Tichem and B. Karpuschewski, Structuring of micro-assembly methods, CD-ROM Proceedings of the 33rd Int. Symp. on Robotics, Stockholm (October 7-11, 2002).
2. R.R.A. Syms, H. Zou, and J. Stagg, Robust latching MEMS translation stages for micro-optical systems, *J. Micromech. Microeng.* **14**, 667-674 (2004).
3. V.A. Henneken, S.P.W. van den Bedem, M. Tichem, B. Karpuschewski, and P.M. Sarro, Design of in-package MST-based actuators for micro-assembly, CD-ROM Proceedings 7th SAFE workshop, Veldhoven, the Netherlands (November 25-26, 2004).
4. Q. Tan, Y.C. Lee, Soldering technology for optoelectronic packaging, Proceedings IEEE ECTC, pp. 26-36 (1996).
5. A.R. Mickelson, N.R. Basavanhally, Y.C. Lee, *Optoelectronic Packaging* (John Wiley & Sons, New York, 1997).
6. R.M. Edge, Flip-chip solder bond mounting of laser diodes, *Electron. Lett.* **27**(6), 499-501 (1991).
7. J. Sasaki, M. Itoh, T. Tamanuki, *et al.*, Multiple-chip precise self-aligned assembly for hybrid integrated optical modules using Au-Sn solder bumps, *IEEE Trans. Adv. Pack.* **24**(4), 569-575 (2001).
8. M. Itoh, Passive alignment on Si optical bench using Au-Sn solder bumps, Proceedings IEEE LEOS 10th Annual Meeting, pp. 126-127 (1997).
9. W. Lin, S.K. Patra, Y.C. Lee, Design of solder joints for self-aligned optoelectronic assemblies, *IEEE Trans. Comp. Pack. Manuf. B* **18**(3), 543-551 (1995).
10. W. Pittroff, J. Barnikow, A. Klein, P. Kurpas, *et al.*, Flip chip mounting of laser diodes with Au/Sn solder bumps: bumping, self-alignment and laser behavior, Proceedings IEEE ECTC, pp. 1235-1241 (1997).
11. C.A. Armiento, A.J. Negri, M.J. Tabaski, R.A. Boudreau, *et al.*, Gigabit transmitter array modules on silicon waferboard, *IEEE Trans. Comp. Hybr. Manuf.* **15**(6), 1072-1080 (1992).
12. R.M. Bostock, J.D. Collier, R-J.E. Jansen, R. Jones, *et al.*, Silicon nitride microclips for the kinematic location of optic fibres in silicon V-shaped grooves, *J. Micromech. Microeng.* **8**, 343-360 (1998).
13. C. Strandman, Y. Backlund, Bulk silicon holding structures for mounting of optical fibers in V-grooves, *J. Microelectromech. Syst.* **6**(1), 35-40 (1997).
14. C. Strandman, Y. Backlund, Passive and fixed alignment of devices using flexible silicon elements formed by selective etching, *J. Micromech. Microeng.* **8**, 39-44 (1998).

THE IMPORTANCE OF CONCEPT AND DESIGN VISUALISATION IN THE PRODUCTION OF AN AUTOMATED ASSEMBLY AND TEST MACHINE

Daniel Smale BEng (Hons) AMIMechE
TQC Ltd, Hooton Street, Nottingham, NG3 2NJ, UK, daniel.smale@tqc.co.uk

Abstract: Producing an automated assembly and test machine is a lengthy process consisting of numerous stages. It can involve several iterations through some or, very occasionally, all of these stages. There is one major similarity linking these stages: the requirement for consistent and accurate information. This unifying requirement makes communication a crucial, but often overlooked, aspect of producing an assembly automation machine.

It is often the case that the simplest and most reliable method of communication is a visual representation or "Visualisation". Visualisation can and does take many forms; anything from sketches on napkins to detailed 3D computer images. All formats have benefits as well as drawbacks.

This paper looks at the role of visualisation and its importance within the production of an automated assembly and test machine. It also further develops the issues and problems faced by the assembly automation industry as well as looking in greater detail at the potential solutions and how these are being developed and expanded on by European research projects.

Key words: Communication, design, visualisation.

1. INTRODUCTION

In the 21st century, communication is affordable, instant and omnipresent. It is because of this that it is easy to forget how essential it can be. Not simply in terms of the physical (or electronic) actions, but rather in terms of the nature of these actions. Everyone is aware of the saying "a picture is worth a thousand words..." something which is probably true except that, in many cases, it under-sells the value of visual representations. Drawings, be they sketches, diagrams or full engineering detailed views, can convey that which words often cannot. This is increasingly the case as international projects become more common where linguistic barriers can result in confusion.

The marketplace today is heavily consumer driven. The increasing consumer expectation and product competition places great pressures on the producers, and this pressure is felt through the whole supply line. A system integrator is generally somewhere in the middle of this line and so feels this pressure not only from the customer who wants his or her new product to be manufactured, assembled and tested as soon as possible, but also from the suppliers who struggle to meet strict lead times and increasing performance expectations.

In this climate, constant and good quality communication is essential for any project to be a success. Visualisation is the engineer's favoured means of communication and thus, as the pressure on the engineers increases, so it does on the visualisation.

2. ROLES AND IMPORTANCE

Engineers have not failed in recognising the value of images; that is why engineering drawings exist. A good engineer tends to be one who can convey ideas and concepts in visual representations, not necessarily one who is good with words, though it is often of benefit to be capable of both. Visualisations reduce the need for complex verbal communications, which are subject to misinterpretation and distortion over time; people generally remember different sequences and outcomes from conversations and meetings. Visualisations, at least good ones, do not allow room for misinterpretation and are traceable.

Meetings and conversations rely upon human interaction and are thus inherently social situations. As a consequence, there is the constant possibility that the ultimate objectives will break down to mundane interactions [1,2]. Visualisations can often be a means of directing conversations and offer a plainly visible objective.

Visualisation has the main purpose of communicating concepts, ideas and detailed information concerning all aspects of the system. This single purpose can be divided into two distinct roles: Inter-company and Intra-company.

Inter-company communication occurs between two or more companies involved in the supply chain. This is where the biggest gains and, by inference, losses can be made.

Intra-company communication occurs internally within a single company and does not involve any outside personnel. Intra-company is much more frequent and generally less formalised than Inter-company communication and so is less crucial for the communication to be right first time as there are more opportunities for corrections. This effect is lessened as the company becomes larger to the extent that, in the case of large multinationals, Intra-company communication only really applies to one department or branch.

3. FORMS OF VISUALISATION – FOUR CLASSIFICATIONS

There are many different types or forms of visualisation. These could be classified as follows: *Sketch, Diagram, Drawing,* and *Detail.* It should be noted that these classes are independent of media, (i.e. just because a visualisation is hand-drawn does not automatically mean that it is a *Sketch*) and also independent of the purpose, or role, of the visualisation.

The above classification list is shown in ascending order of level of detail, degree of prescription and in time to produce (and thus also in ascending order of cost to alter). It is additionally in descending order of the need for additional communication, i.e. a *Sketch* is often most effective when used during a conversation whereas a *Detail* contains all the relevant information needed for manufacture and therefore can be seen as "stand-alone".

It should be noted that, in the following definitions, the intended "audience" or receivers of the information are assumed to have some engineering experience or understanding.

A *Sketch* can be defined as a visualisation that is used to assist verbal communication, not to replace it. A *Sketch* will generally be produced "ad-hoc" and will not be prepared prior to a meeting or conversation. It should be very quick to produce, not dimensionally accurate or even scaled appropriately. Finally, the level of detail should be the minimum necessary to convey the concept to the other person or people involved.

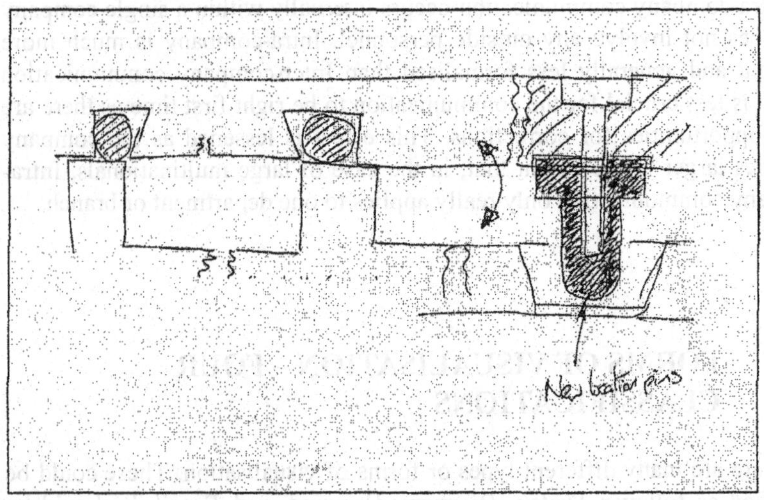

Figure 1. An example of a *Sketch.*

A *Diagram*, essentially, can be defined as a visualisation that is more detailed, accurate and time-consuming to produce than a *Sketch.* However, a more accurate definition would be that a *Diagram* is also used to assist in verbal communication but, unlike a *Sketch,* it is generally prepared in advance of any meeting or conversation. It is likely to consist of a series of boxes which, though approximately correct in scale and dimensions, will generally require labelling so that their nature or purpose is clear.

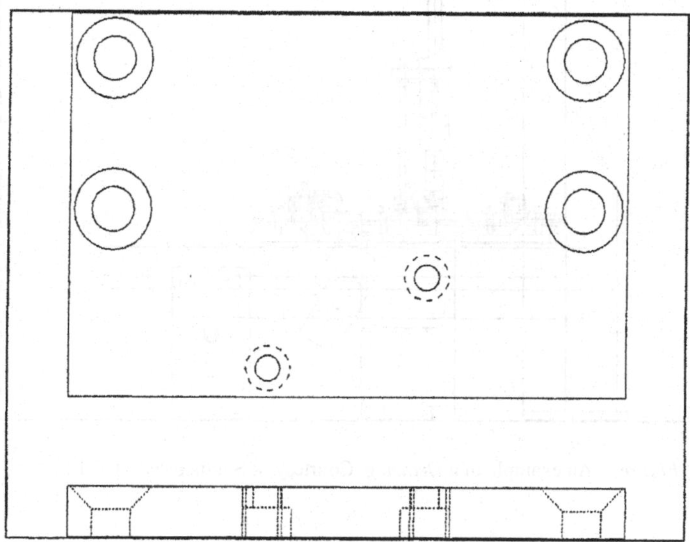

Figure 2. An example of a *Diagram.*

The next most descriptive visualisation is the *Drawing.* This can be defined as a representation that is accurate in both dimensions and scaling. It will be able to accurately convey the product or system in terms of layout, functionality and purpose. The visualisation should require minimal additional communication to be fully understood, though it may be presented as part of a meeting or conversation. The relativity of parts and components to one-another should be clearly visible. An example of this classification could often be the "GA" or General Assembly drawing.

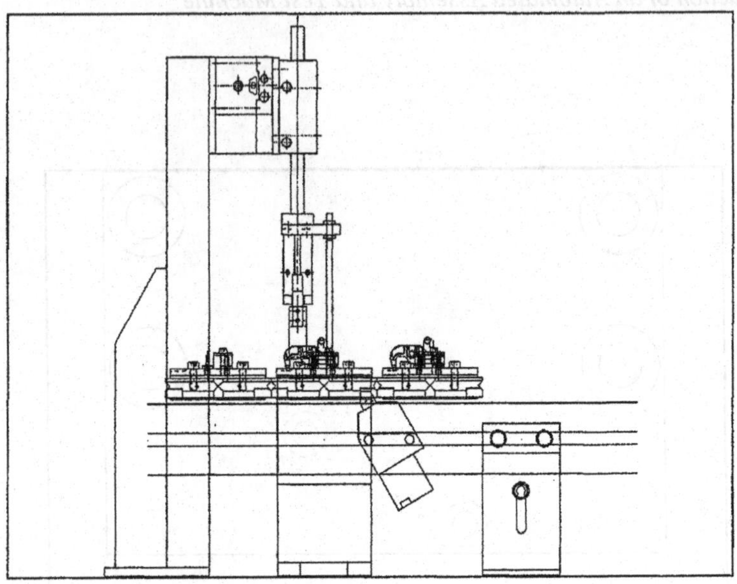

Figure 3. An example of a *Drawing*. Courtesy of Southco and TQC Ltd.

Finally; the definition of a *Detail.* This final classification is, as the name suggests, the most detailed and therefore most time-consuming and costly to produce. The visualisation should contain every piece of information needed for manufacture and/or assembly. It should therefore be entirely self-supporting and not require any commentary: it replaces verbal communication.

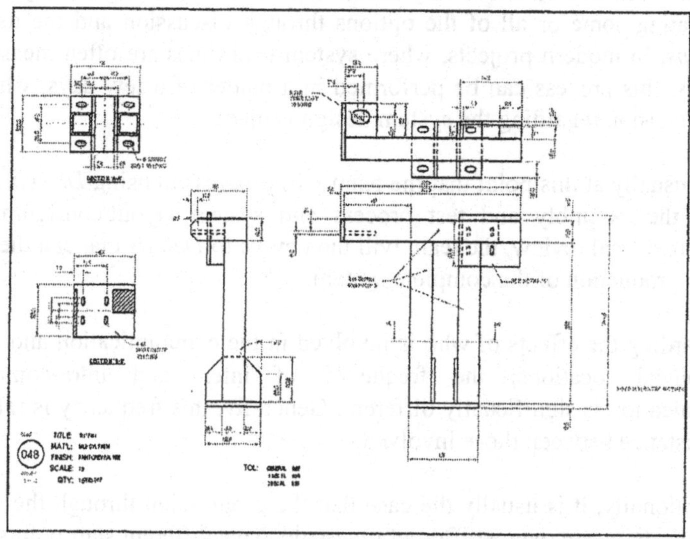

Figure 4. An example of a *Detail.* Courtesy of Southco and TQC Ltd.

Naturally, there are no clear lines that distinguish between the different classifications. However, most visualisations will fall more into one definition than the others.

Any engineering project will involve all of the previously mentioned classes of communication. The class used will vary depending on the phase of the project, though it must be noted that the project will go through numerous iterations of these phases. The class will also depend, to a lesser extent, on between whom the communication is occurring and their geographical proximity.

With regard to the project phase, during the early stages (conceptualisation) communication will generally be in the *Sketch* class. This is because whilst there are several concepts, and thus potential paths for the project team, there is no point in wasting time producing *Details*. This is often aided by the team members working in the same office, or otherwise having frequent physical meetings

As the project progresses in time, and hopefully through the design phases, it is necessary to make decisions that will reduce the number of potential solutions. This is a very iterative process and will generally entail investigating some or all of the options through discussion and the use of *Diagrams*. In modern projects, where system lead times are often measured in weeks, this process can be performed in a matter of a few days, when a "final" decision regarding the system design is made.

It is usually at this point that the team will progress to using *Drawings* to finalise the assembly and test process and system layout/configuration. Subject to a final review, the team will then move to *Details* that can then be used for production of the complete system.

Regarding the effects of who is involved in the communication and their geographical locations; the frequency of inter- and intra-company communication is significantly different. Generally, this frequency is related to the distance between those involved.

Additionally, it is usually the case that the progression through the class of visualisations are over a different timescale, have different start points and are out of phase with one-another. This in itself can be for a number of reasons that will not be covered in this paper.

4. PROBLEMS

The Inter-company communication occurs less frequently than Intra-company communication. For this reason, Inter-company meetings are more structured and there is considerable pressure for them to go well first time, driven by the time and monetary costs associated with them. It is therefore desirable to maximise the relevant detail that can be conveyed. However, it is also desirable to minimise the complexity of the visualisation. This is for two reasons: Firstly, complexity generally requires a trained eye to understand. Secondly, complexity is time-consuming to produce, which in turn is an additional cost that can be ill afforded. Both of these problems are expanded upon below.

The fact that engineering drawings require a trained eye to fully understand them has not been an issue whilst the decisions being made regarding the assembly machine and the associated project were conducted, at least predominantly, by engineers within the company.

This is, however, changing. These decisions are not confined simply to the initial sell of the design to a customer; management and financial considerations come into play at every step, both internally within the company itself and externally. This change has been catalysed by the ever-more competitive market and the subsequent increasing importance of all cost issues. Assembly and test automation machines are usually a very large investment and so, quite understandably, the management of the customer company want to ensure that the investment is financially sound.

The management will not only base their decision on financial matters, the overall business strategy of the company will almost certainly play a role. It is therefore important, from all stakeholders' perspectives, that staff without an engineering background can understand the ideas, concepts and designs.

Equally, the quoting process itself, from the system integrator to the customer and also from equipment suppliers to the system integrator, can be an expensive one to undertake. The current economic climate, as described in the Introduction, forces each member of the supply chain to be efficient and minimise wastage. No company can afford to repeatedly spend tens or even hundreds of hours preparing quotes and in other sales activities, which do not directly result in a return on this investment. However, the dilemma is that without spending this time and effort, there is little chance of securing any business.

The other issue associated with engineering drawings is that they are time consuming to produce. This is especially the case with CAD modelling, which is often the most time-consuming part of the whole process [3]. The acceleration of product evolution, which is driven by several factors, has reduced the time span of product lifecycles. This has impacted upon the design and manufacture aspects of the product. Neither the individual assembly automation projects nor the companies themselves can afford to waste time producing detailed drawings that, for one reason or another, will not be used. The current way of working amongst all of the stakeholders is to continually monitor and balance this work depending on the situation. This requires continual human decision-making and thus, by implication, mistakes are made. These are rarely of a catastrophic nature, but certainly they each cost time and/or money, the sum of which has an impact on both the project and the companies involved.

It becomes clear that reducing the time needed to produce sufficiently detailed visual representations and ensuring that they can be read and understood by non-engineers is crucial to the progression and increased efficiency of the market.

5. SOLUTIONS

The problems that exist with visualisations provide a significant challenge to any engineer, project team or company. These problems are unlikely to result in the immediate collapse of a project or a company, but the inefficiency and subsequent costs cannot, ultimately, be sustained.

The most obvious solution to the cost and time problems that any company can easily implement is to simplify the visualisation. However, this will not give the engineers the required information and over-simplifying can look un-professional, a serious problem in such a competitive market. It is therefore crucial that the right balance of time, cost and detail is struck.

The solution currently employed by engineers in the Assembly Automation and Test field is to make the majority of the conceptual level design decisions at a very early (pre-quote) stage. The sales team, which consists of experienced mechanical and electrical engineers and project managers, will perform the iterations between *Sketches* and *Diagrams* rapidly generating a list of potential solutions and then reducing this to a short list of 2-4 promising solutions. The sales team will then finalise their decision as to which of the concepts to quote on. Once this final decision has been made, the sales team will produce the *Drawings* that will provide the basis for both the quote and the handover of the project to the project team.

The communication between the individuals involved within the sales team is most often a combination of verbal and visual, with both *Sketches* and *Diagrams* being used. This process is constrained by time; it is essential that a quote of sufficient detail is produced quickly, both to minimise cost and to satisfy the potential customer. This results in the sales team presenting *Drawings* to the customer and, assuming a successful quote, to the project team. The project teams are responsible for the project beyond the quotation acceptance and ultimately implement the physical realisation of the design. The project team will then further develop the *Drawings* into *Details*, which are then used to produce the final system.

This method has proved to be successful in enabling projects to move swiftly and operate within budget. However, there are problems associated with this method; it constrains the project team in terms of design freedom and it also places an even greater pressure on the visualisations. The *Sketches* and *Diagrams* are needed to be efficient in communicating within the sales team, whilst also providing adequate "stand-alone" capability that they will be able to be used as archive information. The *Drawings* are under the most pressure as they must not only communicate the design to other engineers within the company, but also sell the idea to the customer.

The ideal solution, from an engineer's point of view, would be a tool that enables him or her to produce a *Drawing*, with the benefits of high detail and reduced requirement for additional information, but in the time-frame and ease of a *Sketch.* This is of course the ideal solution and would be a very ambitious target for any company or project. However, by gaining an understanding of how and why visualisations are used as well as developing and optimising a tool, perhaps even for each of the classes, progress towards this ultimate target can be made.

Achieving this is not straightforward. What must be accepted is that the solution cannot be provided by a single company trying to impose its ideas on the rest of industry. The solution must come from collaborative work, with representatives from each link in the supply chain and a mixture of both industry and academia involved. The most immediate problem with this strategy is that if you ask ten people what is "too complex" or "too simple" you will doubtless get ten different answers. This is not just due to the non-quantifiable nature of "complexity" but is also due to the fact that individual companies have their own standards and practices, which they will require any solution to fit to, not the other way around. Any solution must therefore be flexible, adaptable or customisable to specific user requirements.

The IMechE in the UK also recognise the problem and believe one of the most promising solution routes is in Simulation [4]. The aim of simulation is to produce highly impressive (aimed at non-engineer customers) and detailed (for the engineers in the companies) visualisations.

Another way of solving these problems, which also involves simulation, is to consider producing a library of equipment and assembly cell models, an idea being developed in the E-RACE project [5]. E-RACE utilises the Internet to provide the required user interface, which allows instant communication between any two companies. These models are both visual representations of the physical item and also contain the performance and costing figures. This concept is clearly in tune with current engineering practises; it is very rare, even for a bespoke manufacturer, to design and build an assembly automation machine from scratch. It is far more common to utilise a collection of "modules", either designed in-house or bought-in, and make the necessary adaptations to ensure that a working machine is produced.

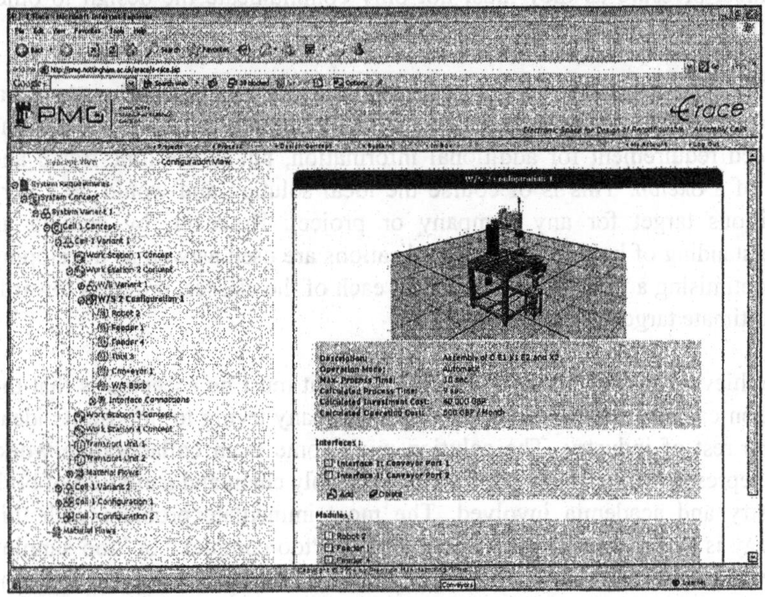

Figure 5. Screen shots from the E-RACE portal. Courtesy of www.e-race.org.uk.

The E-RACE project builds on this, adding a structured user requirement specification and quotation system. This concept can be further developed by designing the machinery to suit this design process. This is being investigated by the EUPASS project [6], which has links to E-RACE.

6. CONCLUSIONS

It is generally accepted that the *Detail* class of visualisation has been greatly improved by the almost universal use of CAD. *Details* are time consuming to produce by nature and necessity; they must enable those involved to convert the raw materials into the final product. Realistically, any further improvements to this class will be the result of software improvements by commercial companies.

It is in the other three classes, for which there is no specific or dedicated tool, that the big improvements can and should be made. Engineers may currently employ various tools, often a CAD tool or even simply MS Office program, to assist them with the earlier classifications, but these are compromises.

Ultimately, an engineer wants a tool or suite of tools that are dedicated to each stage of his or her visualisation needs. The optimum solution would be to combine the benefits of *Sketches* with those of *Drawings* and *Details.*

REFERENCES

1. Boden, D. (1994) *The Business of Talk,* 1st Edition, Polity Press, London.
2. Koskinen, I. Plans, evaluation and accountability. *Social Research Online,* 2000, www.socresonline.org.uk
3. Chua, C K et al. (2003) *Rapid Prototyping: Principles and Application,* 2nd Edition, World Scientific, New Jersey, NJ
4. Rooks, B. 2000, *Winning Ways for Manufacturing,* Assembly Automation, Vol 20, number 1, pp 35-39.
5. UK Eureka Consortium. E-Race, Eureka Project: Eureka Factory E!2851. www.erace.org.uk
6. EUPASS www.eupass.org

DEVELOPMENT OF PASSIVE ALIGNMENT TECHNIQUES FOR THE ASSEMBLY OF HYBRID MICROSYSTEMS

Christian Brecher, Martin Weinzierl and Sven Lange
Fraunhofer Institute for Production Technology

Abstract: Conventional machining methods have been developed to meet the standards of ultra precision machining. Special milling processes utilizing mono-crystalline diamond tools, the so-called fly-cutting processes, are used successfully to manufacture highly precise microstructures with an optical surface finish. In micro assembly often positioning accuracies of only a few micro meters are needed. An approach of the Fraunhofer IPT is to achieve these accuracies using passive alignment strategies. In this paper, the ultra precision machining of the v-groove structures as well as their passive alignment capacities for micro assembly tasks are presented.

Key words: ultra precision machining, passive alignment, assembly, micro components

1. INTRODUCTION

Microsystems are often characterised with a high functional density on an area of only a few millimetres. The combination of optical, mechanical, electronic or fluidic components creates the latest generation of microsystems, the so-called hybrid microsystems. To guarantee the function of these systems, the exact cooperation of their single components is absolutely necessary. In general, this comprises the exact alignment of the components within a range of a few micrometers and below. Therefore, single item production systems are today's standard in the production of hybrid microsystems. These extensive manufacturing and assembly processes increase the cost and complicate line production. In fact, along with the handling and joining operations, the micro assembly process can account for up to 80 % of the total production costs of a micro system[1]. The growing demand on the integration of fluidic and optical components, for instance, does not always allow for the use of silicon wafers. Therefore, existing knowledge from the silicon-wafer technology, which has been established over the last years for applications in microelectronics, as well as conventional production technology, can not only be used for the manufacturing of such micro systems. The production and assembly of hybrid microsystems requires appropriate handling and alignment techniques in combination with precise process control to comply with the specified narrow tolerances. Research is carried out at the Fraunhofer Institute for Production Technology (IPT), to develop cost-efficient methods to support the production, handling and assembly of micro components and -systems. In this paper, the ultra precision manufacturing of passive alignment structures as well as their potential for the passive alignment of micro components is introduced and discussed. The achievable passive alignment accuracies are discussed with the example of passive assembled glass fibre arrays.

2. ULTRA PRECISION TECHNIQUES FOR THE PRODUCTION OF HIGHLY ACCURATE PASSIVE ALIGNMENT STRUCTURES

Passive alignment structures consist of elements that hold the micro components in exactly defined positions. This is accomplished by an in general statically determinate embedding of the micro components in structures with narrow form tolerances and high surface qualities. The manufacturing of these structures requires highly precise machining processes. Lithographic processes in combination with etching have been established in silicone technology. Further developments, such as the LIGA process, combine multiple processes to create microstructures which are precise enough to be used for the passive alignment of micro components. However, the amount of processes involved, increases the cost and effort for the manufacturing of single passive alignment structures. Here, conventional metal cutting processes such as milling or planning offer the possibility for a cheap and fast manufacturing of microstructures. Regarding bulk production, these microstructures may be used for replication processes such as hot embossing of plastics or glass. At the Fraunhofer IPT, the machines and the processes themselves have been designed to meet the requirements of ultra precision machining. The most frequently used ultra precision machining process is fly-cutting. In this milling process, a single tool is rotating on a high precision air bearing spindle with up to 3000 min^{-1} and moved across the surface of the work piece with an accuracy of 0.1 µm/100 mm. Thus, linear groove-structures in the shape of the projected tool geometry are cut into the work piece. The use of monocrystalline diamond as a cutting material enables the manufacturing of tools with cutting edges down to 50 nm. These allow for a chip removal down to a thickness of only a few micrometers which results in very high surface qualities of 10-20 nm Ra.

In order to comply with the narrow form tolerances which are required for highly accurate passive alignment structures, the relative alignment between the tool and the workpiece is decisive. Therefore alignment techniques have been developed to define the tool's position exactly in respect of its six degrees of freedom. During rigging, the tool is prealigned with the precision of an articulated robot which has a repeatable accuracy of ± 30 μm. This repeatability is by far not enough for the required machining precision but is sufficient to position the tool within the measuring range of 480 μm x 360 μm of a measuring device which is integrated in the ultra precision machine tool and actively detects the translational offset between the currently mounted tool and a reference (Fig. 1). The reference is a fixed point which is defined within the measurement range, usually the crosshair of the camera system. Prior to structuring, the position vector between the reference and the work piece is determined with the use of a tactile sensor. The measuring device consists of a CCD-camera combined with a telecentric objective and has a hardware resolution of 0.465 μm. Software is used to measure the tool's position and to calculate the offset to the reference which is then compensated within a submicrometer range by the control system of the machine tool. The accuracy of the offset detection is influenced by the positioning accuracy of the tool within the measurement range and by the measurement uncertainty of the system. The positioning of the tool is mainly influenced by the accuracy of the passive alignment of the spindle rotor, which lies within ± 5 μm. In combination with a misalignment of the camera, that is below 0.011°, deviations of 0.001 μm result in horizontal and vertical direction. Repeated measurements of the tool tip position have revealed a maximum deviation of the measured value of 0.248 μm, which is mainly attributable to the measurement uncertainty of the software. Therefore, the translational offset of the tools is kept well below 0.5 μm. The tool's rotational positions are determined by passive alignment. The standardized design of the tool retainers enables the integration of fitting surfaces to define the tool's roll angle φ by passive alignment (Fig. 1). The tool itself is adjusted appropriately within the retainer. Therefore the measuring device can be swayed by 90° to detect all rotational displacements. Once each tool is adjusted right in its retainer it can be exchanged multiple times with much higher precision than simply mounting it in the spindle without knowing its exact position. The condition of the tool's cutting edge is examined frequently throughout the whole cutting process by use of

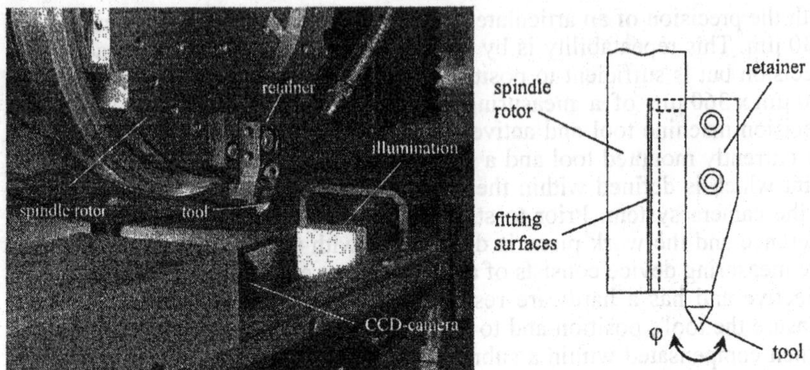

Figure 1. Optical set up (left) for the active alignment and fitting surfaces (right) for the passive alignment of the tool position

the metrology device as well. An automatic tool changer replaces the tool in case of excessive tool wear and is used to switch between different shaped tools to manufacture complex structures. The use of these innovative technologies increases the process stability of ultra precision machining and enables the manufacturing of complex structures with sub micrometer tolerances.

As it is known form conventional cutting processes, high cutting forces can cause tool deflection during the process. This effect causes form deviations and thus increases the work piece tolerances. Concerning the stiffness of the fly-cutting process, the weakest link is the air bearing of the spindle. However, with a working load below 1 N of the ultra precision machining process and a stiffness of the air bearing of 40 N/µm, the maximum tool deflection would amount for 0.025 µm. Therefore, the influence of cutting forces in ultra precision fly-cutting can be neglected regarding the tolerances of microstructures.

3. PASSIVE ALIGNMENT OF GLASS FIBRES

The focal point of the research work at Fraunhofer IPT on the passive alignment of micro components is the analysis of the entire chain of tolerances resulting from the manufacturing and assembly processes. Besides the macroscopic geometry of the alignment structures, the machine and process accuracy during the fabrication of the structures as well as the tolerances of the micro components themselves have an influence on the net accuracy of the passive alignment. For a precise and accurate analysis of the passive alignment structures and the alignment process itself, the use of highly precise micro components is necessary. Single mode glass fibres which are used for the transmission of optical signals have very little form tolerances regarding variations in diameter (± 0.7 µm), roundness (≤ 0.7 %) and core eccentricity (≤ 0.5 µm)[2] . This is the reason, why the experimental research described in the following was carried out on the example of the passive alignment of glass fibres.

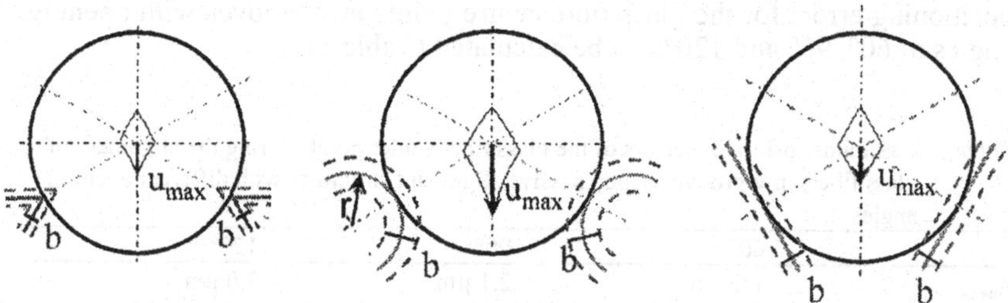

Figure 2. Models for the theoretical analysis of pointed, round and flat passive alignment features

First of all, the theoretical positioning accuracy of glass fibres in passive alignment structures was derived with the help of mathematical models. For this analysis, general tolerance ranges for the glass fibres (a) and the alignment structures (b) were taken from the DIN ISO 1101 standards[3]. In the theoretical analysis, round, pointed and flat passive alignment features were compared regarding their precision in the passive alignment of cylindrical microcomponents (Fig. 2). The flat alignment features were identified to have the best alignment accuracy. Therefore v-groove structures were chosen as passive alignment structures since they can be manufactured easily through ultra precision machining by means of fly-cutting.

Considering the tolerance ranges of the glass fibres (a) and the passive alignment structure (b) a maximum horizontal deviation of the centre points of the glass fibres of

$$u_{max,x} = \frac{a+b}{2 \cdot \cos(\alpha/2)} \tag{1}$$

and a maximum vertical deviation of the centre points of the glass fibres of

$$u_{max,y} = \frac{a+b}{2 \cdot \sin(\alpha/2)} \tag{2}$$

can be derived.

The total tolerances of the glass fibres position add up to approximate 2.0 µm for the centre point positions of the glass fibres. A tolerance range of 1 µm (peak-to-valley) for the diamond machined v-groove structures was derived from an optical profile measurement, using a confocal microscope, which is shown in Fig. 3. With examples (1) and (2), the resulting theoretical positioning errors for the glass fibre centre points in v-grooves with opening angles of 60°, 90° and 120° can be calculated (Table 1).

Table 1. Maximum positioning errors of the glass fiber center points during the alignment of glass fibers in v-groove shaped passive alignment structures with different opening angles

	60°	90°	120°
$u_{max,x}$	1.7 µm	2.1 µm	3.0 µm
$u_{max,y}$	3.0 µm	2,1 µm	1,7 µm

Following the systematic geometric analysis of the resulting tolerance fields for the passive alignment of the glass fibres, the achievable positioning accuracy was experimentally investigated at Fraunhofer IPT. For these experiments, test structures were made of aluminium. First the main structure body was machined with a 5-axis micro milling operation and then 40 v-grooves with varying apex angles (60°, 90°, 120°) and a spacing of 250 µm were introduced via ultra precision machining (fly-cutting) using monocrystalline diamond tools.

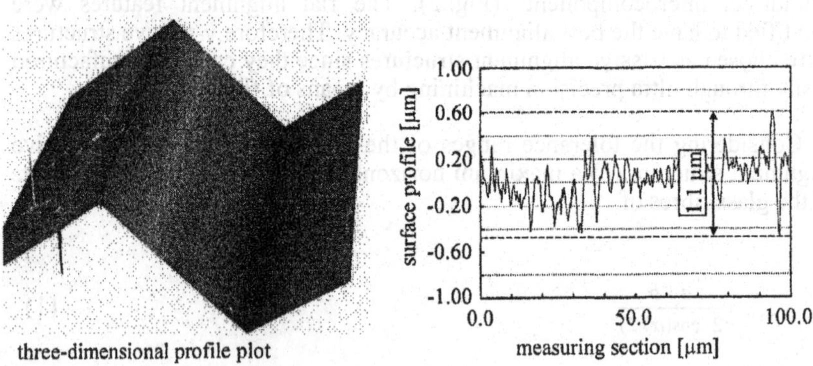

Figure 3. Three-dimensional confocal white light profile plot (left) with the derived two-dimensional profile section (right)

The glass fibres were then manually prealigned within the v-grooves with the help of a stereo microscope and slightly pressed into the alignment structures by means of a high-precision, micro milled clamping device. During this mounting process, the glass fibres were set against a diamond-turned limit block, in order to guarantee that the fibre end surfaces flushed with the end surface of the alignment structure. In order to compensate for the variability in the fibre diameters and to press each of the 40 fibres into the v-grooves with the same force, a clamping block with a 70 µm thick layer of latex was utilised. Figure 4 shows a scanning electron microscope (SEM) image of a section from the one-dimensional glass fibre array.

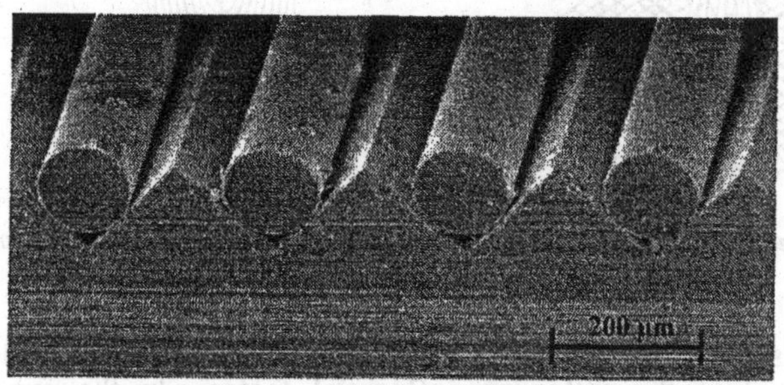

Figure 4. SEM image of passive aligned glass fibers (100x)

The measurement of the relative positioning accuracy of the core centre points of the glass fibres was carried out with a high-resolution image processing system[4-6]. To ensure the highest possible contrast to their surrounding, cold cathode light was fed into the fibres' free ends. In this way the fibre cores were illuminated by the light conducted through the glass fibres only. For the interpretation of the image processed measurement data, the corresponding Gaussian distributions and standard deviations (σ_x, σ_y) of the x-y-pairs of the fibre core centre points were determined. By means of the standard deviations the three-dimensional Gaussian distribution of the fibre centre points can be expressed as follows:

$$P(x,y) = \frac{1}{2\cdot\pi\cdot\sigma_x\sigma_y}\cdot\exp\left(-\frac{x^2}{2\cdot\sigma_x{}^2}-\frac{y^2}{2\cdot\sigma_y{}^2}\right) \tag{3}$$

Figure 5 shows the three-dimensional Gaussian distribution of the fibre centre points positions in the section plane of the passive alignment structures. The z-axis of the Graphs represents the frequency density of the Gaussian distribution. The outline of the top-views represents the 3-σ-line. According to the 3- σ-criteria, 99.7 % of the fibres centre points were positioned with a maximum error of +/- 3.2 µm for apex angles of the grooves of 60°, +/- 3.6 µm for apex angles of 90° and +/-3.9 µm for apex angles of 120°.

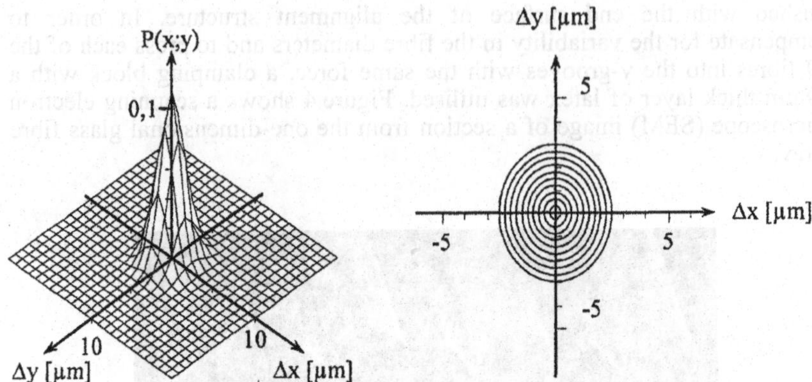

Figure 5. Tree-dimensional Gaussian distribution to evaluate the positioning accuracy of the passive alignment of glass fibers in v-grooves

The measured values lie substantially above the theoretical values which can be traced to the following influential factors. In the experimental research a maximum difference of 2.2 μm was detected to the theoretically determined deviations of the fibre core centre points (see Table 1). The accuracy of the image processing is one of the factors which influence the experimental results. The image processing system provides a resolution of a 0.5 μm per pixel upon the entire field of view. A software interpolation between individual pixels enhances the resolution by a factor of ten during the measurement process. Thereby measurements with a theoretical resolution of up to 0.1 μm can be done. Measurement errors occur here, for example from stray light which hindered the precision of the imaging of the fibre core contour. Additionally, errors are caused by the interpolation of the circular contour of the fibre cores and the centres of the circles. The choice of the threshold values for the binary filter also plays an important role. A further but rather minor (< 100 nm) influence on the experimental results is the positioning accuracy of the z-axis of the positioning system which is used to position the glass fibres within the measurement range of the camera system.

Assuming, that 0.2 μm of the measured difference form the theoretical values can be assigned to the influences described above, the greatest influence on the experimentally determined positioning accuracy is due to the contamination of the glass fibres and alignment structures with coating remains and atmospheric dust. According to the deviations of the fibres core centre points which were determined in the experimental analysis, these account for a total of 2 μm

Recapitulatory it is safe to say, that passive alignment structures which are manufactured by ultra precision machining processes like fly-cutting are well suited for a high-precision passive alignment of glass fibres. With an optimization of the cleaning methods for the stripped glass fibre ends and for the ultra precise microstructures which are used for the passive alignment as well a minimization of the failure influences on the used measuring system, positioning tolerances of down to +/- 1 μm should be attainable within the passive alignment of glass fibres in ultra precision machined v-groove structures.

4. CONCLUSION

For the passive alignment of glass fibres in v-groove structures, the alignment of 40 fibres in a one-dimensional array resulted in a best positioning accuracy between the centre points of the fibre cores of +/- 3.2 μm in v-grooves with an apex angle of 60°. Slight differences were identified between the presented theoretical model and the experimental investigations. These can be attributed to contaminations of the glass fibres and the alignment structures as well as uncertainties in the measuring system. The experimental results show, that the passive alignment of glass fibres in alignment structures which are manufactured by ultra precision processes like fly-cutting using monocrystalline diamond tools can be realized with positioning accuracies of few micro meters. This shows that ultra precise manufactured passive alignment structures are well suited for lots of alignment tasks within micro assembly.

5. OUTLOOK

In the field of passive alignment of micro components, future research work at the Fraunhofer IPT will extend into the experimental investigation of passive alignment of spherical and cubical micro components to validate the theoretical results which have come out so far. Beyond that, theoretical as well as experimental investigations will be made on the simultaneous passive alignment of multiple micro parts and the handling of such arrangements of micro parts. The results form these researches are expected to be suitable for the fully automated mounting of hybrid microsystems.

ACKOWLEDGEMENTS

The achievements presented in this papers are the results from the publicly funded research projects GroßMikro (AiF / BMWA), SFB 440 (DFG) as well as the EC-funded Network of Excellence 4M. The authors would like to thank the AIF, the DFG and the European Commission for their support which enabled the work done in the field of ultraprecision and microsystem technology.

REFERENCES

1. S. Koelemejer Collet, J. Jacot: Cost Efficient Assembly of Microsystems. In: MST News, No. 1 (9), 1999, p. 30-32
2. Corning Incorporated: Corning SMF28 Optical Fiber, Product Information, 2004

3. N.N.: Norm DIN ISO 1101: Form- und Lagetolerierung. Form-, Richtungs-, Orts- und Lauftoleranzen, Deutsches Institut für Normung e.V., 1985
4. C. Demant, B. Streicher, A. Waszkewitz: Industrielle Bildverarbeitung, Springer, 1998
5. B. Jähne: Digitale Bildverarbeitung, Springer, 1989
6. K. Hentschel, M. Wendelstein: Telezentrische Objektive für die industrielle Bildverarbeitung, Stemmer, 2002

PART IV

Modular Assembly Systems and Control Applications

MINIATURE RECONFIGURABLE ASSEMBLY LINE FOR SMALL PRODUCTS

Alain Codourey[1], Sébastien Perroud[1], Yves Mussard[2]
[1]*CSEM Centre Suisse d'Electronique et de Microtechnique SA, Untere Gründlistrasse 1, CH-6055 Alpnach Dorf, Switzerland;* [2]*HTI Biel/Bienne, Quellgasse 21, CH-2501 Biel, Switzerland.*

Abstract: In this paper we present a new concept of miniaturized production line for the assembly of small components. The concept is based on miniature robotic cells that work together. A miniature parallel robot has been developed for this purpose and is presented in this paper.

Key words: Parallel Robot, Microfactory, Miniature Production Line, Miniature Robot

1. INTRODUCTION

Production of microsystems today requires usually bulky and costly machines that have disproportionate size compared to the size of the product they produce. A new concept called microfactory was introduced in the early 90's in Japan to bring solutions to this problem. A first prototype has been presented that is able to produce micro ball bearings[1]. This first prototype includes several machines of small size: a micro-lathe, a milling machine, a micropress, a transfer arm and a two-fingered microhand. The control of the microfactory is simple. No automatic transfer is organized between the machines. The target device is selected with a selector and then manually controlled. Several works have already been done on global concept of the microfactory[2] and on mechanical aspects. Minifactories have also been studied and prototypes have been developed by Hollis[3] and Gaugel[4].

In this paper, we present a new concept for the microfactory based on small production units of 1 dm3. A special emphasis is given to the development of a miniature robot based on the delta parallel structure and on its control system.

2. CONCEPT OF THE MICROFACTORY

A new concept has been established for the design of microfactories in collaboration between CSEM, EPFL and HTI-Biel. This concept has been named pocket factory. It is based on elementary assembly units called microboxes assembled together in series so that they build up a complete factory line. Each microbox has a volume of 100 x 100 x 100 mm^3 and contains a robot, a transfer system for the components, a feeder system and some devices for the process such as a glue dispensing system for example. In certain cases, the microbox is built as a clean room environment of class 100. A possible layout of such an assembly line is shown in figure 1.

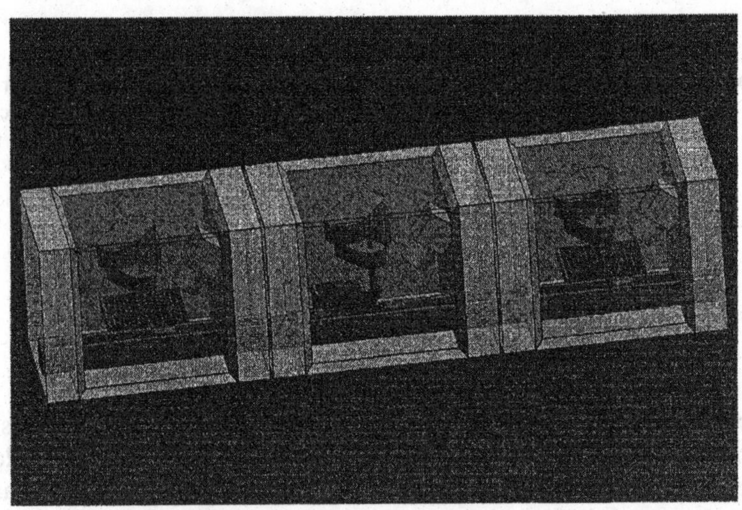

Figure 1. A possible layout of the pocket factory

Based on this idea, a fast robot with parallel structure based on the delta principle[5] has been developed. This robot is described in the next section.

3. MINI-DELTA ROBOT FOR THE POCKET FACTORY

The design of miniature robots is challenging, especially because the scaling laws are not favorable to the miniaturization of motors. On the other hand, miniaturization of mechanics leads, for the same rigidity, to a much lighter structure, giving high potential to high cadency of motion. Because the motors get big in comparison to the mechanics, it is also favorable to use parallel structures, where the motors are fixed onto the base and do not move with the robot.

3.1 Mechanical Design

Small motors usually have very high rotational speed for very little torque, but what is required is exactly the contrary. A gear box with high reduction ratio is thus needed to achieve the desired torques. This implies unfortunately the introduction of high friction in the mechanism and very often also backlash. Thanks to the parallel structure of the delta, the size of the motors is not very important in our case. The chosen motors are placed above the microbox and the transmission of the motion to the arm is realized through a cranted belt with a reduction of 1:2 only. High resolution encoders with interpolation electronics are used to get the desired resolution of 0.5 μm.

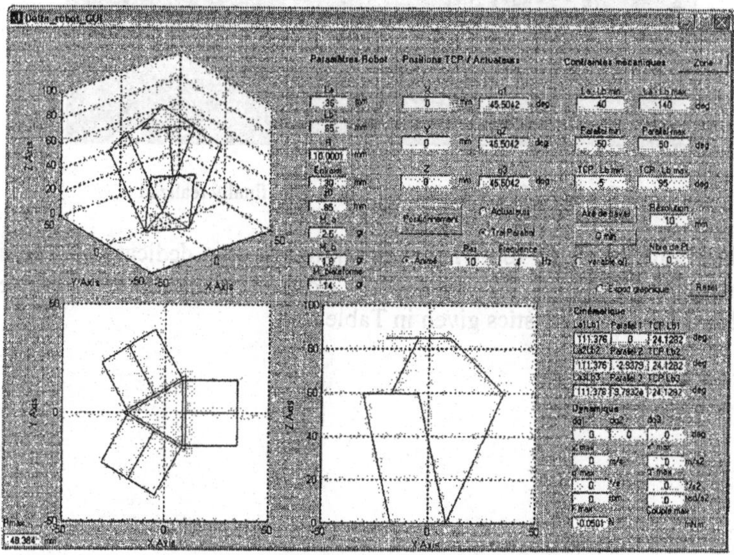

Figure 2. Simulation Environment for Optimization Purposes

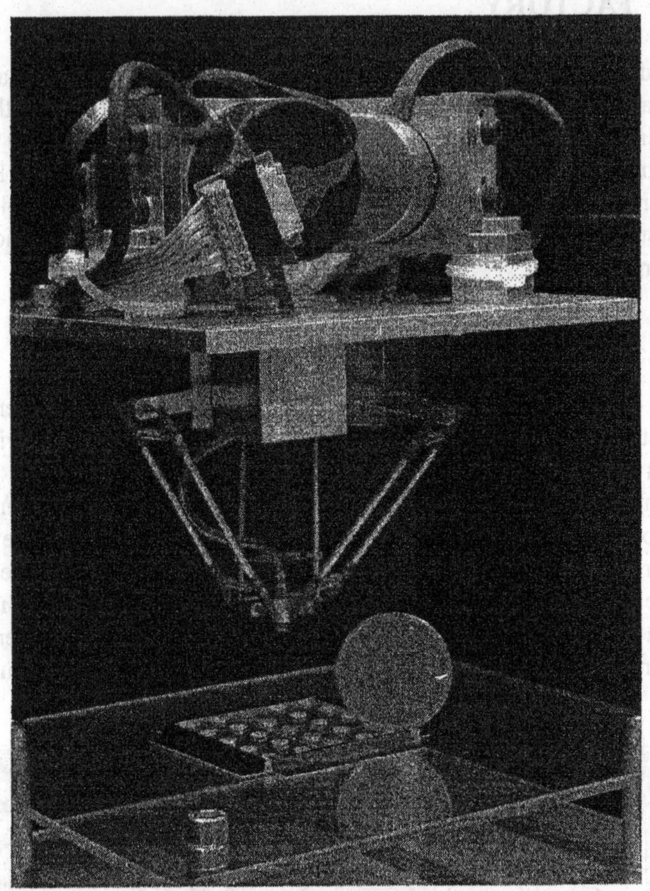

Figure 3. The optimized robot structure finally realized

The geometry of the robot has been optimized in a dedicated simulation environment (Figure 2). The optimization led to the robot shown in Figure 3, which has the characteristics given in Table 1.

Table 1. Main characteristics of the Pocket Delta

Feature	Values
Size	130x130x160 mm
Workspace	60x60x40 mm
Moving mass	23 gr.
Payload	10 gr.
Repeatability	± 2 μm
Cadency	> 3 pick & place /s

3.2 Control Structure

The controller is based on a classical PC with simple interface boards for interfacing the robot hardware (Figure 4). It might be connected to a webcam or another PC through the Ethernet interface. The control software is based on the real-time CSEM Java Framework[6].

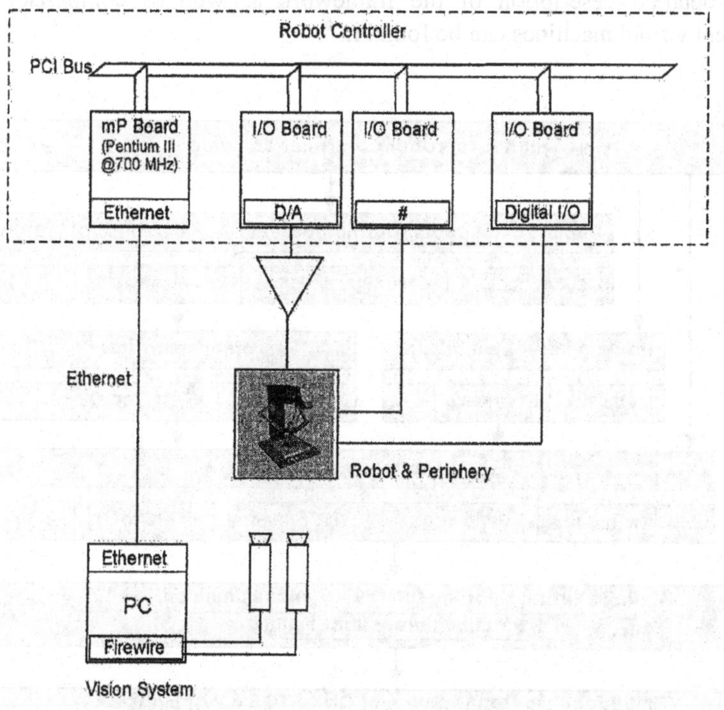

Figure 4. Control structure of the robot

This software framework has been developed at CSEM with an open design that allows extending it in any way, depending on the actual application. It is organized as a set of packages offering classes for low-level functions that may be useful for a wide range of technical software, and specific classes for robotics (see Figure 5). The low-level class library consists of the following components:

- A real-time thread class that can be used to install periodic, high priority tasks for control purposes.
- An embedded web-server that executes software codes within the robot controller through "virtual" cgi scripts. This enables control of the robot over a network.
- A hardware periphery abstraction layer that completely separates the application software from the actual I/O boards. Thus, changes of the I/O periphery only require updating a configuration file and restarting the application.

A detailed description of the framework as well as comparison of different virtual machines can be found in [6].

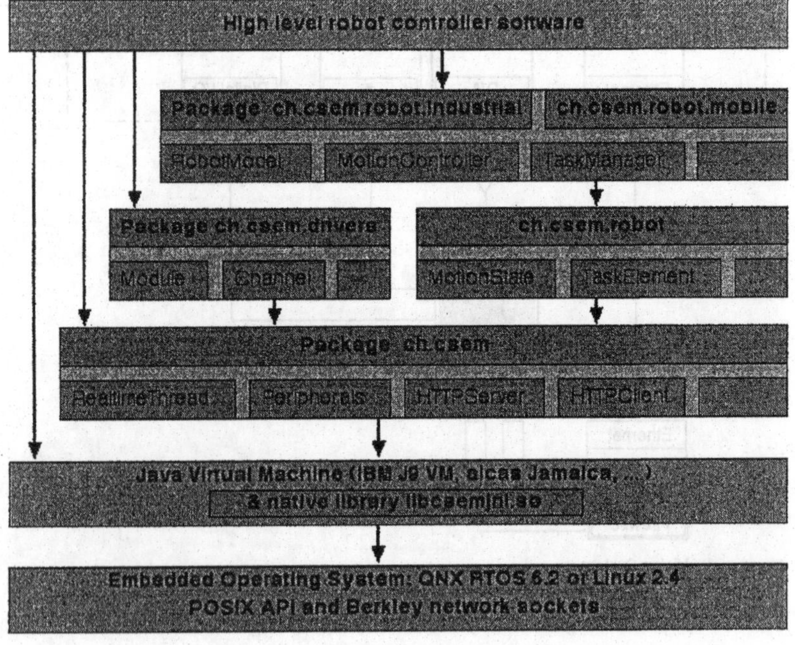

Figure 5. Software layers of the CSEM Java Framework for robot control

4. CONCLUSION

The new miniature robot is fast and precise and is well suited to assembly tasks of small components inside the microfactory concept. As demonstrator, the robot was able to realize conditioning of watch's gearwheels from a vision assisted feeder into a palette. The combination of many of these small robots will lead to new concepts for the production of small parts. A typical assembly line might look like the one in Figure 6.

Figure 6. A pocket factory for the assembly of small electrical components. It is based on 6 PocketDelta robot, each of them realizing a single assembly task.

ACKNOWLEDGMENT

This project is supported by the cantons of central Switzerland and the MCCS Micro Center Central Switzerland. Their support is gratefully acknowledged. Part of this work has been done at EPFL and at HTI Biel. We thank them warmly for their collaboration.

REFERENCES

1. N. Ooyama et al., *Desktop Machining Microfactory*, Proceedings 2nd International Workshop on Microfactories, Fribourg, Switzerland, Oct. 9-10, 2000.
2. T. Hirano et al., *Industrial Impact of the Microfactory*, Proceedings 2nd International Workshop on Microfactories, Fribourg, Switzerland, Oct. 9-10, 2000.

3. R. Hoolis and J. Gowdy, ***Miniature Factories for Precision Assembly,*** Int. Workshop on Microfactories, Tsukuba, Japan, pp. 9-14, November 1998.
4. T. Gaugel et al., ***Advanced modular production concept for miniaturied products***, Proceedings 2nd International Workshop on Microfactories, Fribourg, Switzerland, Oct. 9-10, 2000.
5. R. Clavel, ***Conception d'un robot parallèle rapide à 4 degrés de liberté***, Thèse EPFL No 925, 1991.
6. M. Honegger, "A Java-Based Framework for the Design of Robot Controllers", in ***Proceedings of Msy'02***, Winterthur, Switzerland, October 2002.

CONCEPTION OF A SCALABLE PRODUCTION FOR MICRO-MECHATRONICAL PRODUCTS

Systematics for planning and platform with process modules for the production of micro-mechatronical products

Jürgen Fleischer, Luben Krahtov, Torsten Volkmann
wbk - Institute of Production Science, University of Karlsruhe, Germany

Abstract: Miniaturization and mechatronics are important technological trends of the last years and will be gaining importance in the future. The micro-mechatronical products which have already been put on the market are currently designed to meet the abilities of existing production techniques and are manufactured as mass products in specialized and capital intensive plants. In order to make miniaturized mechatronical systems more attractive for the market, their production must be made possible in small and medium volumes at acceptable costs. A holistic approach for the conception of a scalable micro production which consists of the development of an integrated planning process as well as the creation of a flexible production platform with general process modules is presented and demonstrated considering as example the realization of a scalable production for a micro-mechatronical actuator. The approach is characterized by a high level of flexibility considering joining technologies as well as small space requirements. A reduction of capital costs for production technology and a stepwise extension of the production are achieved.

Key words: Planning Systematics, Micro-Assembly, Micro-Mechatronical Products, Automation

1. INTRODUCTION

In the last years, an important market with considerable growth rates and a large number of different products has developed in the field of micro technology[1,2,3]. The main application areas are the information and communications technology, automation, automotive and medical engineering[4]. The production methods essentially being used for manufacturing of micro-products can be subdivided into two differing procedures. On the one hand, there is micro-systems engineering based on lithographical processes, which developed out of the domain of semiconductor engineering. On the other hand, there is precision engineering where classical manufacturing methods are used to produce parts with increasingly smaller dimensions.

For manufacturing micro-mechatronical products, both the manufacturing methods of the micro-systems technology and the precision engineering are applied[5]. In this context micro-assembly plays an important role in integrating the variously produced components into a complete system. Micro-assembly is characterized by high requirements in accuracy and constraints in package dimension. Numerous and differing joining processes, which are very complex due to the micro-specific requirements, e.g. small influence of gravitational forces in comparison with surface forces, high sensitivity and small tolerances of the components, are applied. The micro-mechatronical products discussed in this paper are composed three dimensionally and the component size is within the range of 400 µm up to 4 mm and the total product size within the range of 2 mm up to 50 mm.

The selection and verification of the micro-assembly processes are very extensive since there is a deficit of a holistic planning systematics. As a result, a long product introduction phase which takes a significant portion of the total product life cycle is caused. At present, most micro-technical products are manufactured as mass products in specialized and capital intensive plants[6]. Production of miniaturized mechatronical systems in small to medium volumes, which is mostly performed manually, is restricted due to complexity and quality requirements. In order to make miniaturized mechatronical systems increasingly more attractive for the market, production must be enabled in small and medium volumes at acceptable prices[7]. Currently, there are only a few solutions which permit an economical access to micro production and suit for a scalable extension or a reconfiguration.

The Institute of Production Science at the University of Karlsruhe is elaborating a holistic approach for the conception of a scalable micro production which consists of the development of an integrated planning process as well as the creation of a flexible production platform with general process modules[8, 9, 10].

2. INTEGRATED PLANNING PROCESS

The input of the integrated planning systematics (see Fig. 1) is the structure of a micro-mechatronical product. The selection of processes for micro-assembly is supported by a feature-based procedure. A micro-assembly precedence graph is derived and contains all technically possible solution alternatives of the assembly for the product. Due to micro-specific requirements, e.g. gripper selection considering adhesion behavior or sensitivity of the components, the micro-assembly precedence graph includes micro-specific information fields. This ensures that assembly sequence as well as information of handling steps and linkage of the single processes is contained. Based on this micro precedence graph, linkage concepts, e.g. small volume manufacturing with a stand alone system, series manufacturing in linked systems, are developed. By means of the simulation tool Tecnomatix eM-Plant, which is adapted, the concepts will be evaluated and improved in an iteration loop. As a result of the planning systematics, an optimized and configured assembly is provided.

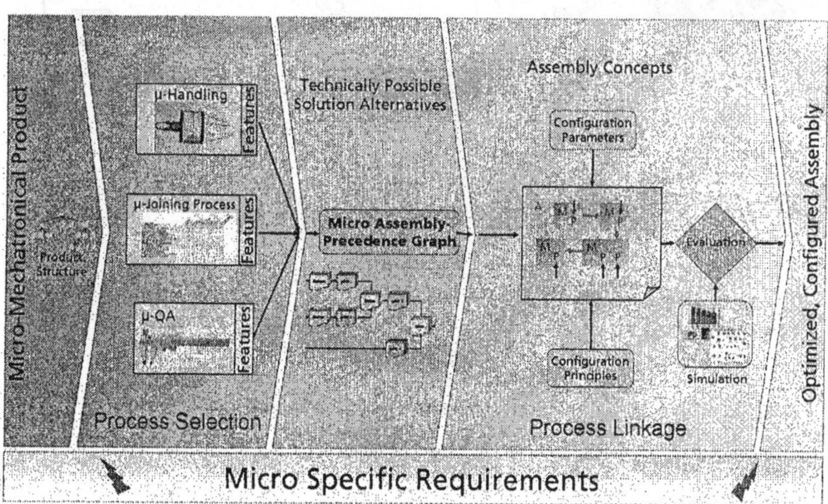

Figure 1. Integrated planning systematics for micro-mechatronical products

2.1 Selection of Processes and Handling Tools

The selection of processes and handling tools plays an important role within the planning of micro-assembly. Micro-assembly can be divided into two basic operations: joining processes as well as the appropriate handling steps. Between each joining process, a handling operation comprising mainly gripping procedures, but also the transport of components from one process step to the next, is applied. These handling operations should be integrated into the planning process like all other process steps. To plan the micro handling processes a method based on feature technology is developed.

Complex handling operations like flipping or joining of miniaturized components must be planned in detail due to the interference effects of the micro world, e.g. adhesive forces. As a consequence of the insufficiently examined effects of micro handling, the planning of these processes is difficult. It is quite impossible to make exact predictions about the handling of micro components until it is tested in empirical studies. Thus, in the earlier phase, when physical micro components are not available, exact planning of micro-assembly processes is impossible.

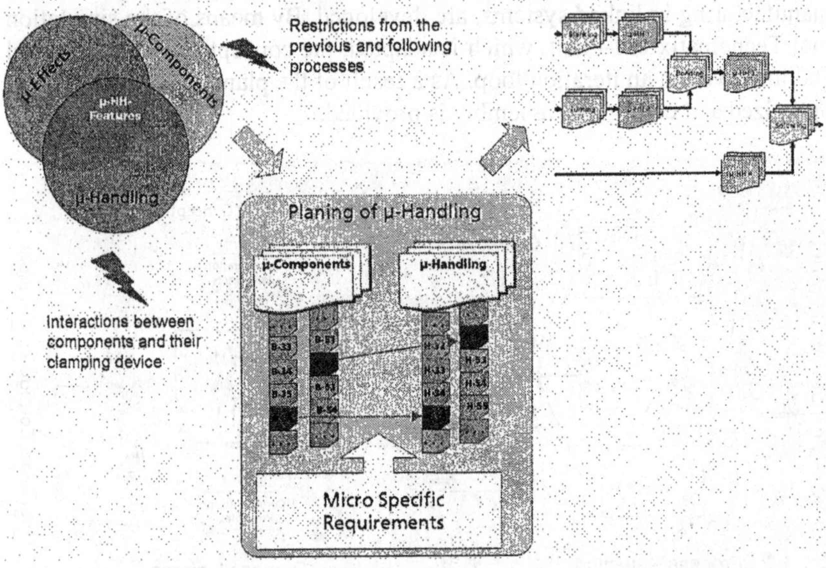

Figure 2. Derivation of the micro-assembly precedence graph with feature technology

An important precondition for the holistic planning process is the removal of these planning deficits by creating a methodology for the description of micro components, micro interference effects and the micro handling tools by feature-technology. The method (see Fig. 2) conjugates features (e.g. sensitivity) to the components and performs a comparison with the corresponding gripper features to determine the compatibility. In addition to geometrical information the micro handling features contain semantic information (e.g. tendency of adhesion), restrictions from the previous and following processes as well as interactions between components and their clamping device. The micro handling features are the logical succession of the construction, manufacturing and quality assurance features[11, 12]. By applying feature technology in the planning phase of the product, the accurate assignment of components to a suitable gripper with regard to the corresponding environmental restrictions becomes possible. As described above a detailed product specific micro-assembly precedence graph is derived in which all assembly processes are planned in detail.

2.2 Process Linkage

Based on the micro-assembly precedence graph, a linkage concept is elaborated. A methodology to provide concepts depending on a rule-based selection is developed. Rules considering technical, economic and logistical design parameters and principals as well as input and output relations and interactions have to be defined. An important requirement on the concepts is an accelerated cost and risk minimized realization of the production with regard to varying production volumes. The stepwise extension is supported by the integrated planning systematics and in each case the optimal configuration is determined. As a basis for the planning systematics, a flexible production platform and standardized process and linkage modules are required to set up a scalable production for micro-mechatronical systems. In research there exist only a few approaches which are related to the manufacturing of sensors and biological chips[13]. However, no integrated planning in terms of a scalable production is taken into consideration.

An important criterion for a scalable production is to use the production equipment of the development and prototype phase also for the production of small to medium volumes. Particularly at a production within this range an unsorted supply of single components with picking and orienting is difficult and expensive to realize or even not applicable. This is caused by the sensitivity against mechanical damage and adhesive problematic of the micro components. Especially in bonding processes the components must be cleaned and pretreated. For this reason, the concept of a work piece carrier is used as a solution approach. A work piece carrier which passes through the entire production can be designed according to different technical and logistical constraints. Numerous functions can be performed by the work piece carrier including storage of raw materials, hybrid 3-dimensional micro-components and tools as well as additional and auxiliary axes or gripping and clamping devices. The work piece carrier integrates tasks of the assembly area, the devices and the transportation medium in itself. A direct assembly on the work piece carrier is easy to realize and leads to shorter assembly time and better achievable accuracies because this process requires less space. The tolerances of the work piece carrier are compensated by an image processing system of the assembly platform and a high joining accuracy is ensured. The transport of the components between different processes or machines is reliable with the work piece carrier and is simplified by providing the once achieved orientation of the components and the standardized interface. In addition, components and assembly relevant data can be easily assigned and traced by a identification tag at the work piece carrier.

3. MODULAR PLATFORM AND PROCESS MODULES

The production of micro-mechatronical products in small to medium volumes plays an important role for the investment in the production equipment. Automated solutions are mostly dimensioned for high quantities and comprise special solutions for one product with several machines and expensive linkage technology. An approach for the economical access to automated micro-assembly is to start automation with a modular platform, process modules and work piece carrier linkage.

The platform makes an economical basic infrastructure available for the micro-assembly. An axis system which can reach all positions within the work space and offers a tool or gripper fitting is provided for. For precise positioning of the axes a high definition image processing system is applied and allows high flexibility. Furthermore, an open control concept makes the integration of the process modules quick and easy to integrate (CAN-Open).

The required processes in micro-assembly are integrated in general process modules, e.g. positioning, rotating, pressing, riveting, adhesive / soldering paste dispensing, heat hardening / UV exposing, ultrasonic / laser welding or quality assurance.

The platform concept enables an easy upgrading and multiplication of the production and a stepwise extension can be fulfilled with relatively equal processes. While within the pilot phase the focus is on the verification of the processes and the determination of the necessary process parameters, the commissioning and the flow of material become increasingly important as the production quantity increases. With this concept, the components can initially be cleaned manually, sorted, and then commissioned on the work piece carrier for a small volume production. In further stages of development, the commissioning can be performed automatically by commissioning devices. Furthermore, the process modules can be distributed on multiple machines and linked by appropriate linkage modules to increase the productivity.

The advantage of this proceeding is that the assembly processes are used from the first phase up to the expansion to a production of large-scale volumes in an automated environment. Thus, it is possible to identify the manufacturing, assembly, material allocation and the logistical problems at an early stage. Beyond that, the modular platform as well as the process modules can be modeled and evaluated with simulation tools.

4. PRODUCTION OF MICRO-MECHATRONICAL SYSTEMS

4.1 Demonstrator Product

An example of a micro-mechatronical product is a flat motor of the company "mymotors" (diameter 12 mm, thickness 3 mm). It is characterized by a complex assembly and has a characteristical mechatronical structure, e.g. mechanical components generate signals and printed circuit boards have mechanical functions.

The product consists of eleven single components and several joining processes are applied for the assembly. As described in chapter 2, the features, e.g. sensitivity of the components were analyzed at first and according to the product structure a micro-assembly precedence graph (see Fig. 3) was derived. It contains all processes, e.g. bonding, soldering, pressing, riveting and describes the linkage between processes, handling and linkage. Furthermore, the micro-specific restrictions are taken into account by designing the handling tools and the work piece carrier.

The layout of the process and linkage modules within the modular production platform is derived out of the micro-assembly precedence graph. As modular platform a micro-assembly machine from the company "Häcker Automation" is applied and described below.

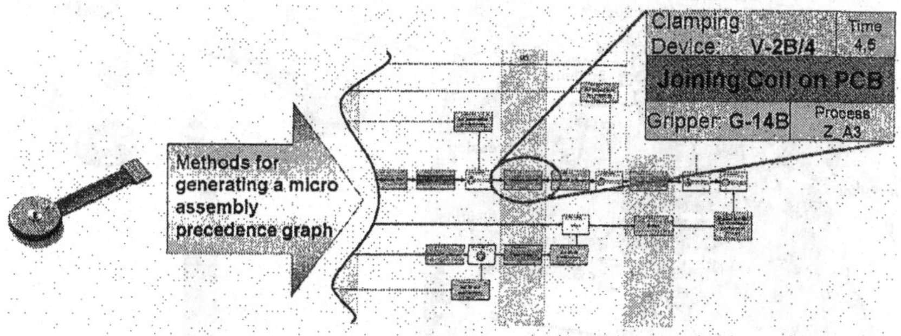

Figure 3. Micro-mechatronical flat motor and corresponding micro-assembly precedence graph

4.2 Modular Production Platform on the Example of a Micro-Assembly Machine

The micro-assembly machine consists of a precise serial kinematics with a repeat accuracy of 20 µm and a flexible and modular configurable working space of 450x450 mm². The placement head integrates an image processing system (IPS) and a double head system.

The IPS is able to percept and analyze objects three dimensionally. Two high-resolution cameras are used and positioned at an angle of approximately 23° in relation to one another. Thus, the system is able to locate the position of the components as well as to detect their height with an accuracy down to 5 µm.

The double head system consists of a placement head and a dispense head. The placement head has a vacuum connection and an integrated gripper change module which enables a fully automated change of various vacuum and mechanical grippers. An innovative measure system (spatio-volumetric flow rate) enables the dispense head to perform dosings of media with different viscosities with an accuracy in the dimension of picoliters. After each dispense procedure a correction of the time-pressure control is done in correlation to the quantity dispensed during the previous procedure. Changes in viscosity of the different media are actively adjusted.

In addition to these permanently installed components, several other automated process modules, which can be placed in the working area, are available (see Fig. 4). These are, e.g. a flipping station or a 3D-station with additional axes for flipping. Furthermore, a heat lamp was developed as a process module to integrate bonding and soldering processes.

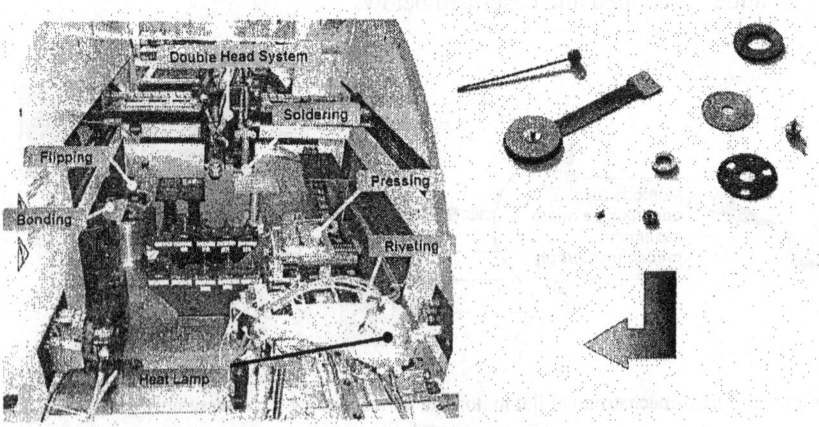

Figure 4. Micro-assembly machine with double head system

4.3 Process Modules on the Example of a combined Soldering, Bonding and Pressing Module

The two joining processes of bonding and soldering are playing a substantial role in the assembly of the flat motor. The rapid curing of adhesive or melting of soldering paste at a defined time within the assembly process is important for the automation of the process. A small and compact heat source which is characterized by a fast selective and local heating area is necessary. As a solution, a heat lamp was developed. It consists of a reflector, a halogen lamp and a case with heat sink and protection glass. The reflector is a bisected ellipsoid mirror and the halogen lamp is positioned in the first focal point. In the second focal point of the ellipsoid mirror the spot of the bundled light is located and heat is generated. The cooling strategy plays an important role when designing of the heat lamp. The interior of the ellipsoid and the heat sink is streamed and cooled by air. Subsequently, the air is conveyed in front of the protection glass towards the light spot to blow away steam from soldering. Visualization and control of the soldering and bonding processes are important for automation. Two laser diodes permit the accurate positioning of the focal point when teaching in the process. Thus, the x-, y- and z-position for the components to be soldered or bonded could be easily determined. Control of the temperature is measured by a laterally attached contactless infrared thermometer.

The z-axis of the heat lamp module is also used for process integration of riveting within this module, due to heat generation and temperature measurement is contactless. The process module can achieve a maximum press force of 1000 N. The pressing force controlled by a force transducer is placed directly behind the replaceable press punch. Thus, three processes with two basic functionalities are implemented within only one module (see Fig. 5).

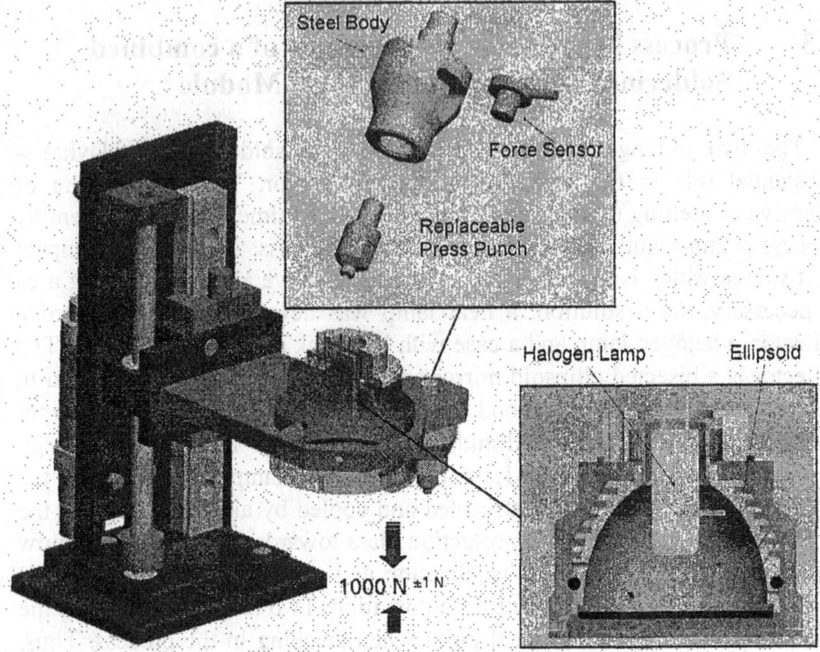

Figure 5. Standardized device with combined soldering, bonding and riveting module

The device is mounted outside the working space of the assembly machine and is supplied by two additional axes. In this way, processes can be executed simultaneously. While performing the soldering process under the heat lamp, the main kinematics of the assembly machine can proceed other operations.

The flexible and modular design of this device offers the implementation of further processes, like e.g. UV-exposure, ultrasonic or laser welding, electronic examination and quality assurance.

5. LINKAGE WITH WORK PIECE CARRIER ON THE EXAMPLE OF A SMALL VOLUME PRODUCTION

A small volume production of a flat motor on the micro-assembly machine by using a work piece carrier was implemented at the Institute of Production Science (see Fig. 6). The layout consists of the process modules for bonding and flipping processes, the tool station for replaceable grippers and a bottom side camera, as well as an x-y-unit for the transport and positioning of the work piece carrier. A multi-functional work piece carrier which is designed according to the German DIN 32561[14] is utilized. At the contacted dispensing of adhesive or soldering paste there is the danger of moving the components out of position or pulling them out of their molding cavity due to strong adhesive forces. A vacuum underneath the components eliminates this problem.

Furthermore, the work piece carrier has the functions of a magazine for the components as well as of the assembly area for all the assembly operations during the manufacturing process. Due to a skillful design of the molding cavities, it is guaranteed that the components remain fixed after each joining and bonding operation. In the phase of a small volume production the components are placed manually on the work piece carrier. With increasing production quantities the work piece carrier can be fitted with components by the assembly system, e.g. by belt feeders. The multi-functional work piece carrier is the central element of the previously described automated micro-assembly and can be adapted to product modifications or changes without having influence on the total assembly system. This guarantees the flexibility and modularity of such a scalable concept.

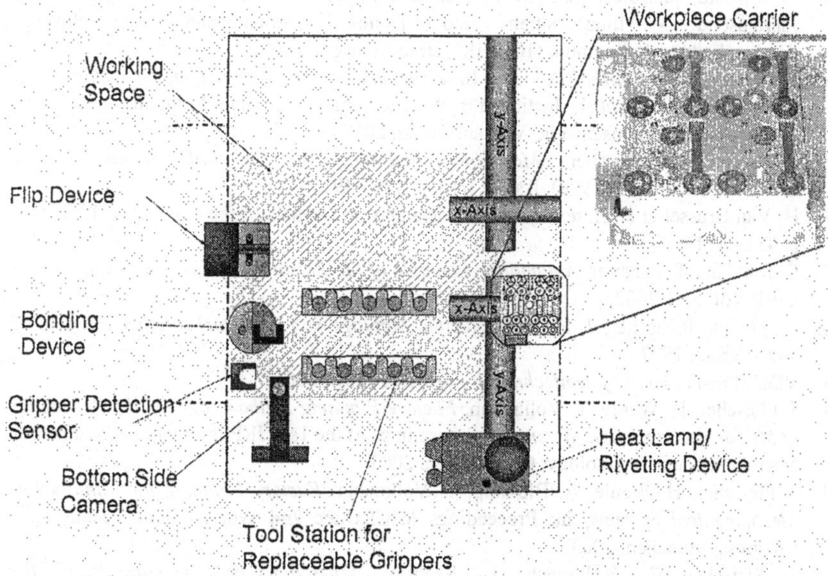

Figure 6. Layout of the automated assembly of the flat motor with multi-functional work piece carrier according to DIN 32561

6. SUMMARY

For the economic manufacturing of micro-mechatronical products, a holistic planning systematics as well as a scalable manufacturing technology plays an increasingly important role. An optimized and configured micro-assembly is enabled by methodical assistance at the process and handling tool selection phase, the deriving of a micro-assembly precedence graph as well as the selection of a linkage concept. As a basis, a modular assembly platform and a construction kit of process and linkage modules is developed. Thus, a high level of flexibility according to joining technologies, variability and reconfigurability of the system is guaranteed. The manufacturing of initially small volumes and the stepwise extension of production quantities is realized with small space requirements. Due to modular design, the reduction of capital costs for the manufacturing technology is possible and the demand for an economic access into micro manufacturing technology can be achieved.

REFERENCES

1. W. Ehrfeld, *Mikrotechnik - Produktion und Produkte*, in: Feinbearbeitung im neuen Jahrtausend - Innovationen und Trends, Internationales Braunschweiger Feinbearbeitungskolloquium (10. FBK), Braunschweig, 2000.
2. J. Hesselbach, *mikroPRO - Untersuchung zum internationalen Stand der Mikroproduktionstechnik*, Institut für Werkzeugmaschinen und Fertigungstechnik TU Braunschweig, Braunschweig, Vulkan Verlag, 2002.
3. M. Weck; S. Fischer, *Maschinenentwicklung für die Mikrotechnik*, in: wt Werkstattstechnik 89 (1999), Issue 11/12, p. 489 - 491.
4. H. Van Brussel et al, *Assembly of Microsystems*, in: Annals of the CIRP Vol. 49/2/2000, p. 451 - 471.
5. G. Reinhart, T. Angerer, *Automated Assembly of Mechatronic Products*, in: Annals of the CIRP Vol. 51/1/2002, p. 1 - 4.
6. D. Zühlke, R.Fischer, J. Hankes, *Flexible Montage von Miniaturbauteilen*, Düsseldorf, VDI-Verlag, 1997.
7. VDI/VDE-IT, *Mikrosystemtechnik in Europa*, Teltow, 2001.
8. J. Fleischer, H. Weule, T. Volkmann, *Factory Planning Methodology for the Production of Micro-Mechatronical Systems*, in Proceedings of the 1st CIRP International Seminar on Micro and Nano Technology, Copenhagen, 2003.
9. J. Fleischer, H. Weule, A. Blessing, T. Volkmann, *Factory Planning for Miniaturized Mechatronical Systems*, in: Proceedings MICRO.tec 2nd World Micro Technologies Congress, München, 2003.
10. J. Fleischer, T. Volkmann, L. Krahtov, *Wirtschaftliche Automatisierung der Mikromontage in der Kleinserienfertigung*, in: wt Werkstattstechnik 94 (2004), Issue 9, p. 390 - 394.
11. J. Elsner, *Informationsmanagement für mehrstufige Mikro-Fertigungsprozesse*, Heinz W. Holler Druck und Verlags GmbH, Karlsruhe, 2003.

12. DI Richtlinie 2118, *Informationsverarbeitung in der Feature-Technologie*, Berlin, Beuth Verlag, 2003.
13. T. Gaugel, M. Bengel et al, *Building a mini-assembly system from a technology construction kit*; in Assembly Automation Band 24, Issue 1; 2004, p. 43 - 48.
14. DIN 32561, *Production equipment for Microsystems - Tray - Dimensions and Tolerances*, Deutsches Institut für Normung e.V., Berlin, 2003.

TOWARDS AN INTEGRATED ASSEMBLY PROCESS DECOMPOSITION AND MODULAR EQUIPMENT CONFIGURATION

A KNOWLEDGE ENHANCED ITERATIVE APPROACH

Niels Lohse[1], Christian Schäfer[2], Svetan Ratchev[1]

[1]University of Nottingham, Precision Manufacturing Group, School of M3, University Park, Nottingham, NG7 2RD, UNITED KINGDOM; [2]Robert Bosch GmbH, Software and System Engineering in Production Automation CR/APA3, Postfach 30 02 40, D-70442 Stuttgart, GERMANY

Abstract: In today's increasingly volatile and dynamic global markets it is increasingly important to react to changing market demands and reduce the time-to-market. The design and re-design of assembly systems has a significant impact on the product development time. This paper reports on the effort that has been put into developing an assembly process decomposition and modular assembly equipment configuration methodology that takes advantage of the current trend towards modular equipment solutions and is expected to reduce design time and improve the design process integration. A general framework for the proposed methodology has been outlined and an ontology for the design of modular assembly systems is being discussed.

Key words: Assembly Process Decomposition, Equipment Configuration, Modular Assembly Systems, Ontology, Agent-based

1. INTRODUCTION

The requirements driven specification of assembly process as well as the selection and configuration of equipment for suitable assembly system solutions are central aspects of the assembly system design and redesign process. The assembly process specification should define the temporally ordered activities from the order in which the different components of a product or product family can be assembled to specific actions that need to be performed to facilitate the actual establishment of the individual liaisons between the components. The equipment selection and configuration process needs to find suitable equipment solutions for the required processes and combine them into a working assembly system.

Both the process decomposition and equipment configuration are highly related. The process definition prescribes the required equipment and the available equipment constrains the process decomposition (Rampersad, 1994). Often however these two aspects of the design process are considered separately and equally important constraints are neglected. Hence a methodology is needed that facilitates the dynamic decomposition of assembly processes and the configuration of equipment solutions within a single integrated framework that makes best uses of available expert knowledge.

Currently there is a strong trend towards Evolvable and Reconfigurable Assembly Systems (RAS) that enable enterprises to rapidly respond to changes in today's increasingly volatile and dynamic global markets without having to commit large investments in advance (EUPASS, 2005; Onori et al., 2002; Koren et al., 1999). One of the enabling factors of RAS is the availability of highly standardized modular equipment solutions that can be rapidly configured to deliver different assembly solutions. This opens the scope and need for higher degree of integration and automation during the design of such systems. Configuration methodologies that have been demonstrated in the computer industry which benefit from a higher degree of modularization can be harnessed to solve the challenges of the assembly system design process. Examples of such configuration methods include XCON (McDermott, 1982), MICON (Birmingham et al., 1988), and COSSACK (Mittal and Frayman, 1989).

A number of different frameworks for the configuration and design of products and system has been proposed (Bley et al., 1994; Lu et al., 2000; Boër et al., 2001; Jin and Lu, 2004). Distributed collaborative design frameworks provide clear advantages for the considered design problem as discussed by Rosenman and Wang (2001) particularly when combined with object and component oriented modelling approaches as are commonly used under the CIM paradigm (Schäfer and López, 1999). We propose to use a distributed knowledge based reasoning approach for the process decomposition and equipment configuration and an agent based framework to facilitate their integration.

The paper provides a more detailed definition of the proposed framework. Furthermore, the underlying knowledge model will be outlined and the approach will be illustrated with an example reconfiguration. To conclude the paper, the whole approach will be critically discussed and further work has been outlined.

2. METHODOLOGY OVERVIEW

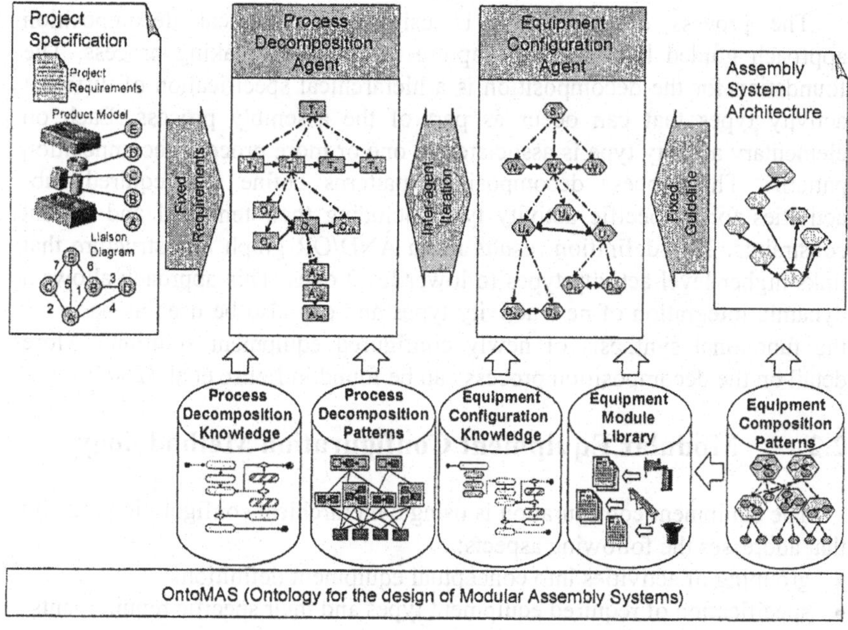

Figure 1. Integrated Assembly Process Decomposition and Equipment Synthesis Framework

Figure 1 shows an overview of the proposed approach. For this work it has been assumed that the product requirements that drive the assembly process decomposition in the first instance are fixed and are none negotiable. Furthermore it was assumed that a set of equipment modules exists that permits at least one viable configuration to fulfil a given set of requirements. However, the framework has been defined to allow for later extensions to include these aspects.

The assembly process decomposition is guided by rule-based patterns that define how different types of assembly activities break down into sub-activities including their temporal and logical relationships. The decomposition is prescribed through the product requirements and influenced by continuous equipment choices. The configuration of suitable equipment solutions is based on a virtual library of available equipment modules. The configuration process is guided by equipment configuration knowledge and constraint by a chosen system architecture for the given product domain. The interaction between the process decomposition and the equipment configuration is defined as an iterative process between cooperating software agents.

2.1 Process Decomposition Methodology

The process decomposition is using a hierarchical decomposition approach guided by rules that capture the decision making process. The foundation for the decomposition is a hierarchical specification of different activity types that can occur as part of the assembly process. Each on elementary activity type is associated to one or more process decomposition patterns. The process decomposition patterns define the required sub-activities for a specific activity type including their temporal and logical constraints. This definition results in an AND/OR graph like structure that links higher level activity types to lower level ones. This approach allows a dynamic integration of new activity types and can also be used as basis for the functional synthesis of newly configured equipment solutions. More detail on the decomposition process can be found in Lohse et al. (2005a).

2.2 Modular Equipment Configuration Methodology

The equipment configuration is using a hierarchical configuration method that addresses the following aspects:

- grouping of activities into conceptual equipment definitions
- specification of required equipment types and their specific requirements
- selection and evaluation of suitable equipment modules
- integration and functional synthesis of selected equipment modules

Each aspect of the configuration process is performed by specific agents presenting domain experts. For example the selection and evaluation of equipment is done by different expert agents for the different types of equipment. There is an agent that provides the capability to select and evaluate grippers, one to do the same for manipulators, etc.

The assembly equipment configuration is guided by predefined module types and interface specifications that are specified as part of a chosen system architecture for a specific product domain. The module specifications defined the required functional capabilities of different module types and their connectivity constraints based on the interface specifications. The system architecture also defines the logical and spatial constraints between the different types of modules. The use of an architecture definition makes the configuration process more effective by reducing the number of possible solutions. This is advantageous as long as there is a mechanism to ensure that the architecture is constantly updated. It is still an open question where the break even point between improved effectiveness and lost advantage due to the restriction of possible solutions is. The approach was designed under the assumption that there exist domain specific architectures that cater for the majority of the needs in their domain.

2.3 Agent-based Interaction/Negotiation

The interaction between the process decomposition and the equipment configuration is defined as an iterative process between cooperating software agents with different design objectives. The agents that facilitate the assembly process decomposition aim to find the best possible fulfilment of the assembly process requirements whilst the agents responsible for the equipment configuration search for the most effective configuration of the assembly equipment. These objectives are naturally contradictive since one is trying to minimise cycle time and process flow and the other cost and space.

Each agent has its own knowledge resources and is associated to a human expert who is responsible for the critical decisions that can not be fully automated. The conflict resolution strategy is based on inter-agent iterative negotiation as suggested by Lu et al. (2000). The interactions between the different agents in the framework are defined in terms of the FIPA interaction protocols (FIPA, 2005). Further detail on the general decision making framework can be found in Lohse et al. (2004).

3. ONTOLOGY FOR THE DESIGN OF MODULAR ASSEMBLY SYSTEM (ONTOMAS)

The decomposition and configuration methods are underlined by an Ontology for the design of Modular Assembly System (OntoMAS) that defines the product, the assembly process, and the assembly equipment domain knowledge models. The ontology is defined based on the general engineering ontology structure suggested by Borst et al. (1997). They suggest a fundamental ontology structure based on mereological, topological, and system theory principles. Their suggested structure has been extended to also include abstraction relationships. This is a knowledge based definition that is closely related to the object-oriented paradigm.

The concepts in the proposed overall ontology are split into three separate domain models; the product, assembly process, and assembly equipment domain models. The product is modelled as assemblies, components, and the liaisons between them; the assembly process as activities and their temporal relationships. The most complex model is used for the assembly equipment. It is modelled as virtual components with a function-behaviour-structure representation (Lohse et al., 2005b).

Functions express the capabilities of an equipment module based on the intentions of the designer and are therefore subjective and domain specific. For example the intended function of a robot is to move end effectors. **Behaviour** characterises how an equipment module reacts to changes in its environment and in turn how its reaction influences the environment based on physical phenomena. For example the high level behaviour of a robot is the transformation of electrical energy into kinetic energy under the guidance of control signals. **Structure** defines the physical aspect of the equipment model with geometric objects and connections. In the case of the robot that would include the links and joint definitions of its structure. The attributes of the three aspect models are all based on a fully parametric model.

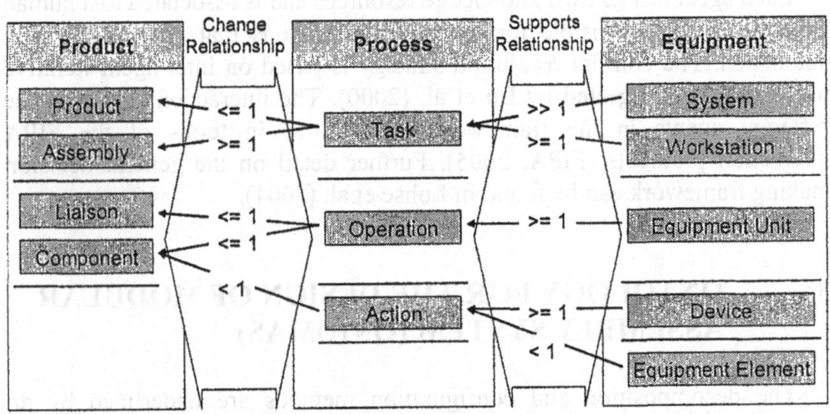

Figure 2. Logical relationships between hierarchical levels of the domain models (extension from Lohse et al., 2005b)

For the hierarchical structure of the three models we suggest that it is advantageous to define a number of distinct levels which have quite rigidly defined relationships. We expect that this will significantly reduce the configuration effort since a straight forward comparison between the required assembly activities and the functional capabilities provided by the synthesised equipment functionally will be possible on each level. Figure 2 shows the proposed levels of hierarchy and how they related to each other.

On the assembly process side, the activities are structured on three levels of hierarchy: task level, operation level, and action level. Actions form the fundamental activities that can be performed by a piece of equipment without the goal to directly influence an object related to the product. Operations are processes that facilitate state changes of entities that are part of a product. Tasks are processes that facilitate clear definable portions of work towards the completion of a product.

The equipment modules are allocated to five hierarchical levels depending on their functional capability. Systems are assembly equipment configurations that perform all the tasks required to assemble the whole product. Workstations are the smallest equipment modules that facilitate the whole assembly of at least two components. Equipment units execute at least one operation and devices at least one action. Equipment elements denote the lowest level of equipment modules that do not have an active function that would enable them to perform an action.

4. ILLUSTRATIVE EXAMPLE

The proposed iterative process decomposition and equipment configuration can best be demonstrated with an illustrative example. The given example here is not entirely based on real data and has been defined for illustration purposes only. Also not all the stages of the design process are shown since this would unnecessarily overcomplicate the example and reduce its explanatory value.

The given example is based on a new design process without any given equipment. The starting point is a set of user requirements for an assembly system including a complete product specification (see Figure 3). The example product is a simple peg-in-hole assembly with a loose fit liaison between component 1 and component 2. From this definition it can be determined that for the required assembly task (T_{A3}) both components need to be supplied, they need to be assembled (O_{A9}), and the completed assembly need to be removed. Furthermore, it can be determined that component 1 needs to be feed (O_{F7}) if component 2 has been defined as base part and also that the assembly operation needs to be an insertion. The classification of the assembly operation as insertion directly entails the specification of a number of required lower level actions.

The first iteration of the configuration can be defined from this initial assembly process specification. An assembly workstation (W_3) is needed to complete the assembly task (T_{A3}). The workstation needs to contain a number of equipment units. In this case an existing feeder unit (U_6) has been found that matches the requirements of the feeding operation (O_{F7}). The assembly operation (O_{A9}) is associated to an assembly unit (U_5) which is configured from a number of lower level devices and elements.

The connection of the feeder unit (U_6) and the assembly unit (U_5) cause the spatial location at the end of the feeding operation (O_{F7}) and at the starting point of the assembly operation (O_{A9}) to be different. This requires an adaptation of the initially defined assembly process which causes new operations to be added. This in turn changes the responsibility of the selected equipment and might cause them to be changed. The process continues until a stable process-equipment solution has been found which can be subjected to performance evaluation.

A re-configuration process would be similar with the only difference that the design process would start with an existing equipment configuration and the required changes need to be established first.

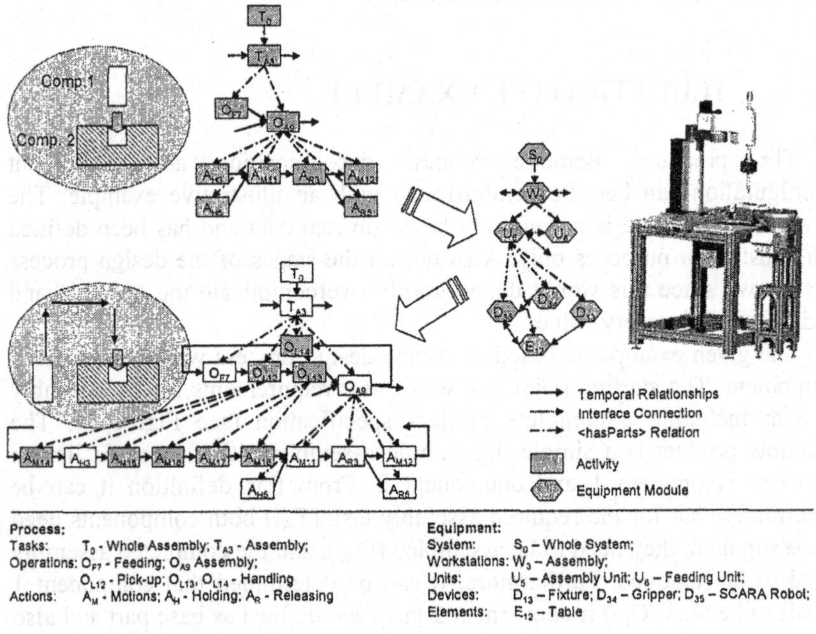

Figure 3. Illustrative Decomposition and Configuration of an Assembly Task

5. CONCLUSIONS AND FUTURE WORK

In this paper we have outlined a new methodology for the integrated specification of assembly processes and modular assembly equipment configurations. The methodology is based on a distributed agent-based reasoning approach that is supported by expert knowledge. The process specification has been defined as a decomposition process using rule-based process specification patterns and the assembly equipment specification as an architecture constraint, hierarchical configuration process. Both methods have been integrated into a common framework with an integrated ontology for the design of modular assembly systems (OntoMAS).

We perceive this methodology to have potentially a significant impact on:

- The reduction of the time-to-market of new products that need automated assembly systems,
- The reduction of the reconfiguration effort permitting more reconfiguration steps
- The improvement of the integration and quality of the design process of modular, automated assembly systems by providing a consistent knowledge infrastructure throughout the whole design process

Future work will be focused on:

- The test of the proposed methodology in industrial use-cases
- Incorporating a wider range of process prototypes and equipment definitions
- Extending the methodology to cover a wider part of the design process to include for example incomplete product definitions

REFERENCES

1. Birmingham, W. P., Brennan, A., Gupta, A. P., and Sieworek, D. P., 1988, MICON: A single board computer synthesis tool, *IEEE Circuits and Devices*, 37-46
2. Bley, H., Dietz, S., Roth, N., and Zintl, G., 1994, Knowledge of Selecting Assembly Cell Components and Its Distribution to CAD and an Expert System for Processing, *Annals of the CIRP*, **43**(1):5-8
3. Boër, C. R., Pedrazzoli, P., Sacco, M., Rinaldi, R., De Pascale, and G., Avai, A., 2001, Integrated Computer Aided Design for Assembly Systems, *Annals of the CIRP*, **50**(1):17-20
4. Borst, P., Akkermans, H., and Top, J., 1997, Engineering Ontologies, *International Journal of Human-Computer Studies*, **46**:365-406
5. EUPASS, 2005, Evolvable Ultra-Precision Assembly SystemS, http://www.eupass.org

6. FIPA, 2005, The Foundation for Intelligent Physical Agents, http://www.fipa.org
7. Jin, Y., and Lu, S. C-Y., 2004, Agent Based Negotiation for Collaborative Design Decision Making, *Annals of the CIRP*, **53**(1):121-124
8. Koren, Y., Heisel, U., Jovane, F., Moriwaki, T., Pritchow, G., Van Brussel, H., and Ulsoy, A. G., 1999, Reconfigurable Manufacturing Systems, *CIRP Annals*, **48**(2)
9. Lohse, N., Hirani, H., and Ratchev, S., 2005b, Equipment ontology for modular reconfigurable assembly systems, in: *Proceedings of the CIRP sponsored 3rd International Conference on Reconfigurable Manufacturing*, Ann Arbor, MI, USA, 10-12 May, 2005
10. Lohse, N., Hirani, H., Ratchev, S., and Turitto, M., 2005a, An Ontology for the Definition and Validation of Assembly Processes for Evolvable Assembly Systems, in: *Proceedings of the 6th IEEE International Symposium on Assembly and Task Planning*, Montréal, Canada, July 19-21, 2005
11. Lohse, N., Ratchev, S., and Valtchanov, G., 2004, Towards Web-enabled design of modular assembly systems, *Assembly Automation*, Emerald Group Publishing Limited, **24**(3):270-279
12. Lu, S. C-Y., Cai, J., Burkett, W., and Udwadia, F., 2000, A Methodology for Collaborative Design Process and Conflict Analysis, *Annals of the CIRP*, **49**(1):69-73
13. McDermott, J., 1982, R1: A Rule-Based Configurer of Computer Systems, *Artificial Intelligence*, **19**:39-88
14. Mittal, S., and Frayman, F., 1989, Towards a generic model of configuration tasks, in: *Proceedings of the Eleventh International Joint Conference on Artificial Intelligence*, San Mateo, CA, USA, 1989, Morgan Kaufmann
15. Onori, M., Barata, J., António, Lastra, J., and Tichem, M., 2002, European Precision Assembly Roadmap 2012, The Assembly-NET Consortium
16. Rampersad, Hubert K., 1994, *Integrated and Simultaneous Design for Robotic Assembly*, John Wiley & Sons Ltd., Chichester, ISBN 0-471-95018-1
17. Rosenman, M., and Wang, F., 2001, A component agent based open CAD system for collaborative design, *Automation in Construction*, Elsevier Science B. V., **10**:383-397
18. Schäfer, C., and López, O., 1999, An Object-Oriented Robot Model and its Integration into Flexible Manufacturing Systems, in: Multiple

Approaches to Intelligent Systems: 12th International Conference on Industrial and Engineering Applications of Artificial Intelligence and Expert Systems, Imam, I. F., Kodratoff, Y., El-Dessouki, A., and Ali, M., ed., Springer, ISBN 3540660763

EVOLVABLE SKILLS FOR ASSEMBLY SYSTEMS

Evolvability by automatic configuration and standardization of control interfaces and state models

Gerd Hoppe

Beckhoff Automation GmbH, Eiserstr. 5, 33415 Verl, Germany, g.hoppe@beckhoff.com

Abstract: The Eupass project (www.eupass.org) proposes to develop assembly systems, especially in the field of ultra precision assembly, with evolvable skills to fit a wide range of products being assembled by applying a "Zero-Engineering" approach for the switch-over from one product to another. The Eupass partners are currently developing a machine line network with new and additional machine and device discovery mechanisms, synchronization and data exchange capabilities. By means of a common behavioral model, machines and devices connected to a network can jointly be discovered, commanded and execute any programmed activity in a collaborative way – developing evolvable skills exceeding the skill set of the singular machines.

Key words: Control, Communication, Assembly, Eupass, Skills, UPnP, Evolvability, Publisher / Subscriber, ADS, EtherCAT

1. INTRODUCTION

In today's manufacturing environments, the return- of- investment (ROI) gap between short product life cycles and expensive fully automated manufacturing systems grows even larger. Whereas there is no alternative to largely automated manufacturing systems due to continuously growing product quality requirements, the traditional way of achieving ROI for an assembly line by extending its life cycle over many product generations results in numerous re-engineering phases for products being manufactured sequentially on the particular assembly line.

The Eupass partners are currently developing an assembly line technology with new and additional machine and device discovery mechanisms, synchronization and data exchange capabilities. By means of a common behavioral (state) model, machines and devices connected to a network can jointly be commanded to start, stop, hold, etc. and execute any programmed activity in a collaborative way to use machining capabilities more freely.

Programmed "skills" of machines and devices are exposed to a planning tool for the assembly sequence via a publication method (blueprint publication), so that assembly steps are being planned within the capabilities of machines being used from a Eupass machine repository. Planning tools are subsequently capable of designing an assembly sequence within the geometrical workspace of the machine modules using the "skill-set" of its devices. As the combination of singular skills is random, new assembly skills evolve even beyond the original capabilities of the set of machinery.

1.1 Engineering is the issue

Assembly processes in manufacturing are characterized by an expensive engineering process spanning a variety of segregated tools for creation of products and subsequently assembly equipment. The lifecycle of the investment in machining lines and the product marketing span are in mismatch, as a result the product cost are higher as needed due to underutilizing investments at the end of the product marketing cycle. To overcome the situation and better utilize the unused investment remaining, it is advantageous to utilize the assembly equipment for ramping up the next product while still producing the fading volume of the previous product.

Therefore, the Eupass project proposes to develop assembly systems, especially in the field of ultra precision assembly, with evolvable skills to fit a wide range of products being assembled by applying a "Zero-Engineering" approach for the assembly line conversion from one product to another.

1.2 Concept

To achieve a Zero-Engineering approach in product conversion, the gap between Automation Equipment (Controls, Networks, Drives, Motors, I/O Sensors and Actuators) and Product Design Tools (CAD, 3D-CAD, Modeling Software etc.) needs to be closed. To avoid multiple recurrent loops in engineering, the product designers should be aware of manufacturing limitations such as dimensional restrictions of assembly machinery for the new product under design.

If design tools would have not only information about limitations of an assembly process, but also have knowledge about the entirety of available assembly process capabilities (skills of the assembly line); the CAD Tool could compile the assembly sequence itself by means of a post processor. Pre-assembling the product in a simulation before the assembly line and even the physical product exists would become reality.

In order to form a connection between the design tool level and the (assembly) machine line, a tri-directional exchange of information is necessary:

- (Assembly) machines must communicate (publish) skill sets and limitations of those,
- Design Tools and postprocessors must communicate the sequence of executing skills and parameters of execution,
- (Assembly) machines must exchange information with the exchange of products to organize the flow of products throughout the line and vertically.

In order to achieve this, the state-of-the-art approach to programming machine lines requires a new structurization and separation of generic capabilities (skills) from product-specific parameters. Also, the organization of machine behavior and communication systems requires separation of organizational program levels and standardization in order to seamlessly function:

- To connect machines in a "Plug-and-play" behavior, a common network, protocol, communication behavioral model (state machine) are required,
- To operate machines in a "cluster", a common operational behavioral model (state machine) is required,
- A separation of "managerial code" from "functional code" is required,
- All machining functions must operate on parameters provided by the engineering tools, so that all functions behave in an abstracted fashion,
- All machining capabilities must be published as skills within a certain operating area and parameter limitations, such as physical parameters etc.,
- Skills of machinery must be accessible by means of a command interface in form of an API (application programmer interface) via network so that
- Tools are capable to command execution of skills,
- Machinery is capable to interact with each other.

This level of abstraction and interaction is achieved in single manufacturer's installations; however, today no standards exist for open connectivity of systems, especially between design & engineering tools and machinery. With EUPASS; the novelty is the standardization of interfaces for an entire industry, the micro-assembly world. As an achievement, an ecosystem of service providers and service consumers may boost the machine builders (OEM) and possibly even manufacturing industries in Europe to gain weight against competitors in this field.

1.3 Interaction of Design and Execution

In order to interface the design layer and the execution layer of a manufacturing system, a bidirectional interface is proposed to create knowledge about abilities on both sides – the design layer and the control layer.

As shown in Figure 1, the Design System Tools include additional layers between the Product Design Tool (CAD) and the production systems to create an Assembly Sequence, simulate such sequence including all steps executed, time spent, materials consumed and geometries and tools utilized during such assembly sequence. The final result will be a "cookbook recipe" of steps and parameters such as geometries, forces applied, tools utilized, separation into machine modules and more.

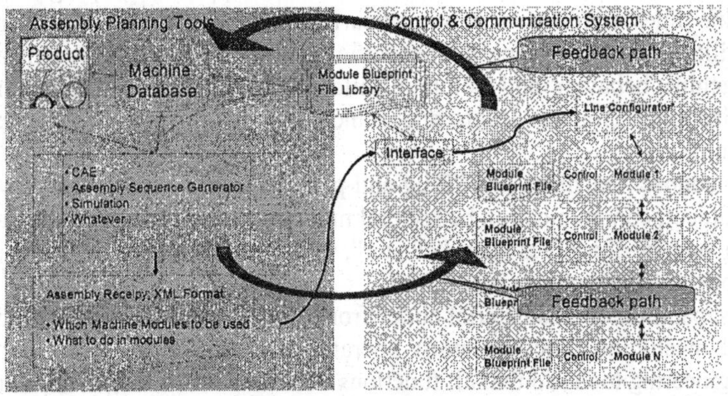

Figure 1 Feedback between Assembly Planning Tools and Assembly Control System

In order to apply useful steps of assembly actions, the Assembly Planning Tools will need to know the specific abilities of machines used to build an assembly line: the "skills". EUPASS compatible machinery will therefore include and publish a "Blueprint File" as electronic datasheet containing - next to other items - the specific skills and their physical limitations for any planning tool. By contacting a storage site, the "Eupass Module Blueprint Library", a virtual repository containing such information plus availability and location of any machinery compatible to such repository information, an assembly line can automatically be planned, simulated, and allocated for commissioning.

On the other side, machinery linked together to an assembly line by mechanics, communication and logic will proceed to form a configuration to a line and then consume the "cookbook recipe", distribute it to module controllers and trigger execution of the assembly recipe step by step and parameter by parameter to invoke assembly of products – so far, a straight forward approach to assembly line control; however, with some structurization, the capabilities of the machinery may develop (evolve) beyond the total sum of singular skills.

1.4 Skills and evolving Skills

Each station or module controller will participate in EUPASS communication via the EUPASS backbone. "Plug & play" capabilities will allow to automatically connect and start up a machine module; additionally, a discovery mechanism will expose an XML scheme to any higher layer computer system, including but not limited to the Line Configurator to publish information about assembly abilities with the Line Configurator and to the Design System Tools.

As additional benefit of the project, the abstraction necessary for this level of standardization allows to combine skills to new "super-skills", creating an evolvement of capabilities of the combination of machinery to a line. Skills may evolve by combining skills of sub-modules; and, beyond the boundaries of a station or module across the assembly line.

This development of features is not only a "Superposition of features" but creates in best evolutionary terms complete new capabilities from interaction of existing (singular) skills; therefore "Evolvability" was chosen for best characterization. In order to create "evolved" skills beyond the combination of two prior existing skills, a planning tool must know the set of all available singular skills for compilation into new skills. Additionally, the planning tools must be capable of combining the published machine skills in any meaningful way for assembly of products within the limitations of the physics of the machinery system.

With advanced software technologies, the planning tools may bridge the gap between a product design tool (CAD, 3D-CAD) and the machinery assembly execution logic. With the ability to know all the requirements for an assembly sequence and the machine skills, algorithmic approaches may solve the problem of combining the skills to other skills that could not be formed by machinery itself by communication: the detailed knowledge about the product to assemble is missing at the production control level.

1.5 Control and Communication

In order to combine skills and utilize machinery for randomly changing products to be assembled down to batch size of quantity 1, the control system must be able to link functionality within a module and across module boundaries within the real-time control environment. A deterministic powerful Publisher/Subscriber technology based on Ethernet was chosen to achieve real-time execution even for complex multi-axis motion control across equipment boundaries.

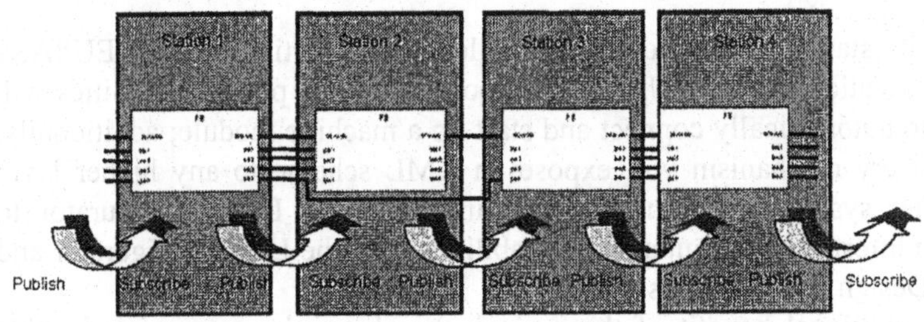

Figure 2 Eupass backbone communication by Publisher/Subscriber Technology

1.5.1 Communication Model

The Eupass communication must support both the hard real-time communication within the machine control environment as well as background, database-level communication of no hard real-time requirements. A model of cyclic and acyclic communication services was chosen to best represent a solution to this requirement: cyclic communication covers axis synchronization and inter-module logic signaling, whereas data transfer and status exchange is handled by the asynchronous or acyclic communication. Based on these two generic types of communication, all other Eupass-specific services may develop by means of the Eupass Communication Library to handle status control, recipe transfer, product tracing services, etc..

1.5.1.1 Interfacing cyclic data

Handling of handshakes and data transfer is established by distributed objects and Network variables in the communication layer by means of cyclic data communication: to transport products between EUPASS stations, a data link between these stations is necessary. Additionally other information, e.g. parameter values, must be exchanged. Because the characteristic of this information can be very dynamic and is used for synchronization tasks, a distributed real-time capable cyclic data transfer is used.

As shown in Figure 2, this is achieved by an open Publisher/Subscriber model on the Ethernet layer that is configured by the EUPASS Line Configurator.

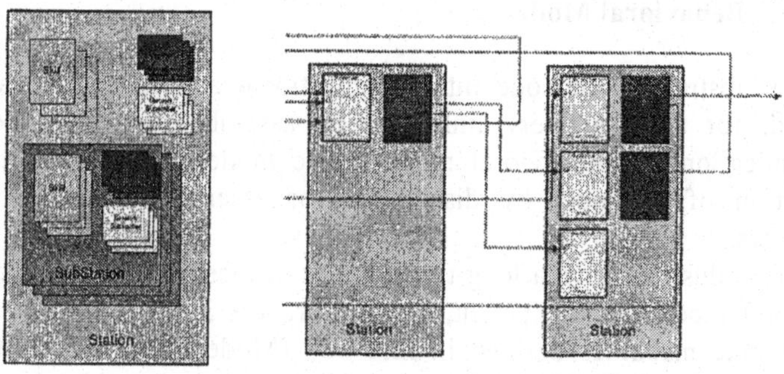

Figure 3. Skills, combined skills within a module (left) and across module boundaries (right) by means of Publisher / Subscriber Technology

Each of the stations (and substations) is capable to publishing and subscribing to distributed objects to and from other stations. Data exchange within a station is not executed over the network, instead, a controller internal Publisher/Subscriber-model data mapping is used to optimize resources.

Synchronization of the independently operating controllers and their network access is accomplished by publishing and subscribing of distributed objects in Publisher and Subscriber Objects.

1.5.1.2 Interfacing acyclic data

All other traffic is either of no real-time nature or no cyclic nature and is mapped to acyclic communication services whenever bandwidth allows, e.g. file transfer, operator commands, HMI display information, communication between User Applications and real-time control environments. etc..

A protocol-layer independent published protocol was chosen as an access and abstraction layer to various fieldbus protocols and for transportation over Ethernet protocols to transparently access different controllers, different processes and devices independently of the underlying protocol: the open ADS (Automation Device Specification) API. A distributed acyclic data exchange service allows communication In-process, In-device, via UDP, TCP/IP or SOAP / Web Service - even via Java.

The EUPASS controller platform already contains the ADS API and is easily accessed by User Applications and/or PLC programs. To facilitate device communication within the Eupass project by communication services, the Control & Communication Workgroup provides a Eupass Communication Library including updates to the partners throughout the project.

1.5.2 Behavioral Model

Also, abstraction of code into organizational and functional layers is required: for all machinery linked to an assembly system, a common communication and behavioral model (state model) should facilitate the application of assembly by distribution of assembly steps and recipe parameters.

Other industries have a long-time history of research and application of behavioral models to equipment, e.g. the process industry with it's batch control state machines defined in ISA S88 (Modes) and S95 (Enterprise Integration) standards. Recent industry initiatives show that these batch state models can be successfully applied to discrete product manufacturing, e.g. by the OMAC Packaging Workgroup. For reasons of convenience, the OMAC PackML state model was chosen for modeling the operational behavior of assembly equipment.

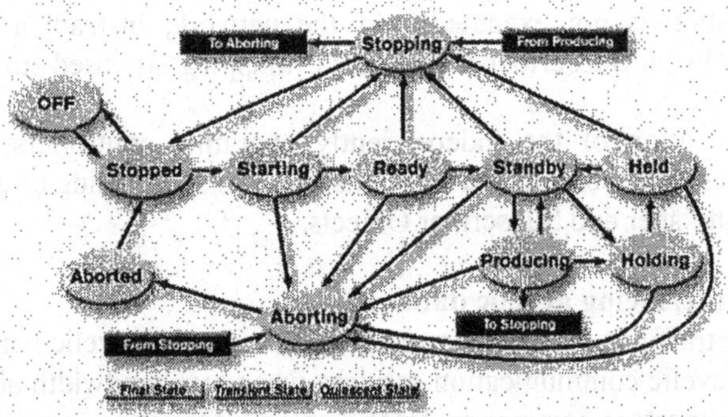

Figure 4. OMAC PackML State Model describes Eupass module behavior and facilitates linking of machinery and common behavior

Other than the machine module, the network communication for each backbone-connected controller itself requires its own state machine to facilitate discovery, operational and non-operational states. The Eupass backbone communication system utilizes the state model of the underlying EtherCAT network technology.

1.5.3 Evolving new Skills

By combining the advantageous design of structured software technology and abstraction of functionality from the behavioral model and communication subsystem, linking of objects can be achieved in a new way to create skills that go beyond the numerical combination of skills of the machines combined to a network. For this achievement, the boundaries of a single skill and a single machine module must become transparent to operate this equipment with a new skill, planned by matching the newly product and its requirements in assembly with the set of machinery available. Figure 5 shows the skill instance with all the elements necessary for skill evolvement: cyclic and acyclic communication interface, state model, recipe interface.

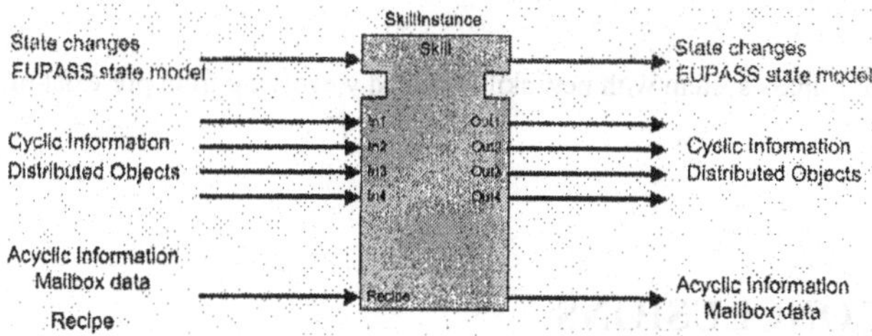

Figure 5. Complete Eupass Skill instance

By linking of skills beyond module boundaries, the project creates a new and open way of linking instances of functionality across hardware boundaries and with exposed software interfaces to allow intelligent planning tools to create new skills.

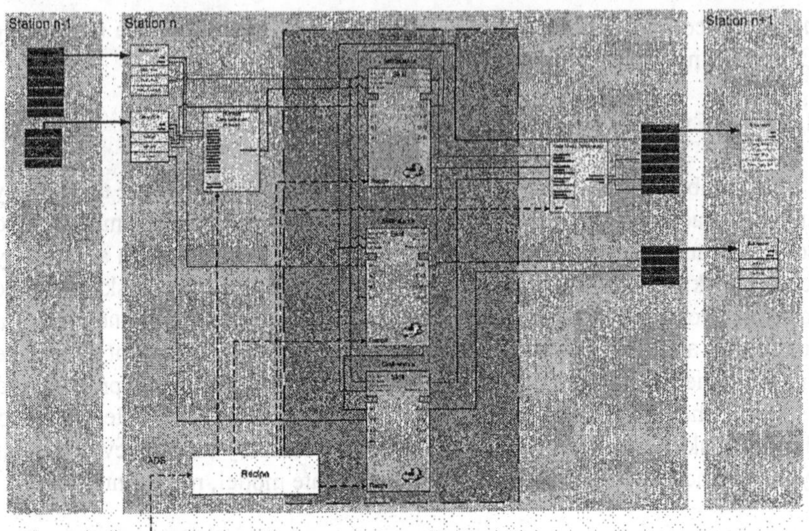

Figure 6. Eupass system with new skills created y evolving from pre-existing skills

2. CONCLUSIONS

The Eupass project targets to implement a new ecosystem of machine module producers and consumers with innovative new distribution models for a user community comprising of Eupass databases, virtual repositories of machinery and their published skill sets. In order to facilitate the deployment of equipment while driving down capital cost for assembly, standardization of equipment and random utilization of skills by automatic assembly line commissioning becomes customary state of the art.

With new technologies on the horizon, the standardization of assembly systems must go far beyond physical interfaces, communication protocols and functions o achieve this goal. Eupass proposes to drive the inventive creation of assembly skills by means of abstraction and layered, object oriented design. The achievements can ultimately lead to a fully computerized workflow from the product design concept to its assembly in an industrial environment with full toolset support to supervise, simulate and finally even create the assembly process from the product design tools to generic machinery elements publishing their capabilities as Eupass skills.

REFERENCES

1. Institute of Electrical and Electronics Engineers, IEEE 802.3 Standard for Information technology - Telecommunications and information exchange between systems - Local and metropolitan area networks - Specific requirements - Part 3: Carrier Sense Multiple Access with Collision Detection (CSMA/CD) Access Method and Physical Layer Specifications
2. International Engineering Consortium (IEC), IEC 61784-2 (Digital data communications for measurement and control - Part 2: Additional profiles for ISO/IEC 8802-3 based communication networks in real-time applications), EtherCAT
3. International Engineering Consortium (IEC), IEC 61158 Digital data communications for measurement and control - Fieldbus for use in industrial control systems; EtherCAT
4. OMAC Open Modular Architecture Controllers, Packaging Working Group, www.omac.org
5. ANSI / ISA S88, Batch Control Part 1: Models and Terminology, Instrument Society of America, www.isa.org
6. ANSI / ISA S95 Enterprise-Control System Integration Part 1: Models and Terminology
7. OMAC Open Modular Architecture Controllers, Packaging Working Group, www.omac.org
8. Universal Plug and Play (UPnP™) Forum, Basic Device Definition Version 1.0 www.upnp.org,
9. Beckhoff Automation, ADS, Automation Device Specification, www.beckhoff.com

TOWARD THE VISION BASED SUPERVISION OF MICROFACTORIES THROUGH IMAGES MOSAICING

J. Bert[1], S. Dembélé[1], N. Lefort-Piat[1]
[1] *Laboratoire d'Automatique de Besançon - UMR CNRS 6596 - ENSMM - UFC*
24 rue Alain Savary, 25000 Besançon, France

Abstract The microfactory paradigm means the miniaturisation of manufacturing systems according to the miniaturisation of products. Some benefits are the saving of material, energy and place. A vision based solution to the problem of supervision of microfactories is proposed. It consists in synthetising a high resolution global view of the work field and real time inlay of local image in this background. The result can be used for micromanipulation monitoring, assistance to the operator, alarms and others useful informations displaying.

Keywords: Supervision; microfactories; mosaicing.

1. INTRODUCTION

Since many years a lot of *microproducts* i.e. in the size range between $10\mu m$ and $10mm$ have been developed: micromechanical parts, microopticals parts, microsensors, microactuators, MEMS, MOEMS, ... For the automatic manufacturing of these elements and, their manipulation to get for example assembled products, the main approach have been the use of precision facilities where no constraint of size is considered. That has led to precision production systems which size is usually very large compared to *micrometric* size of parts, for example in semiconductor industry. Those factories take large space, spend a lot of energy and material. Besides that precision production approach, a micro production approach more known as the microfactory approach has been studied. Its purpose is the achievement of small manufacturing and manipulation systems, the philosophy is to match the size of the production facilities with the size of the products. The potential benefits are the saving of material, energy and place, agility and flexibility. Some experimental microfactories are described in the following references: (Tanaka, 2001; Okazaki et al., 2004).

Because of the impossibility or difficulty to use conventional sensors in a microfactory (lack of resolution and place) vision systems are of great interest. They give images of the work field from which a lot of informations can be retrieved. An image can be used for detecting a micropart or an effector, measuring the position of a part, measuring the force applied on a part, measuring defects of a part, verifying the presence of a part in an assembled product, detecting events, displaying alarms, displaying effector position, ... Those informations cover the control, inspection and supervision functions.

The paper deals with the supervision of microfactories, which is still a few studied. The following points are developed next: the characteristic of the vision system in a microfactory, the scene synthesis by mosaicing, proposition and validation of the dynamic mosaicing solution.

2. CHARACTERIZATION OF THE VISION SYSTEM

Supervision is one of the many functions achieved in a production system specially a microproduction one (Breguet et al., 2000). The supervision of a microfactory is a few studied. (Tanaka, 2001) and (Okazaki et al., 2004) *simply noticed the presence of three miniature cameras for* displaying the image of each machine in the experimental microfactory of Japan AIST. (Kuronita et al., 2001) use a camera to monitor the activity of their swarm robot based drilling system. Actually supervision is more than monitoring i.e. displaying image of the microfactory, it also includes assistance to the operator, detecting events (lost of a microproduct, contact of the effector with the substrat, ...), displaying alarms and informations.

In the paper a vision based supervision paradigm is proposed. The usual vision system used in a microfactory is a camera with an optical (photonic) microscope. The resolution is high (up to $0.25\mu m$ according the law of Raleyght) but the depth of field and the field of view are low, the overall dimensions are important, the images are weakly contrasted (Vikramaditya and Nelson, 1997). Instead of the microscope based vision system, a microfiberscope and a camera also can be used. A fiberscope consists of a bundle of optical fibers for lighting and an other bundle for image transportation, a microoptical set allows the connection with the camera. The system is flexible and not cumbersome (end diameter reaches $0.5mm$). The resolution is less high, the depth of field is high, the field of view is low. That kind of system has been developed by (Tohyama et al., 2000) for stereoscopic observation of the work field. The microcamera with the appropriate microoptical set (endoscope) also could be used.

The common characteristic of the above vision components is their low field of view, the corresponding image represents a local view. Then the vision system of the microfactory must includes a set of local vision components positioned at adequate places. Finally the vision system must be distributed and if possible modular. We propose the reconstruction of the global view of the microfactory by mosaicing the local images of the vision components.

3. OFF-LINE VIEW SYNTHESIS BY MOSAICING

The mosaicing consists in constructing a large image (the mosaic image) from a set of small images. It virtually increases the field of view of vision systems without lost of resolution and with minimum deformation. It has a lot of applications :

- 360° panoramic image achievement that can be view with the Quicktime®VR software (virtual camera) (Chen, 1995),
- video compression (Irani et al., 1996),
- increasing the field of view (Heckbert, 1989; Kumar et al., 1994; Szeliski, 1994; Potsaid et al., 2003),
- digitization of large printed documents (Zappalá et al., 1997; Pilu and Isgro, 2002; Kumar et al., 2004).

Static mosaicing includes two stages.

- In the first stage the registration of the images is performed, they are aligned in the same reference according to their transformation (camera motion) with this reference (Figure 1).
- The second stage is the blending stage. After been registred the images are fused to form the mosaic image.

There are three approaches of mosaicing according to the method used to recovery the camera motion: the calibrated motion approach, the intensity based approach (named also direct method), the feature based approach. Below, we present these approaches by considering only two images.

3.1 Calibrated motion approach

When the camera motion is perfectly know i.e. the transformation between the images is known, the registration is immediately achieved.

If the motion is a translation or a small rotation the problem is easy, no registration is achieved, the images are strips that are directly aligned. This approach was used by (Rousso et al., 1997; Peleg and Herman, 1997), (Blanc et al., 2001) in satellite image mosaicing and (Potsaid et al., 2003) for increasing work field in biological micromanipulation application. The advantage of this approach is the fact it can be used even if the scene contains no texture.

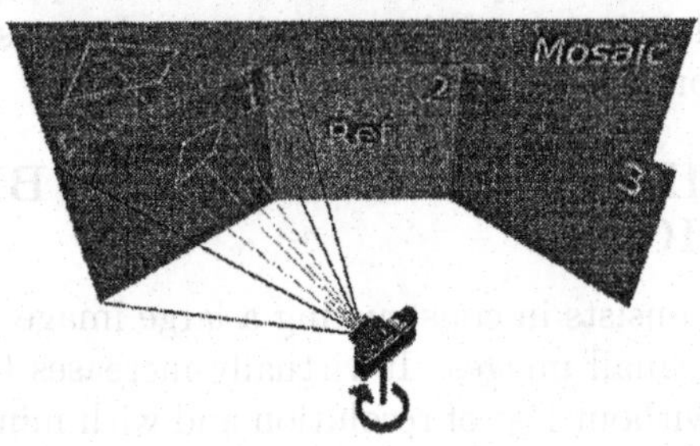

Figure 1. Illustration of images registration: image 1 and 3 are aligned with image 2 (mosaic reference).

3.2 Intensity based approach

The direct approach allows the registration of images with an iterative intensity correlation method. In fact the correlation allows the estimation of the camera motion between the two images. A lot of methods have been developed. (Barnea and Silverman, 1972) used a spatial correlation named L1 Norm. Next, (Kuglin and Hines, 1975) used a phase correlation by FFT that allows the estimation of the translation between two images using the properties of Fourier space. Szeliski and Shum introduced a warp correlation between two images that leads to the colineation matrix (homography matrix in image plan) (Szeliski, 1994; Szeliski and Shum, 1997; Shum and Szeliski, 1997). This colineation matrix integrates translation and rotation of the camera and correspond to a full planar projective motion model (camera pan/tilt):

$$p' \sim Gp \Leftrightarrow \begin{bmatrix} x' \\ y' \\ 1 \end{bmatrix} \sim \begin{bmatrix} g_0 & g_1 & g_2 \\ g_3 & g_4 & g_5 \\ g_6 & g_7 & 1 \end{bmatrix} \begin{bmatrix} x \\ y \\ 1 \end{bmatrix} \tag{1}$$

where $p = (x, y, 1)^T$ and $p' = (x', y', 1)^T$ are respectively pixels in images 1 and 2 represented in homogeneous coordinates (projective coordinates), and $\sim$ indicates equality up to a scale. In translation motion, only the two parameters g_2 and g_5 are used. To compute the 8 parameters G matrix, an iterative method is used. G is first initialized and then updated according to the following expression $G \leftarrow (I + D)G$ where I and D are respectively the identity and incremental matrixes. Image I_2 is iteratively warped by G until the error with image I_1 is less than a defined value, then the problem becomes the minimization of the error $E(d)$ between I_1 and the warped of I_2 ($\tilde{I}_2$). For that, the error is approximated by the first order Taylor series:

$$E(d) = \sum_i \left[\tilde{I}_2(p_i) - I_1(p_i)\right]^2 \approx \sum_i \left[g_i^T J_i^T d + e_i\right]^2 \tag{2}$$

where $e_i = \tilde{I}_2(p_i) - I_1(p_i)$ is the intensity error, $g_i^T = \nabla\tilde{I}_2(p_i)$ is the image gradient of $\tilde{I}_2$ at p_i, $d = (d_0, ..., d_8)$ is the incremental motion vector parameter, and $J_i = J_d(p_i) = \frac{\partial p_i}{\partial d}$ is the Jacobian of the resampled point coordinate p_i with respect to d. This least-squared problem, (eq 2), has a simple solution through the *normal equations*:

$$Ad = -b, \quad A = \sum_i J_i g_i g_i^T J_i^T \quad b = \sum_i e_i J_i g_i \tag{3}$$

A is the *Hessian*, and B is the *accumulated gradient* or *residual*. These equations can be solved using a symmetric positive definite (SPD) solver such as *Cholesky* decomposition. d is solved and G is updated to warp $\tilde{I}_2$ and so on. That 8 parameters projective transformation recovery algorithm works well if initial estimates are close enough to final results. Its contains more free parameters than necessary, it suffers from slow convergence and sometimes gets stuck in local minima. For these reasons, the 3 rotational parameters model is usually preferred. For long images sequences, this approach also suffers from the problem of accumulated misregistration errors. The latter are reduced using a global alignment method next (Szeliski and Shum, 1997; Shum and Szeliski, 1997; Shum and Szeliski, 1998).

3.3 Feature based approach

An alternative solution to estimating the transformation between images by intensity correlation as explained above, consists in the use of invariant feature points. These are points where the intensity changes like corners. The motion estimation follows four stages:

1 find the *interest* points with a corners detector,

2 match points of image 1 with points of image 2,

3 remove outliers i.e. the false matchings,

4 estimate G matrix with at least four pairs of matched points.

The first corners detector algorithm was published by (Moravec, 1977). Today, there are several corners detectors in the literature, but only two are most popular, Susan by (Smith and Brady, 1995) and Harris by (Harris and Stephens, 1988). (Schmid et al., 1998) shows that the Harris detector is the most robust according to illumination changes. This is why, generally Harris detector is often used for features detection. It is based on auto-correlation function since the latter put in light the intensity changes. Actually a bilinear approximation of auto-correlation is used because small shifts are considered. Suppose $[u, v]$ be the shift, the approximation M can be written:

$$M = \sum_{x,y} W(x,y) \begin{bmatrix} I_x^2 & I_x I_y \\ I_x I_y & I_y^2 \end{bmatrix} \tag{4}$$

with $W(x, y)$ a window function (rectangular or Gaussian), I intensity of image, I_x and I_y respectively the gradient along the axes X and Y. The result of the corner detector is a set of points in each image (figure 2).

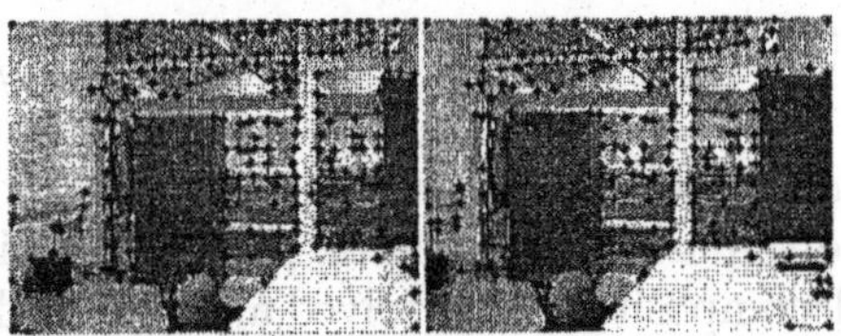

Figure 2. Feature points detected in the images with Harris corner detector. The motion between the two images is a rotation around the camera centre (pan).

The next stage consists in matching the two sets of points. For each point of image 1 a correlation window is defined centered at that point. The latter is used to perform the correlation between that window and the same size region centered at each feature point of image 2. Sum of Squared Difference (SSD) (Smith et al., 1998) correlation is usually used:

$$\begin{aligned} SSD(p_1, p_2) = \tfrac{1}{W} \sum_{k=-K}^{K} \sum_{n=-N}^{N} & [I_1(x_1 + k, y_1 + n) \\ & -I_2(x_2 + k, y_2 + n)]^2 \end{aligned} \tag{5}$$

If that result is greater than a defined threshold the points are supposed matched each other. The SSD gives some erroneous matchings, then in stage 3 it is necessary to remove the bad matchings. The RANSAC algorithm (Random Sample Consensus) is often use for that purpose and for motion (G) estimation. It is an algorithm for robust models fitting, first introduced by (Fischler and Bolles, 1981). It is robust in the sense it has good tolerance to outliers in the experimental data. It is capable of interpreting and smoothing data containing a significant percentage of error. The estimation is only correct with a certain probability, since RANSAC is a randomised estimator. The algorithm has been applied to a wide range of model parameters estimation problems in computer vision, such as registration or detection of geometric primitives.

Figure 3. Before applied RANSAC (left), after applied RANSAC (right). Every line indicates the matching between two points

3.4 Blending

A set of images represented in the same reference are obtained after the registration. These images usually contains some overlapping zones i.e. common zones and the problem is to attribute values to the pixels of these zones in order to smooth the transition between the images (i.e. to make invisible the seams). The blending solves this problem. A weighted averaging is usually used: every pixel is weighted according the kth power of its distance to the image boundary (hat filter):

$$f_{res}(p) = \frac{\sum_{i=1}^{N} f_i(p) d_i^k}{\sum_{i=1}^{N} d_i^k} \tag{6}$$

with d the weighting coefficient. After this stage we obtain the image mosaic.

3.5 Comparison

The calibrated motion approach suppose the precise knowledge of the camera motion between the different successive images. It works even the images are not textured, that is not the case for the two others approaches. In fact, they use correlation methods (for motion estimation) that requires textured images. The success of the intensity and feature based approaches is not guaranteed because respectively of problems of convergence and local minima dead end, and of the feature detection algorithm weakness. Finally the choice of a mosaicing approach is application dependent.

4. RESULT WITH THE BENCHMARK

A small local view of the work field is not sufficient to allow the performing of a micromanipulation or microassembly, a global view is required and this is true even if the vision system is distributed or not. So we propose to use mosaicing for syntheting the global view. It can be noticed that (Potsaid et al., 2003) mosaics optical microscope images to get a view of the scene for biological observation and manipulation. Our solution is quite similar to that of (Kourogi et al., 1999), it consists in an off-line construction of the global image by mosaicing and an on-line inlay of dynamical local images in this background. The following benchmark is used to valid that solution: a microendoscope based vision system (the camera is a cylinder of length $20mm$ and diameter $5mm$, the angle of view is 90°, the CCD sensor resolution is 768x576 pixels, a 8 bits frame grabber), an xyz stage (resolution $2.5\mu m$, travel $55mm$, a stand alone control system) (figure 4). The maximal resolution (for the minimal work distance of about $35mm$) is $50\mu m$/pixels. The small products manually manipulated with a brussel gripper are components of watch, an axis, a gear, a support. Matlab, C++ Builder with OpenCV library environments are used to program the application.

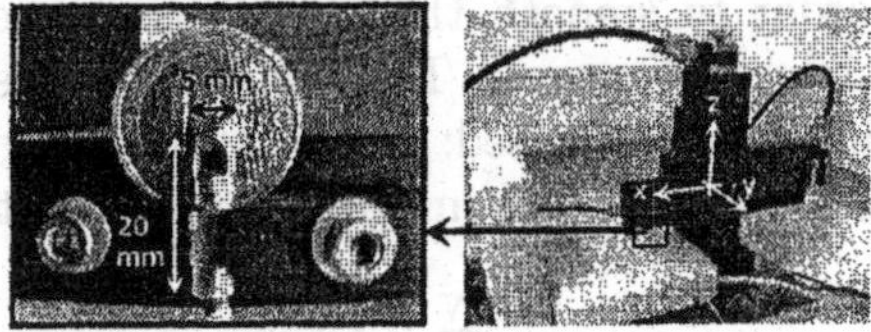

Figure 4. Micro camera compared with a coin of 1 euro (left), xyz stage (right).

4.1 Off-line background construction

Usually, in micromanipulation and microassembly the scenes (work fields) are rarely textured and precise positioning stage are used. So we use the calibrated motion approach.

The positioning stage is equiped with the microcamera and this system is used to scan the work field in order to obtain the set of local images. Calibrated translations are performed. The motion step d_m (metric) corresponds to the image width d_p (pixels), it can be written $d_m = d_p \times S_p$. S_p is the pixel size and depends on the work distance (scene-microcamera) Z_m. We perform partial calibrations of the system by analyzing the image of a $2000\mu m$ diameter circle based template at different work distance. The image of the template is acquired, the diameter of the circle is computed with a resolution of 1/20 pixel using interpolation with B-Spline functions, and the pixel size S_p is computed. By applying a median least squared method, a robust optimisation method, we find the following function:

$$S_p = 1,277.10^{-3} \times Z_m + 90,532 \tag{7}$$

The images delivered by the vision system are very distorted because of the lens of the microcamera (angle of view of 90°). In order to minimize the distorsion in the final mosaic image we crop a small strip I_c around the centre of each local image (distorsion is always smaller around the image centre) then the actual motion step is $d_m = d_{pc} \times S_p$ where d_{pc} is the width of I_c.
The result of the method is presented figure 6 at the minimum work distance, the image size is 1100 × 1100 pixels for a pixel size of $50\mu m$

4.2 On-line local image inlay

Now, we have to perform the real time inlay of the local view in the above global view. That local view define a dynamic zone in the static background. We do not perform a simple overlay which stays visible the seam between images. We perform the fusion of the two images by a symetric fade mask which part is defined by the following equations (8):

$$\begin{array}{ll} f_1(x,y) = 0 & f_2(x,y) = \frac{1}{\beta-\alpha}x + \frac{\alpha}{\alpha-\beta} \\ f_3(x,y) = \frac{(x-\alpha)(y-\alpha)}{(\beta-\alpha)^2} & f_4(x,y) = 1 \end{array} \tag{8}$$

α and β are respectively start and end slopes, (x,y) is the pixel position. Figure 5 shows the 2D and 3D forms of the mask.

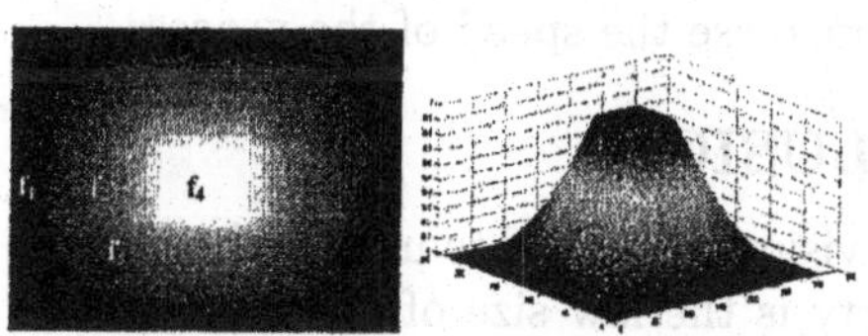

Figure 5. **The fade mask in 2D representation with the functions according to equation 8 (left), the fade mask in 3D representation (right)**

The local image I_l is multiplied by the mask and the background I_b image is multiplied by the mask complemented:

$$I_{ic}(x,y) = I_l(x,y)f(x,y) + I_b(x_{ic}, y_{ic})\left[1 - f(x,y)\right] \tag{9}$$

I_{ic} represents the final dynamic image, (x_{ic}, y_{ic}) is the centre of the dynamic zone in the final image.

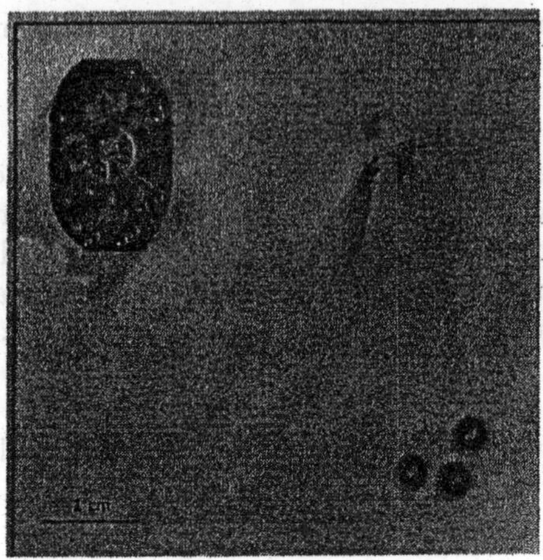

Figure 6. Dynamic mosaic of the benchmark: 1100x1100 pixels, 50μm/pixels, 55mm x 55mm.

Figure 6 shows an image of the final video. In the background we find a watch support (upper left) and gears (lower right). Upper right of that image we can see a gripper manipulating an axis. The user has a global view of the work field with high resolution, he sees the assembly target, products stocks, he also sees the local view showing the manipulation in progress. In addition, the fade mask prevents the gripper to hide the work field, it is visible only in dynamic zone which is updated as often as possible. We obtain a frame rate of 10 Hz, but our code is not optimize and we can easily increase the speed of the process.

5. CONCLUSION

We analyze the vision components used and usable in microfactories. Their main property is the low size of the field of view because of the requirement of high resolution and lower distortion. These local views of the work field are not sufficient to perform the tasks, a global view is required. A solution to perform the latter is static image mosaicing, so we have summarized the three approaches of mosaicing, calibrated motion, intensity based and feature based, and pointed out their advantages and disadvantages. According to the characteristics of micromanipulation (precise positioning stage, non textured work field) we select the calibrated motion approach to reconstruct the global image of the work field with high resolution. That background is dynamically updated by inlay live local images. This dynamic mosaicing was validated with a benchmark including a xyz positioning stage, a microcamera, a non textured work field containing watch components. The proposed dynamic mosaicing defines a step toward the vision based supervision of microfactories. The mosaic gives visual feedback and could be used to assist the operator, to display alarms and informations about the tasks being performed. The solution is also valid for distributed vision systems and can be combined with visual servoing for the tracking of mobile target every where in the mosaic.

References

Barnea, E. I. and Silverman, H. F. (1972). A class of algorithms for fast digital image registration. In *IEEE Transactions on Computers*, volume C-21, pages 179–186.

Blanc, Philippe, Savaria, Eric, and Oudyi, Farid (2001). Le mosaquage d'images satellitales optiques a haute resolution spatiale. In *18e colloque sur le traitement du signal et des images (GRETSI'01)*, volume 2, pages 251–254, Toulouse, France.

Breguet, Jean-Marc, Schmitt, Carl, and Clavel, Reymond (2000). Micro/nanofactory: Concept and state of the art. In *SPIE Proceedings of the Microrobotics and Microassembly II*, volume 4194, pages 1–12.

Chen, S. (1995). Quicktime vr - an image-based approach to virtual environment navigation. In *Computer Graphics (SIGGRAPH'95 Proceedings)*, pages 29–38.

Fischler, M. A. and Bolles, R. C. (1981). Random sample consensus: A paradigm for model fitting with applications to image analysis and automated cartography. *Communications of the Association for Computing Machinery (ACM)*, 24(6):381–395.

Harris, C. and Stephens, M. (1988). A combined corner and edge detector. In *Proceeding of 4th Alvey Vision Conference*, pages 147–151.

Heckbert, P. (1989). Fundamentals of texture mapping and image warping. Master's thesis, The University of California at Berkeley, USA.

Irani, M., Anandan, P., Bergen, J., Kumar, R., and Hsu, S. (1996). Mosaic representations of video sequences and their applications. In *Signal Processing: Image Communication, special issue on Image and Video Semantics: Processing, Analysis, and Application*, volume 8(4).

Kourogi, Masakatsu, Kuratay, Takeshi, Hoshinoz, Jun'ichi, and Muraoka, Yoichi (1999). Real-time image mosaicing from a video sequence. In *IEEE International Conference on Image Processing*.

Kuglin, C. D. and Hines, D. C. (1975). The phase correlation image alignement method. In *Proceeding IEE International Conference on Cybernetic Society*, pages 163–165, New York, USA.

Kumar, G. Hemantha, Shivakumara, P., Guru, D. S., and Nagabhushan, P. (2004). Document image mosaicing: A novel approach. In *SADHANA - Academy Proceedings in Engineering Sciences*, volume 29(3), pages 329–341, India.

Kumar, R., Anandan, P., and Hanna, K. (1994). Shape recovery from multiple views : a parallax based approach. In Publishers, Morgan Kaufmann, editor, *In Image Understanding Workshop*, pages 947–955, Monterey, CA.

Kuronita, Tokuji, Tadokoro, Shigeru, and Aoyama, Hisayuki (2001). Swarm control for automatic drilling operation by multiple micro robots. In *Australian Conference on Robotics & Automation*, pages 7–12, Sydney.

Moravec, H. P. (1977). Towards automatic visual obstacle avoidance. In *Proceeding of the 5th International Joint Conference on Artificial Intelligent*, page 584.

Okazaki, Yuichi, Mishima, Nozomu, and Ashida, Kiwamu (2004). Microfactory - concept, history, and developments. *Journal of Manufacturing Science and Engineering*, 126:837–844.

Peleg, S. and Herman, J. (1997). Panoramic mosaics by manifold projection. In *IEEE Computer Vision and Pattern Recognition*, pages 338–343.

Pilu, M. and Isgro, F. (2002). A fast and reliable planar registration method with applications to document stitching. In *Proceeding of the British Machine Vision Conference*, pages 688–697, Cardiff.

Potsaid, Benjamin, Bellouard, Yves, and Wen, John T. (2003). Scanning optical mosaic scope for dynamic biological observations and manipulation. In *IEEE Intelligent Robotics Systems (IROS)*, Las Vegas, NV, USA.

Rousso, B., Peleg, S., and Finci, I. (1997). Mosaicing with generalized strips. In *DARPA Image Understanding Workshop*, pages 255–260, New Orleans, Louisiana, USA.

Schmid, C., Mohr, R., and Bauckhage, C. (1998). Comparing and evaluating interest points. In *IEEE International Conference on Computer Vision*, pages 230–235.

Shum, Heung-Yeung and Szeliski, Richard (1997). Panoramic image mosaics. Technical Report MSR-TR-97-23, Microsoft Research.

Shum, Heung-Yeung and Szeliski, Richard (1998). Construction and refinement of panoramic mosaics with global and local alignment. In *IEEE International Conference on Computer Vision*, pages 953–956, Bombay , India.

Smith, P., Sinclaif, D., Cipolla, R., and Wood, K. (1998). Effective corner matching. In *Proceeding of the British Machine Vision Conference*, volume 2, pages 545–556.

Smith, S.M. and Brady, J.M. (1995). Susan - a new approach to low level image processing. Internal Technical Report TR95SMS1c, Defence Research Agency, Chobham Lane, Chertsey, Surrey, UK.

Szeliski, Richard (1994). Image mosaicing for tele-reality applications. Technical Report CRL 94/2, Digital Equipment Corporation, Cambridge Research Lab.

Szeliski, Richard and Shum, Heung-Yeung (1997). Creating full view panoramic image mosaics and environment maps. In *Computer Graphics (SIGGRAPH'97 Proceedings)*, volume 31, pages 251–258.

Tanaka, Makoto (2001). Development of desktop machining microfactory. In *RIKEN Review*, volume 34, pages 46–49.

Tohyama, Osamu, Maeda, Shigeo, Abe, Kazuhiro, and Murayama, Manabu (2000). Fiber-optic sensors and actuators for environmental recognition devices. *IEEE Trans. Electron.*, E83-C(3):475–480.

Vikramaditya, Barmeshwar and Nelson, Bradley J. (1997). Visually guided microassembly using optical microscopes and active vision techniques. In *IEEE International Conference on Robotics and Automation*, pages 3172–3177, Albuquerque, New Mexico.

Zappalá, Anthony, Gee, Andrew, and Taylor, Michael (1997). Document mosaicing. In *Proceeding of the British Machine Vision Conference*, volume 17, pages 589–595.

PRECISION MULTI-DEGREES-OF-FREEDOM POSITIONING SYSTEMS

Modular Design For Assembly Applications

[1]Gheorghe Olea, [1]Kiyoshi Takamasu and [2]Benoit Raucent
[1]The University of Tokyo, 7-3-1 Hongo, Bunkyo-ku, 113-8656 Tokyo, Japan; [2]Université Catholique de Louvain, Place de levant 2, B-1349 Louvain-la-Neuve, Belgium*

Abstract: The paper is dealing with the development of a new class of multipurpose *multi-degrees-o-freedom* (MDOF) *Positioning Systems* (PS) to be used as *precision devices* in a cell manufacturing desktop environment. Based on a structural investigation and systematization with imposed criteria, PM with 2dof actuators on the base was proposed. In its first modular design concept, each of the members–*Positioning units*(Pu), having at least 2DOF motion capabilities over a large workspaces, are symmetric parallel mechanisms (PM) architectures, fully actuated, by in-parallel linear-linear (L-L) *actuation modules*(aM). The kinematics and design aspects for the representative cases – 2, 4 and 6 DOF Pu, by using planar motors and scissors (and/or, parallelograms) as *transfer of motion modules(TofM)* are presented, as they exhibit, both, an increased stiffness and speed features, respectively.

Key words: parallel mechanisms; positioning, desktop assembly; precision; multi-degrees-of-freedom; 2dof actuators; modular design; parallelograms; planar motors.

1. INTRODUCTION

In the *positioning technology*[1], for example, a new system, should exhibit not only an increased level of accuracy (as the main index of the performance), but some times, a higher speed and/or bigger payload capacity over a large work spaces. They must, generally, to be able to perform complex tasks motions with multi degrees of freedoms capabilities. A compact, small size is always welcomed (if, possible) and the modular

[*] Now working with Université Catholique de Louvain as Guest Researcher

aspects taken in to account, for the flexibility of production. A development of such *advanced systems* to cope with all (or, some) of these requirements are to be (strong) innovative, including the latest theoretical results from the specific field(s), and the actual practical progresses in the technology domain (ex., linear direct drive technology, piezo,..).

This is also a case, with a positioning/manipulation system (P/MS) inside of a *assembly cell* from the semiconductor, optics, or more generally, precision manufacturing industry. Today, the assembly processes are still preponderant at the low level of automation, comparing with machining, for example. And, a lot of work is now focused on the manipulation at the micro/nano level for very small parts. But, there are still a lot of work to do, and gaps at the desktop level: desktop factory, automation cell, etc. Excepting simple task (transfer) which requires one DOF, most of the tasks here require more than two DOF(>2DOF) motion capability. They are to be (highly) precise(<100 μm), fast enough (at least 1m/s) and in a flexible manner (modules) done on a working table (1000mm).

As a Positioning System (PS) always play an important role in an ensemble of a machine, equipment, device, etc. the choice of an adequate *mechanical structure* for a Positioning System (PS) will be the first step. The main *concept* in the *design* of standard multi degrees of freedom (MDOF) positioning systems, previously used for many years, have been based on the *serial chain structures,* when individual units (*stages*), with standard number of DOF (1) in translation or rotation, having a variable low accuracy (generally, ball screws) were chosen to be added by stacking one after another, to fulfill with the required total (multi)number of DOF. *Motorized(robotic) axes* on belt, rack/pinion or air transmission power systems have a good speed (>1m/s) and large strokes(>500mm) but, are at the limit of the accuracy required in a precision assembly(>0.1mm). With an appreciable load, the positioning tasks become difficult, increasing their drawbacks. Besides of the advantages of these two concepts - low cost and feasibility, and their high standardization in modules, they can not overcome with their drawbacks because of the vicious circle accuracy-speed-force-stiffness, characteristic of the in-serried structures (presented elsewhere) and their consequences (low stiffness, and poor dynamics behavior as the vibrations is increasing). In such case, the design concept on serial structures based cannot be seen as a solutions, and the PS not properly work.

To accomplish with these, *parallel structures* - have to come and solve the problem as they are appearing to be a good alternative to solve some of the insurmountable problems presented before. As a result PMs have been studied and researched intensively, in the last period of time, in this context and every time an increased accuracy, stiffness, and dynamics, were main requirements. Unfortunately, they have small workspaces and complicated kinematics/dynamics as the DOF increase. On the other hand, intelligent machines (like, SCARA robots) using the transfer line[2] and/or planar motors[3] as auxiliary devices systems in cooperation with other axes, prove to be a productive and adequate solution for manufacturing (assembly) in mass production of products. However, if the type and volume of the products is changing, less automated (robotized) system increased more in attention by their possibilities to be highly productive, in using their flexibility. Equipped with local and global intelligent sensors (like, CCD video camera) the *new environment* together with the "assistants" can manage various tasks inside. Especially, when workers are with different abilities and/or are subjected to constraints due the physical conditions this system should exhibits the usefulness. These assistants could be small robots working on a table or other new adequate systems with manipulation/positioning capabilities[4].

In our paper an investigative work was conducted on the PMs general class with imposed design conditions (constrains) from above, in a systematic manner, in order to find a suitable solution to the previous problems. The study has had as a consequence the selection of PMs symmetric structures class, fully actuated which is using 2dof actuators on the base, to be further developed, using the planar: a) in-parallel kinematics 2dof actuators and b) simple and/or double scissors/ parallelogram transfer of motion mechanisms. With these improvements and others in the design solutions, or practical implementations, some of the members exhibit new and/or improved features. Their general kinematics and (pre)design considerations are presented as a single *modular concept*.

2. PRECISION MULTI-DEGREES-OF-FREEDOM POSITIONING SYSTEMS (P/MDOF-PS)

2.1 Conceptual Design

The conceptual design stage is the most important one, in any process of creation. New innovative products are based on this, and it is the crucial one. The final conclusions in the short (investigative) analysis from the introductory part were: a) *PM structures* could be a good (and, single alternative) solution to the serial structure design concept for PS, and b) The *linear direct-drive technology* is able and capable today to deliver advanced practical solutions for a variety of purposes (including, the planar one).

We adopted them as *fundamentals principles* of the conceptual design process. Then, with more deeply investigative actions, some systematization, and based on our own work and ideas, to be able to deliver the right solution at our design. We did it, and now shortly we are presenting the results.

Parallel Mechanisms as Positioning Systems(short investigation)
PMs were from the beginning[5] studied as mechanical systems to be used in manipulation, involving the positioning ones. In the recent years, an appreciable effort was done towards their investigation and development to be potential used in industry, based on the previous successful works and results. But, in the assembly, famous implementations still have to wait.

The first design as a manipulator has been done by MacCallion[6] even for an assembly station. It has 6DOF, three points actuated and used three pantographs to perform the insertion tasks. After that, some other researchers dealt with this aspect[7,8], too. The PM as a compliant mechanism was studied at INRIA[5], and one of the prototype shown their high capability to accurate sense an external force by accurate mapping the force. But, a first commercial parallel manipulator (GADFLY) was sold in assembly(IC-integrated circuit) by Marconi, only in 1992. Apart of the well-known yet structure, *Gough-Stewart platform*[9], an enormous amount of new structures and design improvements are now on the "market" and were (or, are to be) proved as practical PS solutions. Their number is increasing dramatically, being difficult to be counted, and a data base maintained (and, updated). Occasionally, groups or individuals belonging to famous research centers or laboratories in the frame of small or large projects (ROBOTOOL, ..etc) tried and made good efforts to investigate as much as possible structures for some of their purposes (scientific, didactic, etc). An individual, but with constant and rigorous selection updates on INTERNET was doing by Merlet and the results published[5] in a book which stand as a design guide. Most of the PMs structures have generally, only one(1) dof actuators, and with all the *fixed points* at the base. (Gough-Stewart PM in its general well-known design includes six linear actuators with all the six(6) points from one end, fixed on the base). In a normal (standard) design concept-motors, ball screws, open loop control, etc it not reached an impressive level of accuracy, and the small workspace is a result of a maximum number of joints used (6) and their complexity. Taken in to account the errors from this complicated actuation technology(guides, wear,..) the control becomes difficult. For tasks involving less than all six DOF, they are not well suited structures for manipulation and the coefficient WS/foot print is disadvantageous. Even if, PM seems to be highly desired in the industry, as micro/nano manipulation system (semiconductor, assembly,..), because of the specific features (high accuracy, small space, good stiffness,..) in our case of requirements (table top motion systems) the G-S PM will be not the first solution. Another well-known PM famous structures, the DELTA[10] and HEXA has as principle of actuation initially, the fixed points, but latter, depending on the applications these were moved in one direction (ex., LINAPOD,..) or, others(ex. TOYODA,..) by using the linear (ball screw, or, linear motor technology).

One of the first solutions of PM with all *unfixed points* of the actuation, with a motion parallel with the base was done by Han and Tahmasebi[11], using four and five (pantograph) bar mechanisms, respectively, for each of their six actuators. By these solutions, and with fixed lengths links, the ws of the actuators in a plane was increased, but total WS not significantly. (Note, the actuators have a closed loop linkage, and in-parallel actuated structure, respectively.)

As it was stated in the claims[11], the 6DOF parallel (mini)manipulator with 2 dof actuation concept, for fine position and force control in a hybrid serial-parallel system and with fixed lengths (limbs) was designed to provide high accuracy and stiffness, but it was not designed to provide very large displacements. Latter a possible combination of passive (SR) and actuation (stepper motor X-Y stages) joints was proposed.

PM with 2dof was used in two fundamental prototypes, by Kohli(R-L)[12] and Behi(L/L) disclosing the advantages derived from a symmetrical simplified structure. (Note, the first one gave a practical solution for in-parallel actuation actuators, when the last was serial one.) First experimental implementation using planar motors was done with a kinematics different from previous ones by using S-R joints in one chain[13]. From then, other developments in this direction was done, too.

The work presented in this paper here is in somewhat an extension of what was partially done before. A special class of spatial symmetrically PM structures with 2dof actuators (R-L and L-L) and 6DOF capabilities, labeled as-type "A" was studied[14] towards the development of new and innovative mechanisms with an improved potential and capabilities as positioning, carry load, speed,.., no necessary all together. Two of them, in horizontal and vertical linear actuation: 6-3[(PP)RS]-H and -V, respectively were from the positioning point of view analyzed, because of its appealing potential to be implemented as mechanisms for precision manipulations, by using (then) the new direct-drive linear technology. The actuated joints were conceived to work in three limited rectangular planes. As a result, the PM exhibited a workspace, insides only of the triangular base. Other technical solutions, by using the planar actuations, were proposed from the design pov, and for an immediate implementation, by using standard Japanese mechatronic products, components and technologies[15] latter. Finally, a Parallel Positioning Device[16] was developed.

Structural Systematization and Developments. Using the traditional tool of investigation applied on the actual PM data base (references) of structures, and based on our own experience and work, a process of systematization and development was conducted. The results are shown below, and will be discussed now shortly.

Even from the beginning, we imposed several criteria to be followed in this systematization:

a) *actuators should be on the base* (from many good considerations, but mainly from the static and dynamics pov-stiffness, speed, force,..),

b) to use as much as possible the *similar components* (links and pair, opening the way towards the modularization, and flexibility of their manufacturing and assembly),and

c) *few parts in motion* (for dynamics and precision, diminishing the chance to difficult manage the manufacturing errors and/or the calibration).

As a *first result* (not shown here, for the space reason), we find that a closed loop PM structures with MDOF capability (2-6 DOF) and with 2dof actuation (from the level I and/or II), symmetrical, which have a reduced number of elements-links and pairs, (compared with that using 1dof actuators) could be the first solution. The number of closed loops(branches) were the minimum for this PM class -1(2) -5(6) but, from the entire class, finally we selected a subclass which shown the minimum number of components on a branch. For these, the actuators (all) are located only at one level (I), and should have incorporated at least two degree-of-freedom.

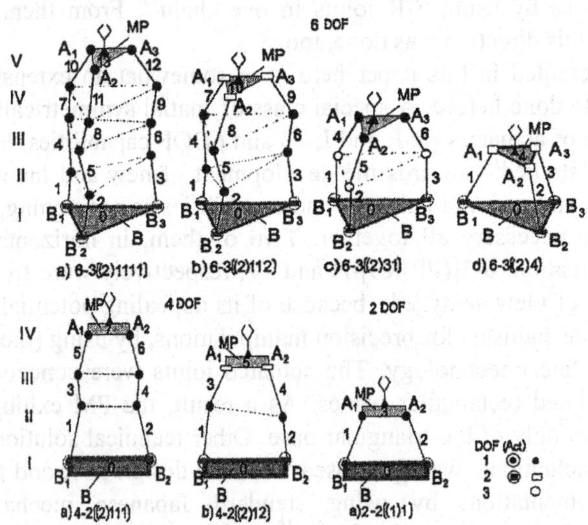

Figure 1. 2dof actuated PM class(fundamental structures)-6, 4 and 2DOF (actuation)

This class we selected to be analyzed and discussed, and further developed. It has as components, two *families* - PMs with 4 and 6 DOFs, Fig.1. Several representative members as a standard combination of the pairs with 1, 2, 3 and 4dof between levels(II–V) are shown and all the possible combination and their specific features synthesized and included in Tab.1. The structures belong to the symmetrical ones, general class. We should point out, even from now, that based on the results included in the figure (and, the PM definition – "PM should have at least 2 branches") for mechanisms with (2DOF) seems to be that this class doesn't have a representative. This is not so true(or only, partially), probably because of the lack of graphical representation. If we think at a planar motor(slider)-with 2dof, its kinematic(mechanical) scheme would be seen, for instance, as a combination of two (serial) orthogonal Prismatic(P) joints. But, in fact, the best (true?) representation could be as in Fig. 1, by using 4 pairs (Prismatic, see – *Actuation*). In the same time it can be seen as a (fully) 2dof (actuated) planar mechanism with a single element (and, 2DOF actuator), and we should introduce it in the class. True, or not, we adopted it as a full member with all the rights and treated in accordance. And, moreover this will be exploited further in the benefit of the *modularity*. (It will be as a standard module inside of the each of the individual next superior structures (with 4, 6 DOF) from the family, and the class. (And by itself as, the simplest structure for a positioning unit.) Between the structures some are new members and others were the basis of the architectures already mentioned previously, as potentially ones[11,13].

Table 1. 2DOF PM family actuated from the base (structural systematization&development)

DOF	Structure	KC	k	n	N	c_5	c_4	c_3	*** c_2	C
	DOF-k[KC]	(I)* II III IV			kΣn+mP	P,R	U, C, (Pl)**	S	S_c, C_s	$k\Sigma c_i$
2	2-2(1)1	(1) 1	2	1	2	2		-	-	4
	2(2)**	(2)	-	-	1	-	(1)	-		1
4	4-2[(2)111]	(2) 1 1 1	2	3	7	3	(1)	-	-	8
	4-2[(2)12]	(2) 1 2		2	5	1	(1)+1	-	-	6
	4-2[(2)31]	(2) 3 1				1	(1)	1	1	
6	6-3[(2)1111]	(2) 1 1 1 1	3	6	19	4	(1)	-	-	15
	6-3[(2)112]	(2) 1 1 2		3	10	2	(1)	-	-	12
	6-3[(2)121]	(2) 1 2 1					(1)+1	-	-	
	6-3[(2)211]	(2) 2 1 1					(1)+1	-	-	
	6-3[(2)13]	(2) 1 3		2	7	1	(1)	1	-	9
	6-3[(2)31]	(2) 3 1					(1)		-	
	6-3[(2)4]	(2) 4		1	4	-	(1)	-	1	6

* Actuated, ** Proposal, *** ex. Sphere-Curve, Curve-Sphere

The members have a minimum number of kinematics chains, the number of joints and links. And, as a direct result, the direct and inverse (used in) kinematics will be more friendly (calibration, control,..), and the design and manufacturing conduct towards the low cost manipulator. Between the specific individual members (families) features- number of links(n) and joints(c_i), or their standard type (P,R,U,...) in a chain(k), or totally in the mechanism (N) and (C), the class exhibits a common and general feature–all *actuators are at the base*, and moreover, each members are *in-parallel actuated,* by a combination of one(1)), two(2) or three(3) from 2 dof actuation, as previously discussed.

Modularity. Generally, a Multi DOF PM (redundant, hybrid,...) is by its intrinsic structure a system susceptible to be designed in a modular way. This aspect is more evident if the PMs are with a symmetrical structure, as our selected structures are. The role and functional principle will be explained for every selected module.

Actuation Module (aM). There are several possibilities to introduce the motion in a planar or spatial parallel mechanism. In any cases, from the base or near by, we have clearly some advantages comparing with other more exotic, and as a result, less exploited solutions.

By the previous systematic structural systematization, the simplest structure of the class (with 2dof) could act as an standard actuator of the entire family(and, class), and by this, designed in accordance. (See the latter design issues for 2DOF module(Pu)). From kinematics point of view, and based on the scheme included in Fig.1/Tab.1, a parallel mechanism in this case should have at least two(2) kinematics chain KCs (from the definition) to be indeed with a parallel structure. We intend to use an in-parallel actuation device for actuators of all our PM structures selected as candidates.

As one of the fundamental design principles was to use the *linear technology*, not only to be sure that an adequate (high) precision and speed is assured, but to pass/encompass with the other requirement related with the work over appreciable long orthogonal distances (table top), a Prismatic-Prismatic (PP) *in-parallel* (P-P) scheme was chosen, as a mechanism for *input motion* (Fig.2). It has four pair (P) and [2-2(P)P)] structure, to produce a planar actuated *output motion* at the end-effector (mobile platform).

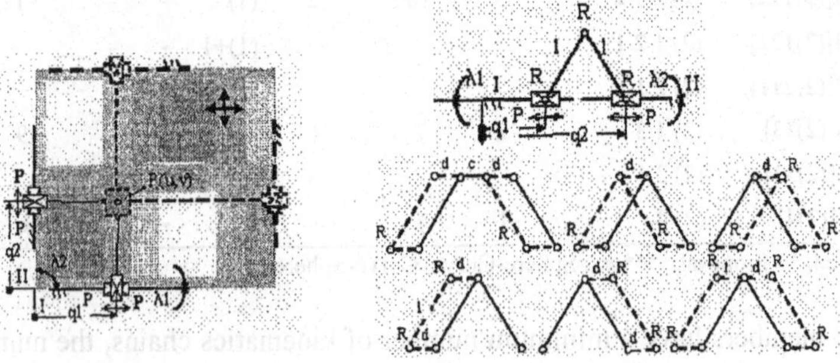

Figure 2. PM Actuation module(aM) and Transfer of Motion(TOfM) modules (kinematics)

Transfer of Motion (TofM). Scissor mechanisms was from a long time used as a fast and simple 2 dof mechanism to transform a linear *input motion* (from one axis) to in plane(*output*) one, but more recently, an increased interest is seen with the development of machines with parallel kinematics(PKM) or, the auxiliary devices in (precision) assembly[17,18]

A considerable amount of study was allocated to 4-bars mechanism in the past, but an important theoretical and practical work has been done in the connection with more recently, parallel mechanisms developments. *Four bar linkages* group are closed loop mechanisms with a single dof *input* and the characteristic element can perform an *output* constrained motion in a plan, but between them, the particular case- *parallelogram (Pa)* was the most widely used as generator of motion in translations, from several years ago. Indeed, the (output) motion of the last element is the same with that of a body in planar translation. This can be from the real use in generating ample motions, as in our case, or for a variety of others purposes, clearly shown[18].

The concept of mathematical Group Theory was used even from the beginning for the (analysis) and synthesis of PMs structures, and by this to find new structures (STAR). As was suggested there[19], for 3dof spatial translations, the parallelograms(Pa) can be used to produce the expected results. DELTA use them as main principle and taken the advantages. Other researchers (re)discovered it within its planar configuration[20].

In our design, we used these two generators of motions transformers - *scissor mechanisms(Sc) and parallelograms(Pa)*, to increase the speed and stiffness for our class of the PS, and to have by this, ample motions capabilities inside of some new PM structures. Their fundamental principle applied in our case, as a combination of them are used as is shown in Fig.3 to develop new PSs architectures, and/or to improve the features of others. This concept can be extended with an adequate design at all the structures which have at least one rotation (R) pairs in their structure.

Mobile Platform (mP) module is based mainly on the end-effector component, which holds the necessary attachments – devices, in doing the scheduled assembly operations in the desktop cell with the required parameters. The shape is in accordance with these operations and its interactions. It could be as a tray for transfer/transportation, probe for control, grasping for manipulation, tools for small adjustments, screwing, etc and designed in accordance with the specific parameters.

2.2 Positioning units (Pu)

The modular concept of design PS as a unique PM structure was previously described. It consist from three modules: 1) actuation (aM), 2) Transfer of Motion (TofM) and mobile Platform (mP). This concept can be extended behind, only to be applied, at an individual structure. It can be a (general) concept for the entire family or, class, which (virtually) includes other members (serial connected, redundant, hybrids). This would be sure a serious advantage in the design and manufacturing costs of multi-agent, multipurpose PSs.

2DOF Pu. Based on kinematics of in-parallel actuation module (Fig.2), an *independent unit* with built-in functions-proper force and guidance can be designed. This integrated multi-degree-of freedom element (platform) as a *direct drives unit* –planar motors(Fig.3), is moving simultaneously through forces coming from a magnetic field, in two orthogonal axis(X-Y), instead of two-one axis (electro) mechanical modules. They are proved to be a good choice for planar high speed precise manipulations/actuations- no friction and no wear, heat or dust, as a result of a air bearing suspension, being suitable for clean room environments. On the other hand, the concentration of the actuated degrees of freedoms in one unit has other advantages compared with more classical ones, by using selectively only one simple dof. Linear motion technology(direct-drive) now is evolving, and their advantages converted in compact, precise and fast *planar modules.* In our design[21], a *slider* (80x80x28mm) which, hold the coils, is moving suspended by the balance between the attracting forces and an air-slide bearing pushing force and moves on a base (desktop table)- *platen*(760x700x15mm,) as a interaction of the coils with the magnetic (steel) environment based on stepper motor open loop principle (pitch-2mm) and. By the combinations (in a standard or, variable way- with the auxiliary contact or, not) this *standard* positioning unit(s) can be configured forming simple or complex planar shapes, or for the members of the class-*Positioning units(Pu)* with 4, 6 or more DOF caring materials, tool, components, finished products,…

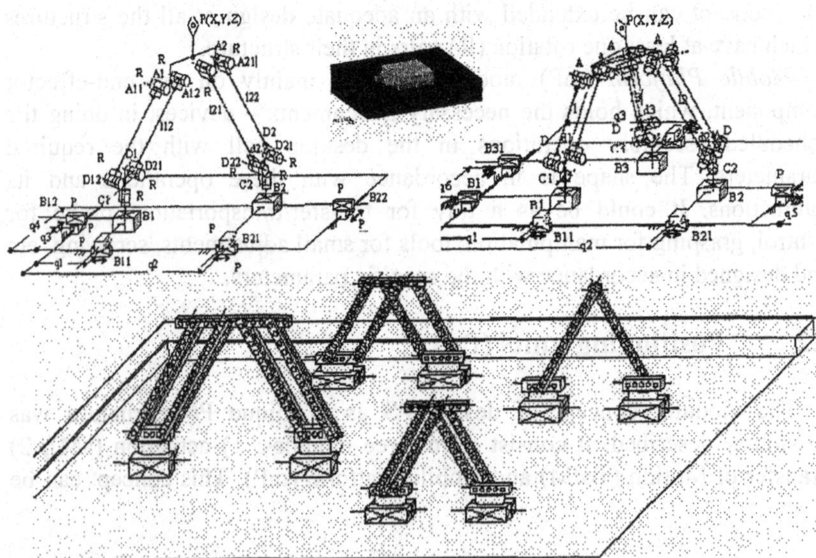

Fig.3. MDOF Pu –2, 4 and 6 DOF, kinematics: 2-(Pl), 4-2[(Pl)RPa and 6-3[(Pl)SPa] & design

4DOF Pu. Based on the previous four structures included in Tab.1/ Fig.1 having 4DOF and a least one (1) dof pairs on a level, we want to draw the attention on 4-2[(2)111] structure. Taken in to account the possibilities to be designed as a combination of only simple common pairs (P and/or R) and their reciprocal spatial axis disposal (0° and/or 90°) one of the architectures could have the kinematics as the Fig.3, the kinematics scheme shows.

The architecture of the mechanism is based on the previous kinematics design principles for the actuation/motion transfer mechanisms, and an additional two R pairs, with the axes perpendicular on the actuation plane. Between the mP and these pairs(R) two (for, symmetry only) planar vertical Pa or (RRRR) is moving. The result is that the end-effector (a rod) can perform, all three translations (X, Y, Z) and one rotation (around vertical axis (Z)). As in most of the PM structures, every motion is generated and resulted from the cooperative actions(motions) of the actuation (sub) systems - the mechanisms has a high index of efficient use of actuators. With other words, an increased force(and/or torque) capabilities is seeing for the mechanics of motion in x-axis, y- axis and z-axis (including the rotation) .

Depending on the mechanisms constructive parameters (link and base radius), an additional higher stiffness can be seen in the axial axis (z) and longitudinal axis (x) of the mechanisms, which can reduces the deflection and provides a higher-level natural frequency.

6DOF Pu. Based on the 6-3[(2)31] structure from Tab.1, a kinematics proposal for a mechanism with full DOF motion, 6-3(Pl)SPa, by using the same (Pa) principles(s) is proposed, too.(Note, symmetrical counter parts, are to be taken into account, also: 4-2[(Pl)PaR and 6-3[(Pl)PaS], respectively.)

3. CONCLUSIONS

A challenging tasks today, in the precision manipulation, is to design flexible systems, for process like cell assembly, and with an increased accuracy for large moving volumes with micrometer accuracy. A selected class of PM was developed by using advanced actuation and high efficient motion transformers (mechanisms) in their kinematics. Several representative members, designed on the modular concept of 2dof in-parallel actuation, and for a practical implementation as positioning/manipulation precision system in a highly environment controlled conditions(multi-agents, multi-assembly, multi-cells) have some specific features. An increased potential of use in term of stiffness and the capabilities to perform multi degrees of freedom motions tasks with accuracy required in some precision assembly operations (transport, manipulation, screwing, ...) from other fields of applications, like (microfactory, micromachining, microscope manipulations,..) can be further expected. For the experimental purpose, the system was successfully tested in an Attentive Work Bench (AWB) solution[4] - component of an assembly cell, proving the kinematic and design concept. This work is indeed in progress, because we intend to study the system from the kinematic and dynamic point of view, and to identify more precise the members' potential, by using a systematic analytical approach and the chance to use an automatic computational algorithm for their analysis and simulation-ROBOTRAN[22] (Multi Body Dynamics symbolic software).

REFERENCES

1. CPT , 2002, "Int. Conf. on New Technological Innovation for Positioning" *Proc. of 1st KOREAN-JAPAN Conf. on Positioning Technology*, Daejon, Korea.
2. V. Scheinman, 1994, Robot World: a multiple robot vision guided assembly system, *Rob. Res*, Proc.4th Int. Symp. MIT Press, pp. 23-27.
3. A. E Quaid., and Hollis R., 1996, Cooperative 2-DOF Robots for Precision Assembly, *Proc. Int. Conf. Rob. Aut.*, Mineapolis, US, pp.
4. H. Suzuki et all, 2003, An overview of Attentive Work Bench(AWB), *Proc. Symp. Real World Information Systems*, 21 Cent. COE , Tokyo, Japan, pp. 65-66.
5. J.P. Merlet, 2000,"Parallel Robots, KLUWER Academic Publisher, ISBN 0-7923-630
6. H. Mac Callion, and D.T. Pham, 1979, The analysis of a six degree of freedom work station for mechanized assembly, *Proc. 5th World Cong. Th. Mach.&Mech.*, Montreal.
9. D. Stewart, 1965,"A platf. with 6 deg. of freedom",*Proc.Inst.Mch.eng.*,180,15,pp.371-386
10. R. Clavel,1988,DELTA,a fast Robot with Parallel Geometry, *Proc.Int.SymRob.*,pp.99-109.
11. F. Tahmasebi et all., 1994, Six degrees of freedom "minimanipulator" with three inextensible limbs, *US Pat. 5279176.*
12. D. Kohli et all., 1998, Manipulator Config. Based on Rotary Linear (R-L) Actuators and their Direct and Inverse Kinematics, ASME J. Mech.Transm. Aut. **110**, 4, pp.397-404.
13. R. Ben Horin, 1994, A Six Degrees of Freedom Parallel Robot with Three Planarly Actuated Links, *Ms.Th.*, Technion, Israel.
14. G. Olea, 1999, Contribution to the Kinematics, Dynamics and CAD of "A" Type Spatial Parallel Mechanisms, in Three Points Guided, *Ph.D ,Th.*, Tech. Univ. Cluj, Romania.
15. G. Olea, 2000, Parallel Mechanisms with Planar Actuators -The general constructive features, *Tech. Rep.*(internal use), Univ. Of Tokyo, Japan.
16. G. Olea, et all., 2001, Spatial Parallel Mechanism with 2DOF L-L Actuators, *32-nd Int. Symp. Rob.*, Seoul, Korea, pp.1129-1134.
17. G. Olea, et all., 2003, Development o Parallel Positioning Systems for Precise Micro/Mini Applications, *Proc. IPAS' 2003,* Bad Hofgastein, Austria, pp.95-101
18. SFB 562, Robotersysteme fur Hanhabung und Montage,www.tu-braunschweig.de/sfb562

19. M. Herve, 1991, Structural Synthesis of "Parallel" Robots Generating Spatial Translations, Fifth Int. Conf. On Ad. Rob. ICAR91, June 19-22, Pisa, Italy, pp.808-813
20. X-J. Liu and J. Wang, 2003, Some New Parallel Mechanisms Containing the Planar Four-
Bar Parallelogram, Int. J. Rob. Res. Vol. 22, pp. 717-732.
21. X. Chen, et all, 2004, Dev. of Planar Linear Motors System, *Proc. of the JSPE Annual Meeting*, September 15-17, , Fukui, Shimane, JAPAN, B18(CD).
22. ROBOTRAN, http://www.prm.ucl.ac.be/robotran/

PART V

Economic Aspects of Microassembly

WHAT IS THE BEST WAY TO INCREASE EFFICIENCY IN PRECISION ASSEMBLY?

Sandra Koelemeijer, Fabien Bourgeois and Jacques Jacot

Laboratoire de Production Microtechnique, Institut de Production Microtechnique, Ecole Polytechnique Fédérale de Lausanne, 1015 Lausanne, Switzerland

Abstract: Assembly of high precision products is often done manually. The main reasons are the complexity of automation and the production volumes that often remain small to medium. Watches, medical devices and sensors are some examples of products requiring high precision assembly: often expensive products with high margins. It is interesting to notice that achieving higher assembly yield allows for relative quick pay-back of equipment. This is also the reason why western European countries remain competitive in this field. In this paper, the important points to remember when selecting a solution to efficiently assist operators in their assembly tasks are highlighted. Good assistance should lead to higher yields, higher throughputs and better quality. One should take into account assembly processes and their difficulties, as well as production volume and economic profitability.

Key words: micro-assembly; precision assembly; man-machine cooperation; semi-automatic assembly

1. INTRODUCTION

Precision assembly is often done manually due to difficult processes, high equipment cost and short product life cycle. But manually assembly results in low yields and thus high assembly cost, which drives to find better solutions. The issue is to provide efficient assistance to assembly operators with automatic or semi-automatic devices: some examples are high precision axis with manual actuation like the Tresky cell's, semi-automatic assembly cells, telemanipulation with haptic devices (Ferreira 2003), intuitive programming, and vision enhancing.

It is a priori difficult to decide how to assist the operator, and to choose the part of the assembly process to automate. This decision shouldn't rely on some kind of dogmatic principle, but on a thorough analysis and on long and short term economic considerations.

2. HISTORY AND MACRO ASSEMBLY

In the past, all assembly operations were entirely manual. Mechanization was introduced to achieve lower cycle times, as well as robotics some decades later. Machines achieved simple feeding operations, and simple assembly operations with straight movements where men could easily be replaced. Operators did the more complex operations, such as orienting, adjusting, tuning, and inspecting.

Design for Assembly rules (Boothroyd, 1991) were introduced in the 80^thies^ by classifying assembly processes from the easiest (or fastest) to the most difficult for automatic assembly. Soon people understood that easy for automatic assembly also meant easy for manual assembly. From that point, the design rule for new products was: make them easy to assemble automatically, even if you do it by hand.

The easiest assembly operation is a simple straight insertion along one axis, the preferred one being vertical. It is the well known peg-in-hole situation. Assembly with two or more axis, in hidden positions becomes difficult for machines, but also trying for operators.

A problem occurs in simple insertion when there is a slight misalignment between the insertion position and the component. This is easily corrected by a human operator. He holds the parts between his fingers, touches the receptor, senses the forces transmitted to his fingers and fine tunes the position. Attempts were made to do the same with robots, and a lot of compliant solutions were proposed: active compliance, compliant grippers, etc. Due to the forces that where transmitted to the gripper (or to the robot) when the two parts to assemble met, the axis of the gripper was modified in order to adapt to the real insertion axis.

But in the end, industries retained simpler ways. One of the easiest solutions with a cartesian or a SCARA robot, is to switch off the control of the x and y axis during insertion, so that the robot can simply adapt its x-y position. As long as the z axis of the gripper and the axis of the receptor are well aligned, insertions are then very easy.

The kinematics the human uses is extremely complex. A multitude of "sensors" are used: touch, multidimensional force feed back, very high resolution force sensors, vision. On the other hand, the best results achieved with robots or mechanical devices are based on precise straight movements and precise positioning. These are two features the operator lacks completely. So the insertion of a peg can be achieved either by an operator using intuitive complex sensing and kinematics, either by an apparatus using straight and precise movements. Copying the human with a robot is not the best solution to achieve high yields and short cycle times. It is better combine the strong features of both.

3. PRECISION ASSEMBLY

3.1 What are the constraints in precision assembly?

Today, products become smaller and micro assembly or high precision assembly is the challenge. Microsystems and other fine mechanical systems are often produced in small to medium batches. Full automation is than not the right solution as the investment costs are too high and pay back can not be reached on such small volumes. On the other hand, human operators are clearly not efficient enough in this domain. Operations become very tedious, resulting in fatigue, low yields and thus high costs. Precision components are often expensive, which reveals the importance of maximizing yield by providing the right tools to ease the operator's task.

The best approach is than semi-automatic assembly. An important step is to define which tasks, processes or part of processes should be carried out by the operator, and which by an automatic device.

Therefore it is important to identify the assembly processes that are needed and the corresponding requirements, to identify the operator and the machine's strong and weak points, and than to allocate the operations accordingly.

3.2 Strong and weak points

Machines and operators don't function the same way. In order to combine the strong points of both (and to avoid the weak points) it is important to identify them with precision. They are described in the following paragraphs.

The strong points of the human operator are the ability to learn, to react to unforeseen situations, high flexibility in case of product change, multi sensorial detection, a very large sensing (10 mN to 100 N) and actuating range (from 10 µm to several meters, more when walking).

The weak points are the impossibility to stop at a predefined position, no force limitation, no straight movements, subjectivity, and the difficulty to move hands with precision without force feed back.

The strong points of an automated device are repeatability and precision, straight movements, force and position control. The weak points are limited motion range, low flexibility to product changes, no reaction to unknown situations, and the fact that it has to be programmed.

The points we surely want to combine are the precision, straight movements, and repeatability with the ability to learn, to react on unforeseen situations, and the ability to make small and very long displacements

3.3 Most frequent micro-assembly processes and their specificities

3.3.1 Component placing

In macro assembly, the peg in hole situation (or simply insertion) is relatively generic of most assemblies. In precision assembly, "peg in hole" also occurs. The difficulties are then:

- To identify the peg and the hole, their exact position, and sometimes to define the hole axis!
- To grip the peg
- To align the peg axis with the hole axis
- To insert the peg in the hole while limiting the constraints on the components
- To control the position.

But most assemblies are of another kind: mainly plane on plane with visual references on both parts (for example die bonding). This is due to the fact that most components are 2 dimensional, manufactured with silicon technologies. Assembly references are often not visible during assembly (for example flip chip, or the reference hidden by a drop of glue).

3.3.2 Specificities of placing

The main difference compared to assembly in the macro-world is that parts can not be positioned on a mechanical reference. Tolerances on parts are either less precise than the precision required, or of the same order, some μm. The functional dimensions and characteristics of the parts have to be identified and measured. The positioning is done by placing or aligning one functional element relatively to another.

3.3.3 Attachment an other processes

The most frequent attachment processes are: press-fitting, glue application, wire bonding and laser welding. A further major process is control and inspection: visual, measurement of forces or positions.

3.4 Difficulties for the human operator

In micro assembly, difficulties are of two kinds: sensing and actuating. Most operations require both simultaneously.

3.4.1 Gripping

First of all, the operator wears gloves and has to use tweezers to grip the component. The multi dimensional force feedback he had in macroassembly through direct contact with his fingers lowers tremendously. Forces are transmitted through the tweezers, resulting in a drop down of tactile information. Furthermore, forces are also smaller, making gripping a real delicate task.

3.4.2 Releasing

Releasing of components is especially difficult in the micro-world due to adhesion forces becoming dominant. Although this problem is not related to the operator but a global one, it has to be taken into account.

3.4.3 Force Sensing

Forces are very small making it hazardous to manipulate small and fragile parts. The level of the force that may create damage increases with the size reduction of the component. The weight and the inertia of the parts are so small, that feeling them with tweezers is very difficult. As stated above, force sensing through tools is much less precise than with the bare fingers.

3.4.4 Visual Sensing

Human vision is not sufficient to control tiny details, and work under a microscope is the rule. The operator only sees a fraction of the component at a time, the depth of field being limited. Training is needed to be able to quickly position the interesting part of the component under the microscope.

It becomes particularly risky to identify functional assembly references that often can not be seen with the bare eye, like on chips.

3.4.5 Actuating

Using mechanical references is the natural way to position a part to another: a book on a table, a key in a key-hole, a washer on a screw. In all those cases, we use one of the components to guide the other one, and positioning is achieved using the dimensions of each part. It is really difficult to position two references to one other without making contact, such as a mass between to parts (first accelerometers), a visual reference on a part at a given distance of a hole, or to align a flipped chip on the corresponding solder balls. A major difficulty in microassembly is to align and to make match the assembly references of two components without mechanical references.

3.4.6 Actuating with force control

An other problem is to position a part without exceeding a given force. This is the case for fragile components such as IC pressure sensors, or fragile mechanical parts, like watch needles. The assembly operator places the needle on the shaft and applies a force; when he feels a resistance, he increases the force to reach the right position. This may damage the inside of the watch movement.

3.4.7 Yield

Yield is always an important factor in assembly, and becomes really significant in precision and micro assembly. Yield consequences are of two kinds:

- Bad components or processes conduct to stops in the assembly installation. Throughput is then lower than expected, resulting in production delays and/or slow pay-backs of the installation. A detailed description is given by Oulevey (2006).
- The assembly of a bad component leads to a bad product, resulting in rework and/or the loss of all assembled components. When they are expensive, which is often the case with precision products, very high losses occur.

The main attention when automating or providing tools should be on the reliability of the assembly processes.

4. WHERE DO THE SOLUTIONS LIE? – PROPOSED METHODOLOGY

Operators have to be assisted where precision is needed. Several approaches are proposed, some academic, some industrial. Our objective is to propose a method to define a suited technology for each particular situation, and to provide technologies to enable precision assembly. Three main thoughts guide our approach:

1. To be economically interesting, the cost of the equipment to assist the operator shouldn't exceed the manual assembly cost, and a quick payback of the investment is necessary. This means: high productivity rate, thus few down-time. A thorough cost calculation is necessary to evaluate the total assembly cost of the products to be produced, and to estimate of how much may be invested. Losses due to human mistakes because of very difficult work have to be taken into account in this calculation.
2. To assure precision assembly, we have to guarantee the processes: this means high yield. This can be achieved only by the precise identification of the processes, of the assembly functions needed, and of the functional references on the parts.
3. For each process or operation, we have to consider the best way to fulfill the function: manual or automatic. Automation should be restricted to where it is really needed in order to control costs; full automation only is interesting for big production volumes.

5. CASE STUDIES

Two case studies will illustrate our approach: the development of a flexible semi-automatic assembly cell at the LPM with the collaboration of the firm Sysmelec SA (Koelemeijer 2002, 2003, 2005), and the press-fitting of a watch jewel in a hole.

5.1 Case study 1 – Flexible micro assembly cell

As the investment for a flexible assembly cell is quite high, it is important to achieve a high throughput. Operations have to be allocated either to the robot, either to the operator, according to the requirements. It is important to restrict the man-machine interaction time for a better efficiency. Their speed is different, the operator shoudn't have to wait on the machine, and the machine should'nt have to wait on the operator.

The preferred solution is then :

- To use the operator for programming, and to avoid teleoperation. Programming time for a new product shouldn't exceed 10 minutes.
- To use the operator for feeding. Parts are very small, an operator can easely carry thousands of parts prepared on palets. As batches are small, this is a very cost effective solution, while automatic feeding would be very problematic (Koelemeijer, 1999, 2001).
- To use the robot (a high precision cartesian Sysmelec robot, equipped with a high resolution camera) for precise positionning, with straight and fast movements. Functionnal refrences on the components are identified by image processing. The robot is also equipped with a gripper mounted on a force control device, and a gluing unit. This ensures that the assembly cell can carry out most of pick and place operations that occur in micro-assembly: insertions, alignments, force controlled placing.

This type of collaboration is a good combination of strenghts. The precision is ensured by the robot and the image processing system. The operator hasn't have to manipulate the components during difficult assembly operations. But he uses his skills and know-how for the programming of the assembly sequence, to choose and to parameterize generic operations and to define references.

5.2 Case study 2 - Press-fitting of a jewel in a hole

A jewel is manually positioned in a hole of a watch bar and then inserted with force to a given position. This jewel is the bearing of a shaft in a mechanical watch. Several jewels are inserted in the same watch bar. A tool is needed: either a hand operated press, either an automated press with position and force feedback. The functional requirement of this assembly is that the shaft should have a play of 20 µm. The capability of the hand operated press and the servo-controlled press are different, resulting in different assembly yields. Details can be found in (Bourgeois, 2005). If the resulting play is outside the tolerance range, the watch bar has to be removed, the jewels pressed out, and then reassembled.

The costs with the hand operated press C_m and the numerical controlled press C_n are:

$$C_m = C_{pm} / Y_m + (1 - Y_m) \,.\, C_{rwk} \,.\, N + c_h \,.\, T_{pm} \,.\, N$$

$$C_n = C_{pn} / Y_n + (1 - Y_n) \,.\, C_{rwk} \,.\, N + c_h \,.\, T_{pn} \,.\, N$$

The break even point is reached for the assembly of N jewels:

$$N = [C_{rwk} \cdot (Y_m - Y_n) - c_h \cdot (T_{pn} - T_{pn})]^{-1} \cdot [C_{pn} / Y_n - C_{pm} / Y_m]$$

The following values are used:

C_{rwk}	=	Cost of rework = 8 €/part
C_{pn}	=	Cost of numerical controlled press = 200'000 €
C_{pm}	=	Cost of manual press = 10'000 €
Y_n	=	Yield of numerical controlled press = 93.7%
Y_m	=	Yield of manual press = 36.7%
c_h	=	Cost of operator = 80 €/hour
T_{pn}	=	Cycle time of numerical controlled press = 15 s
T_{pm}	=	Cycle time of manual press = 10 s

Rework being long (10 minutes) and parts very expensive (high end watches), the servo-controlled press is more interesting as soon as the break even point of 42'000 jewels is reached. As a watch has about 5 to 10 jewels, a production of 4 to 8'000 watches a year allows for a pay back of an expensive servo controlled press in one year.

6. CONCLUSION AND FURTHER WORK

To achieve high precision assembly, operators have to be assisted by some tools or automatic devices. The question of what to automate remains difficult, but always has to respond to economic efficiency. The three following points are important guidelines:

1. To achieve high throughput and a high productivity rate of both machines and operators, tasks should be well separated and collaboration periods limited.
2. Low assembly yield is often what makes precision assembly very expensive. Difficult processes should be automated to achieve better capabilities and higher yields. This quickly results in cost reduction, even when some investment is necessary. This can't be achieved without a very good understanding of the process.
3. Each operation or process should be analyzed in terms of precision, cycle time, range, and capacity, and allocated to the most suited of either the automatic tool, either the operator.

Further work at the LPM-EPFL is done on the identification of micro-assembly processes and their in depth understanding. High process yields can only be achieved through the comprehension of the role of each parameter. We are especially active in laser heating and welding (Seigneur, 2005, 2006), and micro-press fitting (Bourgeois). Another research topic is the use of surface forces for gripping and positioning of small components (Lambert, 2005).

BIBLIOGRAPHY

1. G. Boothroyd and P. Dewhurst, 1991, Product Design for Assembly, Boothroyd Dewhurst, Inc., Wakefield.
2. F. Bourgeois, Y. L. de Meneses and J. Jacot, 2005, Routes & Déroutes - Sur les traces d'un jeune ingénieur qui se lance dans la microtechnique, Revue Polytechnique, Suisse, Novembre 2005
3. A. Ferreira, 2003, Strategies of Human-Robot interaction for Automatic Microassembly, proceedings of the 2003 IEEE International Conference on Robotics & Automation, Taipei, Taiwan
4. S. Koelemeijer, L. Benmayor, J.-M. Uehlinger and J. Jacot, 1999, Cost Effective Micro-System Assembly Automation, Proceedings of the IEEE Emerging Technologies and Factory Automation
5. S. Koelemeijer, 2001, Méthodologie pour la conception de micro-systèmes et de leur équipement d'assemblage, Thèse EPFL
6. S. Koelemeijer, 2002, Small lot size micro-system assembly: The Flexible Micro-Assembly Cell, International Symposium on Robotics, Stockholm 2002
7. S. Koelemeijer, F. Bourgeois, J. Jacot, 2003 "Economical justification of flexible microassembly cells", Proceeding of International Symposium on Assembly and Task Planning, ISATP 2003, Besançon, France.
8. S. Koelemeijer, C. Wulliens, J. Jacot, 2005, Programming of assembly-cells by shop floor operators: a solution for short reconfiguration, CARV 05, International Conference on Changeable, Agile, Reconfigurable and Virtual Production, München.
9. P. Lambert, S. Koelemeijer and J. Jacot, 2006, Design of a capillary gripper for a submillimetric application, IPAS'06, Bad Hofgastein
10. M. Oulevey, S. Koelemeijer, J. Jacot, 2006 Impact of bad components on cost and productivity in automatic assembly, IPAS'06, Bad Hofgastein
11. F. Seigneur and P. Lambert, 2005, Polymérisation de colle epoxy par laser, Actes de la première journée sur la modélisation et l'analyse dimensionnelle,, Lausanne
12. F. Seigneur and J. Jacot, 2006, Laser sealed packaging for microsystems, IPAS'06, Bad Hofgastein

LIFE CYCLE AND COST ANALYSIS FOR MODULAR RE-CONFIGURABLE FINAL ASSEMBLY SYSTEMS

Juhani Heilala, Kaj Helin, Jari Montonen, Otso Väätäinen
VTT Industrial Systems, P.O.Box 1702, FIN-02044 VTT, Finland

Abstract: This article presents a case study in the design of a modular semi-automated reconfigurable assembly system using life cycle cost analysis methodology. To ensure that an assembly system is appropriately designed, system measurement schemes should be established for determining and understanding design effectiveness. Understanding life cycle costs is the first step toward increasing profits. The authors are developing an analysis tool that integrates Overall Equipment Efficiency (OEE), Cost of Ownership (COO), and other analysis methods to improve the design of flexible, modular reconfigurable assembly systems. The development is based on selected industrial standards and the authors' own experience in modular assembly system design and simulation. The developed TCO (Total Cost of Ownership) methodology is useful in system supplier and end-user communication and helps in trade-off analysis of the system concepts.

Key words: Modular Assembly System Design, Cost of Ownership Analysis, Overall Equipment Efficiency

1. INTRODUCTION

The objective of modern assembly processes is to produce high quality customized products at low cost. To ensure that an assembly system is appropriately designed, system measurement schemes should be established for determining and understanding design effectiveness. Measurements can be classed in two categories: cost and performance. Understanding manufacturing costs already in the system design phase is the first step to increasing profits. Throughput, utilization, and cycle time continue to be emphasized as key performance indicators for existing operations and for the planning of new assembly systems, and they have an influence on the cost efficiency of the system. All life cycle related cost issues should be identified and analyzed before making investment decisions, as early as possible in the system design phase.

The authors are developing an analysis tool that integrates Overall Equipment Efficiency (OEE), Cost Of Ownership (COO) and other analysis methods to improve designs of flexible, modular re-configurable assembly systems. The Total Cost of Ownership (TCO) analysis tool development is based on selected semiconductor industry standards[1,2,3] and is applied now for electronics final assemblies and the authors' own experience from assembly system design and simulation. The TCO method is useful in system supplier and end-user communication and helps in trade-off analysis of the system concepts (Figure 1).

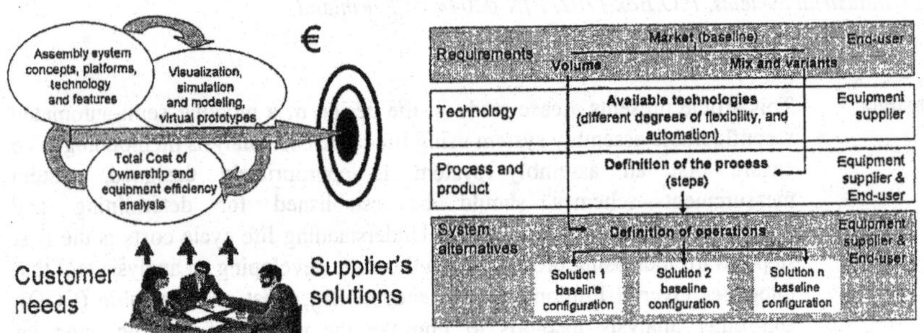

Figure 1. Collaborative assembly system design between end-user and system supplier.

2. DEVELOPMENT OF ASSEMBLY SYSTEM DESIGN AND ANALYSIS METHODS

Cost issues, especially reduction of cost, are key issues when investing new assembly hardware or processes. This was clearly shown in the assembly survey done by Assembly Magazine[4] in 2004. The purchase cost of the system is just one parameter to consider when performing a cost of ownership analysis. Different cost estimation methods have been devised; a few of them measure intangible costs such as flexibility, product yield, parts quality, process time variation, system modularity, re-use value, and so on. Although not all of these intangibles are easily understood, their costs may be measured by indirect methods. In many cases, a cost estimation method can be derived from performance measurements. For example, flexibility affects the capital investment plan. Yield and quality are related to capacity and material handling costs. Process time variation may cause problems with workstation utilization or in-process inventories[5].

2.1 Modular re-configurable assembly systems

Modular structure and reconfiguration is needed in the current market climate, where system changes occur at ever shorter intervals (Table 1).

Table 1. Typical Life-cycle parameters in electronics industry[6]

Parameters	Life-cycle parameter values
Typical aggregate annual volume	Less than 10 000 More than 1 million, up to 8 million
Product Mix - a logical family	Typically 2-8 variants
Typical life of products in production	6 months - 3 years , up to 15 years
Frequency of new product introductions	Computers: 1-2 months, up to 10 variants, others typically from 4 months to 2 year
Assembly rate objective good/units/hour	Starting 2-25/hour high-end products, 250-1000/hour, consumer products, as high as 3000/hour

The use of a modular structure in the architecture of an assembly system has many advantages. It facilitates standardization in that at least the selected suppliers' modules are compatible and the system is scalable. Design of a modular system is just like selecting suitable modules from a catalog and placing them in the right order to achieve the correct process flow and system layout. The end-user and system integrator can more easily configure the system and later reconfigure it to meet the customer's future needs.

Modularity is also a cost-efficient solution; it supports step-by-step investment, and later upgrades or modifications to the system are also easier. Standardized building blocks also help in calculating the cost of investment. Most of the modules should be standard, with known catalogue prices. Thus, product-related special customization is minimized. Typically, some equipment vendors estimate that 85% of the final assembly system equipment is re-usable in electronics final assembly[6].

Reconfigurable and modular solutions for final assembly systems need equally modular design tools. Each modular building block of the real system needs to have a digital image to be used in simulation model building, reconfiguration and analysis. Component-based simulation software with 3D capabilities is ideal for the design and configuration of modular reconfigurable systems.

2.2 Life-cycle consideration

As mentioned earlier, we need to calculate all the costs arising during the lifetime of the equipment. Some ideas for figuring out the frequency of change of assembly systems are given in Table 1. Typically the fastest changes occur in computer manufacturing and consumer goods. A change could occur at 6-month intervals. The life cycle is longer in the automobile industry, and especially in military or medical applications.

To be cost efficient the assembly system needs to outlive the product it was originally designed for, and here there is a need to analyze the life cycle of the planned system. The life cycle estimations of the system in the design phase are based on scenarios created by the end-user. Usually end-users have product roadmaps and estimations for new variant or product family introduction. Thus engineers can estimate the different products and variant life in production and also estimate the changes needed for the assembly system. If the basic assembly process is the same, only the product-specific system parts need to be changed, such as gripper fingers, part feeding, etc.

The scenarios can be modeled with a simulation tool that supports modularity. Each planned change of the system is modeled, and by comparing models it is possible to estimate the needed equipment changes to the system. This approach is useful in the new product introduction phase, when engineers are adapting the new product to the existing assembly line.

Currently there are requirements for common processes worldwide, and engineers need to analyze different country locations and assembly concepts. The cost of labor is but one parameter; there are others affecting unit costs in different locations. The number of good products produced depends on the efficiency and quality performance of the planned assembly system.

2.3 E-Race analysis toolkit theory, COO and OEE

The basics of COO are simple — it is all of the costs during the system life-cycle divided by the number of good-quality units produced[1,7,8,9]. Thus COO depends on the production throughput rate, equipment acquisition cost, equipment reliability, maintenance, equipment utilization, throughput, yield, rework and scrap cost and useful life-time of the system. The basic COO is given by the following equation:

COO per unit = total cost/number of good-quality products.

$$COO = (FC + VC + YC) / (L \times THP \times Y \times U) \quad (1)$$

Where:

FC = Fixed costs (amortized for the period under consideration)

VC = Operating costs (variable or recurring costs, labor costs)
YC = Yield loss costs (scrap and rework)
L = Life of equipment
THP = Throughput rate
Y = Composite yield
U = Utilization

The basic equation for calculating the COO was originally developed for wafer fabrication tools and has become a common reference between equipment suppliers and equipment users in the semiconductor industry. In the arena of electromechanical assembly, it is virtually unknown even though similar calculations are used. Use of the COO is an implementation of Activity-Based Costing (ABC) that helps in understanding all costs associated with a decision. It improves decisions by relating costs to the products, processes, and services that drive cost. Without such a linkage, it is difficult for organizations to understand the full impact of their decisions on their operating cost structure. With this linkage, COO provides a consistent data-driven method for arriving at important strategic and operational decisions. The difficulty is how to evaluate system flexibility, product mix, modularity and the re-use value of the system.

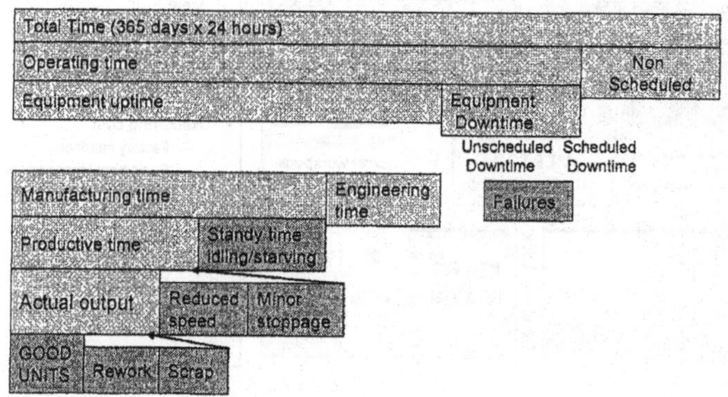

Figure 2. OEE helps to calculate the number of good units produced.

Authors are using OEE analysis to calculate numbers of good-quality products (Figure 2). The OEE (Overall Equipment Efficiency) is commonly used for monitoring the running performance of shop floor equipment. It was developed as an equipment effectiveness metric in Japan to measure the effectiveness of a manufacturing technique called Total Productive Maintenance (TPM). Originally, it was called Overall Equipment Effectiveness. The Semiconductor Equipment and Materials International (SEMI) Metrics Committee changed it to Overall Equipment Efficiency.

OEE is a key performance indicator of how machines, production lines or processes are performing in terms of equipment availability: reliability (MTBF), maintainability (MTTR), utilization, throughput performance or speed and quality produced. It identifies losses due to equipment failure, set-ups and adjustments, idling and minor stops, reduced speed, process defects and start up. All the above factors are grouped under the following three sub-metrics of equipment efficiency:

1. Availability
2. Performance efficiency
3. Rate of quality

The three sub-metrics and OEE are mathematically related as follows:
OEE, % = availability x performance efficiency x rate of quality x 100

General information about OEE can also be obtained from www.oee.com[10]. There are different opinions on how to calculate OEE. The developed Overall Equipment Efficiency analysis method used by the authors is based on a standard[2,3]. There is a systematic way to classify and study equipment efficiency and time losses.

2.4 E-Race VTT analysis toolkit prototype

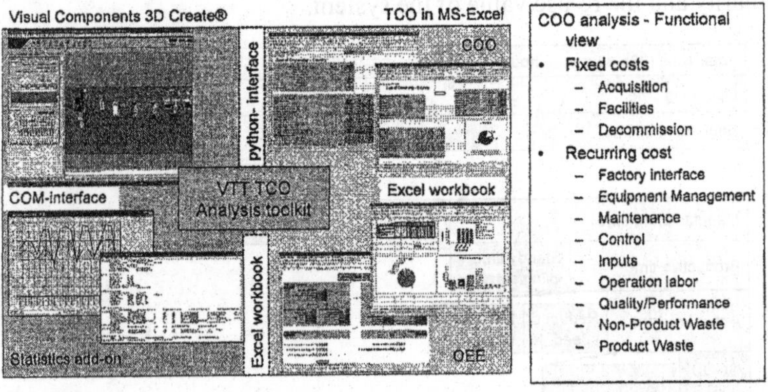

Figure 3. TCO VTT Analysis toolkit integrated into component-based simulation

Based on the standardized methodology presented here, the authors have developed a TCO, Total Cost of Ownership analysis Excel workbook[8]. Only suitable calculation formulas and definitions from selected standards[1,2,3] are used and the method is adapted for electromechanical final assembly system design evaluation. The users can input the parameters to the Excel sheets and analyze TCO, COO and OEE values.

In the second prototype tool[9], the authors have also integrated commercial component-based simulation (www.visualcomponents.com[11]) into TCO Excel analysis workbooks. An overview of the integration is shown in Figure 3. Each time an engineer selects a component from eCatalogue and places it on a simulation model layout, the cost functions start adding equipment cost, and adding an operator to the model adds to the labor cost function. The current version uses a COM interface, Python scripts, and Excel-internal links. Integration of Total Cost of Ownership (TCO) analysis into the simulation provides an effective method for evaluating system alternatives from a cost standpoint. This is clearly a tool for the system sales engineer. Adding easy-to-use cost analysis features to the simulation improves the quality of decisions during a sales meeting.

3. A CASE STUDY

The scenario presented here is based on discussions with industrial partners. The initial data is briefly the following: study of the final assembly line, making Product B (Basic) and later on Product A (Advanced). In all scenarios the production country is Finland, the worker cost of a year is 49 504 €, the cost of floor space is 200 € / m2 /year, and the needed floor space is 200 m2. The cost of rework is estimated to be 20 €. Cycle time of the bottleneck machine is in all cases 7 s. The calculated volume of good-quality units with one shift, 5 working days per week, is 728 676 5 units/year. The OEE factors are: Availability Efficiency 21.19%, Performance Efficiency 77.73%, and Quality Efficiency 98.45%.

The first scenario is to calculate what happens if two separate dedicated assembly lines are built. Figure 4 shows layouts and Tables 2 and 3 list some input data and major results.

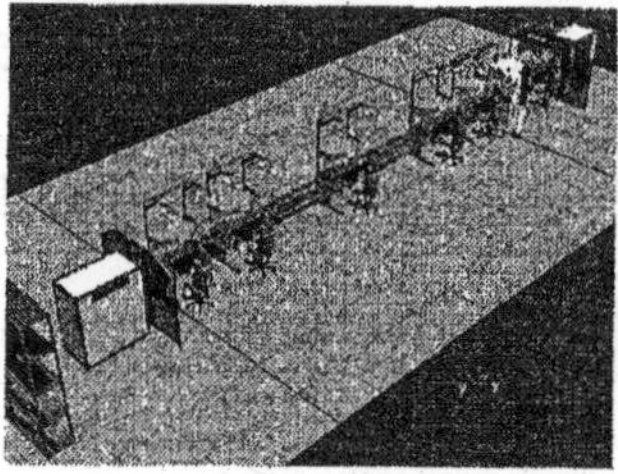

Layout 1, Product **B**, Basic

Layout 2, Product **A**, Advanced

Figure 4. Layouts to be analyzed were modeled with a simulation tool.

Table 2. Comparison Product B (Basic) and Product A (Advanced), 5 years production

	Product B (Basic)	Product A (Advanced)
Number of workstations	7	9
Number of workers	6	7
Number of support workers	1	1
System price	156 000 €	222 000€
Cost of product components	10.00€	12.50 €
Fixed Costs (all costs)	515 424 €	581 424 €
Recurring Cost (components, labor, etc.)	41 216 928 €	50 572 898 €
Yield loss cost (scrap, rework)	821 313 €	898 651 €
Total costs	42 553 665 €	52 052 973 €
COO	11.68 € /unit	14.29 € /unit

Table 3. Life-cycle cost for Product B (Basic) for 5 years

	Year 1	Year 2	Year 3	Year 4	Year 5
Fixed	267 808	59 904	59 904	59 904	67 904
Recurring	9 312 672	7 976 064	7 976 064	7 976 064	7 976 064
Yield loss	164 263	164 263	164 263	164 263	164 263
Total costs	9 744 743	8 200 231	8 200 231	8 200 231	8 208 231
Cumulative COO	13.37	12.31	11.96	11.78	11.68

Table 4. Work time (availability), volume effect on cumulative COO

Yearly volume	Year 1	Year 2	Year 3	Year 4	Year 5
5 shifts/728 676 unit	11.37	12.31	11.96	11.78	11.68
10 shifts/1 550 952 unit	11.92	11.42	11.25	11.17	11.12
15 shifts/2 393 040 unit	11.46	11.13	11.03	10.97	10.94
21 shifts/3 364 712 unit	11.30	11.07	10.99	10.96	10.93

Different work time arrangements show the capacity flexibility of the system and the effect on the cumulative COO (Table 4). The effects of maintenance costs were not evaluated in detail. Value-added cost can be calculated by subtracting component costs from the COO value. Similar results can be calculated for each layout and selected product and working time scenario. There are many variable parameters available in the TCO analysis workbook.

To continue the scenario, the next step would be to analyze a flexible assembly line capable of assembling two products. This is one way to justify flexibility or automation or at least find acceptable investment threshold values.

4. CONCLUSION

This article presents a case study in the design of a modular reconfigurable final assembly system using simulation, system life cycle and cost analysis methodology. The theory behind the analysis is also briefly explained. The developed TCO prototype tools are currently proof of the concept. There is a similar commercial tool available, dedicated for the semiconductor industry, at www.wwk.com[12].

Cost of Ownership (COO) provides an objective analysis method for evaluating decisions. COO provides an estimate of the life-cycle costs. The analysis highlights details that might be overlooked, thus reducing decision risk. COO can also be used for evaluation of processing and design decisions. Finally, COO allows communication between suppliers and users. They are able to speak the same language, comparing similar data and costs using the same analysis methods. Both suppliers and manufacturers can work from verifiable data to support a purchase or implementation plan.

The lifetime cost of ownership per manufactured unit is generally sensitive to production throughput rates, overall reliability, and yield. In many cases, it is relatively insensitive to initial purchase price, as shown by the example. With correct parameters an engineer can justify investments to flexibility and automated equipment or at least determine threshold values.

Overall Equipment Efficiency is usually a measurement of single-machine performance. In the example presented, the calculations are used for a bottleneck machine, and in practice the Overall Throughput Efficiency of the assembly line is calculated. With a serial line and single product, this can be quite simple. The analysis is more complex with mixed production and layout with parallel operations. Simulation studies can pinpoint bottleneck equipment. One of the limitations using OEE analysis is that analysis is process or equipment centric and the material flow or work in process (WIP) is not analyzed, another reason to use factory simulation.

For future development, real case studies with industrial partners are in progress in a new research project. The aims are to improve integration of the analysis into the component-based simulation software, to obtain analysis data from simulation input data files, and from the results of simulation runs with minimum effort from the users.

Users should remember that, as with all simulation analysis, this kind of simulation is sensitive to input data, and that input of false information does not produce the right results. The challenge is getting correct data. Knowing this, the authors are not aiming at absolute results in the design phase but, rather, at obtaining data for comparison of design alternatives. Later on, real factory data and accounting data can be used to verify the models and thus improve the results in the next evaluation round and new system designs.

The authors believe that COO and OEE are becoming increasingly important in high-tech decision-making processes. The challenge is to bring system reconfiguration to the analysis automatically; the idea exists at the conceptual level. Now, reconfiguration and system modularity cost efficiency analysis requires a lot of manual work.

ACKNOWLEDGEMENTS

The development presented here is part of the Eureka Factory E!-2851 E-Race project[13,14]. The authors wish to acknowledge the financial support received from the National Technology Agency of Finland (Tekes), VTT, and Finnish industry. The national project consortium in Finland, research institutes, technology providers, and end-users are working together to enhance assembly systems design methodology and to create next-generation assembly systems.

REFERENCES

1. SEMI E35-0701, Cost of Ownership for Semiconductor Manufacturing Equipment Metrics, SEMI International Standard, http://www.semi.org
2. SEMI E10-0304, Specification for Definition and Measurement of Equipment Reliability, Availability, and Maintainability (RAM), SEMI International Standard, http://www.semi.org.
3. SEMI E79-0304, Specification for Definition and Measurement of Equipment Productivity, SEMI International Standard, http://www.semi.org.
4. Assembly Survey 2004. Assembly Magazine, December 2004, p36 or (www.assemblymag.com).
5. Chow (1990): We-Min Chow. Assembly Line Design, Methodology and Applications. Marcel Dekker, Inc., New York and Basel, 1990.
6. iNEMI Technology Roadmaps 2004 Edition. Final Assembly. International Electronics Manufacturing Initiative. December 2004.
7. Ragona, Sid (2002). Cost of Ownership (COO) for Optoelectronic Manufacturing Equipment. 2002 Microsystems Conference. Rochester, New York, p. 200.
8. J. Heilala, K. Helin, J. Montonen. Total Cost Of Ownership Analysis For Modular Final Assembly Systems. 18th International Conference on Production Research. 30.7.-4.8.2005, Italy, Salerno.
9. J. Heilala, K. Helin, J. Montonen, P. Voho, M. Anttila. Integrating Cost of Ownership Analysis into Component-Based Simulation. 1st International Conference on Changeable, Agile, Reconfigurable and Virtual Production 22-23.9.2005. Munich, Germany
10. www.oee.com
11. www.visualcomponents.com.
12. www.wwk.com
13. E-Race Finland 2002-2006, project brochure.
14. www.e-race.info, international project Web site.

IMPACT OF BAD COMPONENTS ON COSTS AND PRODUCTIVITY IN AUTOMATIC ASSEMBLY

Mathieu Oulevey, Sandra Koelemeijer, Jacques Jacot
Laboratoire de Production Microtechnique (LPM)
Ecole Polytechnique Fédérale de Lausanne (EPFL)
CH-1015 Lausanne, Switzerland
mathieu.oulevey@epfl.ch

Abstract This article shows the consequences of using parts with scrap in an automatic assembly system. The scrap is the percentage of parts that do not reach the required conformity and will cause either a defective assembly or a jam in the system. The two consequences considered here are extra cost and loss of productivity. To show this, we will focus on one part assembly operation anywhere in the process, which consists of feeding, assembling the part and inspecting the resulting sub-assembly.

We will show that the effect of scrap can be huge and depends strongly on how the bad part is detected. 4 cases are considered: 1.The parts are sorted before feeding 2.The bad parts jam in the feeder 3.The bad parts jam while assembling 4.The sub-assembly with bad parts is ejected after inspection. By evaluating each case in 3 examples, we will show that case 3 is often the worst. We will also define the conditions where it is better to sort the parts first (1), let them jam in the feeder (2) or sort the sub-assemblies (4).

Keywords: Micro-assembly, cost-modeling

1. INTRODUCTION

When designing or choosing an assembly installation, it is very important to have a good cost estimation, therefore predictive cost models are of very high interest. A good understanding of where the major production costs lie is essential for a good cost model.

In macro-assembly, a productivity rate (cycle time and setup times) is often enough to describe the line throughput and evaluate the costs (). As products become smaller (high precision and micro-systems assembly), processes are more complex to control, often resulting in unexpected prohibitive assembly costs. Not only assembly processes are more difficult, but also the manufacturing processes of the components, whose yield become very low (Hong et al.,

2005). Scrap is then major source of costs, and is not taken into account in primary cost estimations.

In this paper, we show that the way of specifying components and their tolerances is vital not only for the product's functionality but also for the productivity rate of the assembly installation. We furthermore show that the way to deal with yields and scrap is essential for the good functioning of an assembly installation. Those elements are decisive for the cost of the final product, and this model should be useful to specify the scrap when ordering parts and evaluate the consequences on the product cost. We take the example of defaults in screws on an automatic screw installation to illustrate this.

2. MODEL DESCRIPTION

General cost model of product assembly. Assembly costs depend on the cost of ownership of the the installation. This cost is divided into *investment cost* and *running costs*. To evaluate the *investment costs*, we have shown (Oulevey et al., 2003) that the cost of an assembly installation mainly depends on the production cycle time, the components characteristics and the type of processes needed to assemble them. From tables of available processes and feeding systems, we computed the installation cost by estimating a difficulty factor for each component and process. We have also evaluated the *running costs*, including the building rental, production operators, electricity,... With these elements,we can evaluate the assembly cost of one product *as long as the production rate corresponds to the technical cycle-time*. If the installation is jammed due to bad parts or produces bad assemblies, the cost of ownership per part and the yield of We will show that scrap has a wide impact on running costs.

Consequences of bad parts on costs. The main causes of *non-quality* in assembly are components that do not meet specifications (tolerances, functionality,...) and assembly processes that are not perfect (which can generate bad assemblies even it the components were good). The main consequences of this non-quality are

- Cost of bad components
- Cost of good components used in a bad assembly
- Loss of time due to bad assemblies
- Loss of time due to jams in feeders and assembly processes
- Operator cost to unjam

Here, we will focus only on non-quality due to bad parts and assume that the processes will always make good assemblies from good parts.

Methodology and parameters used to estimate costs. Let us consider an assembly system for a product made of n different parts. The assembly system will be composed of n feeding systems, n assembly processes and usually n quality inspections. Here, we will compute the cost due to the scrap of *one* part on *one* process. If more than one part has a scrap, we can use this model recursively. If we focus on one of these n assembly process, we will have 2 inputs (a sub-assembly and a part) and one output (a sub-assembly).

We will compute the assembly extra-cost of one part on a receptor using the parameters in Table 1. In order to have good comparison, we assume that:

- The receptors (sub-assemblies on which the part is assembled) have no defaults.
- The assembly process is zero-default.
- All bad parts will be detected or cause a problem in the same way (see the cases below). If this is not true, the cases can be superposed.

T_c	**Cycle time [s].** Mean time to assemble one sub-assembly when the machine is on. Excludes all down time (jams, setup, ...). ↪*The hourly productivity is given by* $3600/T_c$ [parts/h].
T_{jf}	**Jam duration in feeder [s].** Mean down time due to a part jam in the feeding system.
T_{jp}	**Jam duration in assembly [s].** Mean down time due to a part jam in the assembly system.
k	**Bad parts.** Proportion of bad parts.
V_p	**Part value [€].** Value of one part (including the bad ones).
V_s	**Sub-assembly value [€].** Value of the sub-system on which the part will be assembled. If a rework after a bad assembly is cheaper, use this value instead.
C_{line}	**Assembly line cost [€/s].** Includes running costs (operators, air and electricity supplies, . . .).and costs of ownership (depreciation or rent, used area, . . .). ↪*The hourly cost is given by* $3600 \cdot C_{line}$ [€/h].
C_{op}	**Operator cost [€/s].** Cost of the operator who will deal with jams due to defective parts. ↪*The hourly cost is given by* $3600 \cdot C_{op}$ [€/h].

Table 1. Model parameters

3. CASES DESCRIPTION

The extra cost due to bad parts will widely depend on where they are detected. 5 cases are considered:

0 The bad parts have no effect on assembly, as if they were good;
1 The bad parts are removed before feeding (sorting by an operator);
2 The bad parts jam in the feeder;

3 The bad parts jam on the assembly process;
4 The bad parts cause no jam, but the assembly is detected as bad and ejected during further inspection.

Cases 0 and 1 are trivial but useful for comparison purposes, whereas cases 2 to 4 correspond to the situations we want to analyze. Each case will be explained in the next sections.

Case 0 : No scrap. Let us consider the case where there is no scrap in parts and all sub-assembly are good. We suppose that no other factor will disrupt the production, in order to use the results as a reference for comparison purposes. The production costs are trivial: the hourly productivity P_0 of the line is directly given by the cycle time, since the machine never stops:

$$P_0 = 3600/T_c$$

and the assembly cost C_0 is given by the machine cost during one cycle:

$$C_0 = C_{line}/P_0$$

No operators are required in this case.

Case 1 : Sort the parts first. If we decide to sort the parts before feeding them, we have to consider an operating time T_i for inspection. This case is not pertinent, because nobody would affect operators to sort the bad parts, but it could give an idea of the part extra cost to have zero-default:

$$C_1 = \frac{C_{line}}{P_0} + \frac{1}{1-k} \cdot T_i \cdot C_{op}$$

and the hourly productivity does not change since only good parts are fed:

$$P_1 = 3600/T_c$$

Case 2 : Bad parts jam in feeder. If the bad parts jam in the feeder, usually an alarm occurs T_a seconds before the production line is stopped. If the operator is close and efficient (i.e. the time to arrive and fix the jam T_j is smaller than T_a), the jam will have no incidence on the production. If the operator is far or the jam is difficult to fix ($T_j > T_a$), the machine will be stopped for a time $T_j - T_a$. Since the average down time requires the probability distributions of T_j and T_a to be computed, we will use

$$P_2 = 3600/(T_c + k \cdot T_{jf})$$

$$C_2 = C_{line}[T_c + k \cdot T_{jf}] + C_{op}[k \cdot T_{jf}]$$

Case 3 : Bad parts jam in assembly process. This case is similar to case 2 but worst: firstly because the receptor is damaged and is either trashed or reworked (which costs V_s), secondly because the jam duration T_{jp} is usually greater than T_{jf} in case 2. The productivity rate is similar to case 2

$$P_3 = 3600/(T_c + k \cdot T_{jp})$$

but the cost takes into account the damaged receptor:

$$C_3 = C_{line}[T_c + k \cdot T_{jp}] + C_{op}[k \cdot T_{jp}] + k \cdot V_s$$

Case 4 : Bad parts are detected after assembly (inspection). The sub-assembly is damaged like in case 3, but it is ejected without jamming. In this case, The productivity is better, as no jam occur:

$P_4 = 3600/(\frac{1}{1-k}T_c)$

and the cost is similar to case 3, but without time loss due to jams (no operator required):

$C_4 = C_{line}[\frac{1}{1-k}T_c] + k \cdot V_s$

4. EXAMPLES

Example 1 : single screw. A typical example of extra-costs due to bad parts is a screwing process. Screws are subject to various manufacturing defaults (malformed thread, missing slot, remaining burr, bad length, ...). Let us consider a basic screwing process where one screw is inserted in a sub-assembly. We will neglect the value of the sub-assembly. Here are the values for all the parameters:

- 1% of the screws are malformed ($k = 0.01$).
- We neglect the sub-assembly value ($V_s = 0$).
- 1000 screws cost 10€ ($V_p = 0.001$).
- The screwing machine can screw 1000 screws per hour ($T_c = 3.6$)
- The feeder jams during 10s ($T_{jf} = 10$), and the screwing machine jams during 60s ($T_{jp} = 60$).
- The hourly costs are 30€ /h for the screwing machine ($C_{line} = 0.083$) and 36€ /h for an operator ($C_{op} = 0.01$).

With these figures, the cost for assembling one screw is given by:

$$C_0 = 0.03; C_1 = 0.043; C_2 = 0.0319; C_3 = 0.0411; C_4 = 0.0301$$

In conclusion, if the screws jam in the screwing machine, it is not worth to sort the screws before feeding ($C_1 > C_3$), but we would accept to pay 20 € instead of 10 € for 1000 screws without any scrap.

Example 2 : plastic caps assembly. Plastic caps are glued on carton packages for juices and milk. They are made of 2 parts: the neck and the cap. The cap is inserted on the neck by press-fitting. Then, this sub-assembly is glued on the juice brick. Sometimes, the assembly is detected as bad because either the cap or the neck has an injection failure. The question is : Is it worth sorting the parts before assembly ?

Here, we have the following parameters : $k = 0.001, Tc = 0.2, V_p = 0.001, V_s = 0.002, C_{line} = 30$ If the line can sort the parts first, then $C_{line} = 35$ We obtain then the costs:

$$C_0 = 0.0022; C_1 = 0.0025; C_2 = 0.0024; C_4 = 0.0022$$

C_3 is not evaluated because the press-fit will not stop if the parts are defective. C_2 is given for information, even if all bad parts would not jam in the feeder. The answer to our question is that it is not worth sorting the bad parts before feeding, since the cycle-time is small and the part value very low.

Example 3 : closing an aluminum housing with 6 screws. Let us consider a sensor packaged in an aluminum housing closed with 6 screws. 0.2% of the screws are defective (which is good) and will cause a jam and damage the housing. As we use 6 screws on each assembly, we use an new value of k which is the probability to have at least one defective screw on an assembly $k_{6screws}$. If $k_{screw} = 1\%$, then

$$k_{6screws} = 1 - (1 - k_{screw})^6 = 1.2\%$$

If a jam occurs due to a malformed screw, the sub-assembly has to be reworked, with an average cost of 3.5 €, including operator time and spare parts. The other parameters are the same as in example 1, and the model gives these results:

$$C_0 = 0.050; C_1 = 0.086; C_3 = 0.112; C_4 = 0.100$$

Here, we can evaluate the catastrophic consequences of using screws even with a small scrap in these conditions. 0.2% scrap that jam in assembly will double the assembly cost ($C_3/C_0 = 2.25$) and cause 10% down time. It is far better to inspect and sort the screws before feeding them (C_1), even if the bad screws do not jam the screwing-machine (C_4).

5. DISCUSSION AND CONCLUSION

As the examples show, a small scrap in the parts can generate very large costs mainly due to the time loss during jams (example 1) and the loss of valuable good parts in bad assemblies (example 3). The model presented here allows a fast estimation of the costs.

This model should also be be useful to design new production lines and choose the inspection policy which minimizes the costs and down-time. It also provides a technique to evaluate the amount that is worth being spent for parts with a lower scraps.

This model is quite simple and does not include the following aspects:

- The model takes into account the hourly cost of operators, but the cost is biased, firstly because you must have an integer number of operators, secondly because it strongly depends on the operator policy: either he stands in front of the machine and can quickly fix a jam when it occurs, or he can perform other production tasks (filling the feeders for instance) and cause a delay in fixing the jam. However, there is no other solution than evaluating every assembly system depending on the tasks affected to its operators.

- The loss of productivity is evaluated for each assembly process of the line. This is useful to identify the bottleneck, but the overall productivity must be evaluated using analytical techniques (Buzacott and Shanthikumar, 1993; Hongler, 1994; Medhi, 2003) or discrete-events simulation (Heinzelmann, 1995)

ACKNOWLEDGMENTS

This work has been partially supported by the EUPASS project (funded by the European Commission), whose aim is to design and evaluate the performances of reconfigurable precision assembly systems.

REFERENCES

1. Buzacott, J.-A. and Shanthikumar, J.-G. (1993). *Stochastic models of manufacturing systems*. Prentice Hall.
2. Dubois, D. and Forestier, J. (1981). Productivité et en-cours moyens d'un ensemble de deux machines séparées par une zone de stockage. *R.A.I.R.O. Autom. Syst. Analysis and Control*, 16:105–132.
3. Gershwin, S. B. (1994). *Manufacturing Systems Engineering*. PTR Prentice Hall.
4. Han, M.-S. and Park, D.-J. (2002). Performance analysis and optimization of cyclic production lines. *IIE Transactions*, 34:411–422.
5. Heinzelmann, E. (1995). Simas ii - ein simulationswerkzeug für kmu. *Technische Rundschau Transfer*, 3:24–29.
6. Hong, C., Milor, L., Choi, M., and Lin, T. (2005). Study of area scaling effect on integrated circuit reliability based on yield models. *Microelectronics reliability*, 45(9-11):1305–1310.
7. Hongler, M. O. (1994). *Chaotic and Stochastic Behaviour in Automatic Production Lines*. Springer Verlag.
8. Jacot, J. (2004). *Industrialisation, Notes de cours*. EPFL.
9. Medhi, J. (2003). *Stochastic Models in Queuing Theory*. Academic Press.
10. Oulevey, M., Roduit, P., Koelemeijer, S., and Jacot, J. (2003). A cost model for flexible high speed assembly lines (erace project).
11. Terracol, C. and David, R. (1987). Performance d'une ligne composée de machines et de stocks intermédiaire. *APII*, 21:239.

PART VI

Microassembly – Solutions and Applications

THE RELIABLE APPLICATION OF AVERAGE AND HIGHLY VISCOUS MEDIA

Assembly Net

Thomas A. Lenz and Nabih M. Othman

Fraunhofer IPA, Department of Cleanroom Manufacturing, Nobelstrasse 12, 70569 Stuttgart, Germany

Abstract: The use of bonding as a joining technique is gaining more and more importance in micro manufacturing. There is a demand for systems which are capable of reliably dispensing tiny amounts of adhesive, especially under industrial conditions. By integrating sensors into the dispensing system or checking process results there is the possibility to become aware of the system specific problems and options of improvement. Methods and systems of investigation have been developed and realized at the Fraunhofer IPA.

Key words: Process control, process monitoring, sensor integration, dispensing, adhesive bonding, micro assemble, vision system.

1. INTRODUCTION

New demands are being placed on manufacturing engineering as a result of increasingly miniaturized components, frequent product changes and alterations in the batch sizes of life style products such as mobile phones, PDAs, cameras, etc. For the manufacturer, this means having to adapt production systems to new circumstances and ensuring adequate process stability for the required piece numbers within a very short space of time (Hennemann,1992).

Bonding techniques are becoming more and more popular as a joining technology for fixing and contacting components (Luchs, 1998). Ideally, it should be possible to adjust dispensing systems simply and accurately to produce and maintain a high volume constancy and reproducibility of adhesive application (Gaugel, 2003).

Although current dispensing systems are capable of reliably creating tiny structures both of unfilled adhesive or filled conductive adhesive under laboratory conditions, in industry changes in operating conditions such as fluctuations in material temperature due to warming, variations in batches and many other factors lead to deviations in feeding volumes and thus to the application of faulty structures. Figure 1 shows the results of such changes when the feeding volume at the tip of the needle varies in an uncontrolled manner or is not noticed (Othman, 2004).

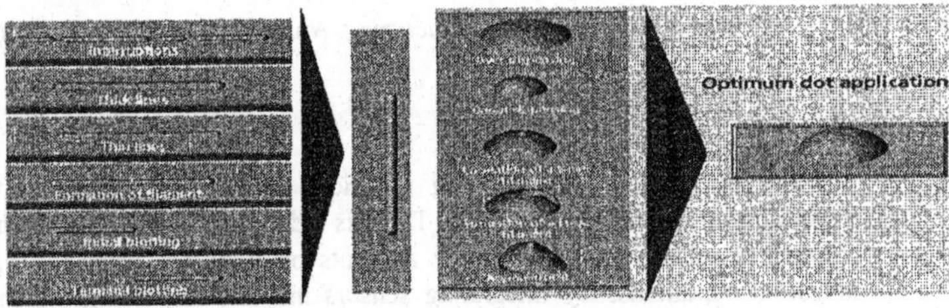

Figure 1. Faults in the application of adhesive due to fluctuations in flow rates.

Most of the dispensing systems currently available on the market still require optimization as they do not possess sensors which can measure fluctuations in parameters relevant to the manufacturing process and thus react accordingly. Therefore, with a view to increased productivity, flexibility and reliability, controlled systems for the accurate application of dots and lines of highly-viscous adhesives (especially conductive adhesives) are required. Methods for investigating dispensing systems, targeted parameter studies and mathematical models are needed to pave the way for more efficient dispensing systems.

2. FACTORS INFLUENCING THE DISPENSING PROCESS

Experiments carried out at the Fraunhofer institute concerned with application of conductive adhesives have shown that the dispensing process is not only dependent upon the correct selection of the application method or system but much more upon the correct process parameters for the application in question. As well as the dispensing system itself with its reservoir (e.g. cartouche), the dispensing needle, adhesive, substrate and handling system also need to be taken into consideration (Figure 2.).

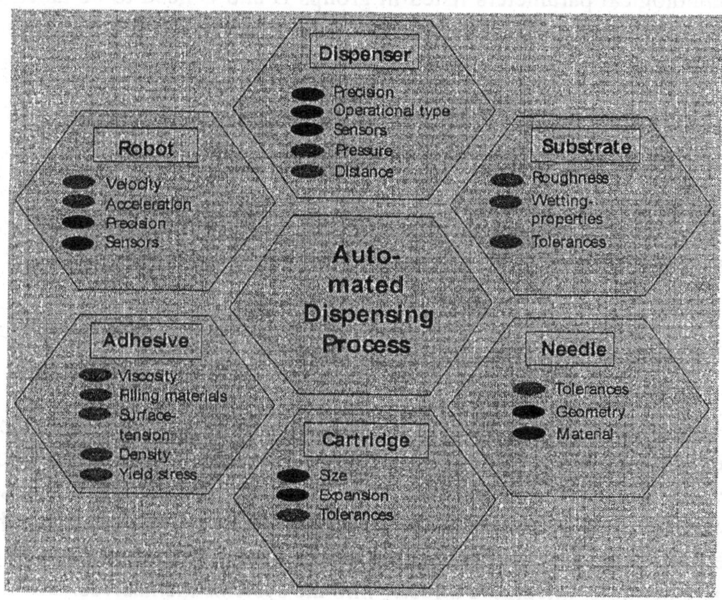

Figure 2. Parameters influencing the dispensing result:
Parameter Group A (not alterable),
Parameter Group B (alterable before commencement of dispensing process),
Parameter Group C (alterable even during dispensing process)

Parameters influencing the dispensing result can be divided into three groups:

- The influencing factors listed in **Parameter Group A** are parameters which cannot be altered or influenced by the user. The viscosity, dropping time and conductivity of an adhesive are all examples of such parameters.
- The factors listed in **Parameter Group B** include parameters which can be altered before the actual dispensing process but not during it. An example of this is expansion of the cartouche which can be minimized if the user takes appropriate measures beforehand.

- **Parameter Group C** includes parameters which can be altered during the dispensing process. One example here is the traveling velocity of the robot.

When defining the requirements of a reliable overall system, especially the technological parameters listed in groups B and C need to be taken into account. Many of these parameters have been investigated at the Fraunhofer IPA (Othman, 2004).

3. INVESTIGATION METHODS

Once possible deficits and influencing factors have been defined, current analysis methods and a general procedure need to be determined.

3.1 Monitoring the process

In order to achieve targeted process monitoring, sensors which have been adapted to meet the measurement task requirements in question have to be integrated. As the main issue here is not the development of special sensors, miniaturized sensors currently available on the market have been used which have either been integrated into sensor adapters for universal implementation or, where possible, integrated directly into the systems themselves.

In the field of miniaturized sensors, the choice of available sensor systems is relatively narrow, the price usually extremely high and delivery times invariably considerably lengthy. These circumstances mean that compromises have to be made regarding the construction of the individual systems in order to achieve a high degree of flexibility and universal implementation. Figure 3 shows the first prototype utilized. A special optical monitoring device for microfluidic systems is presented in another paper (Güttler, S.; Lenz, Th. A., 2005).

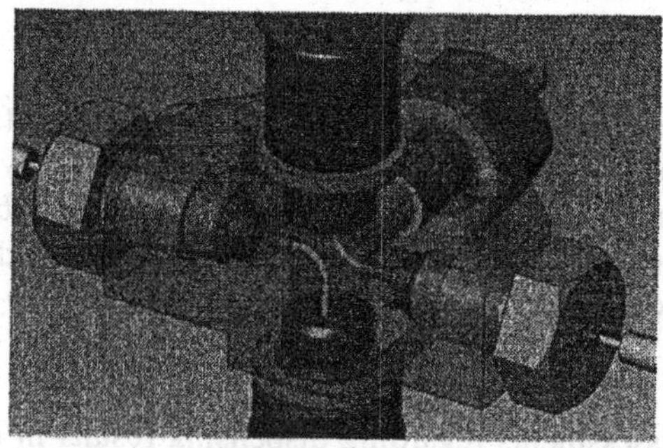

Figure 3. CAD model of the first prototype of a sensor adapter with integrated pressure and temperature sensors.

After further experiments concerned with construction and joining technologies, sealing techniques and the integrated sensor technology of this adapter, it was possible to create optimized systems for special applications as shown in Figure 4.

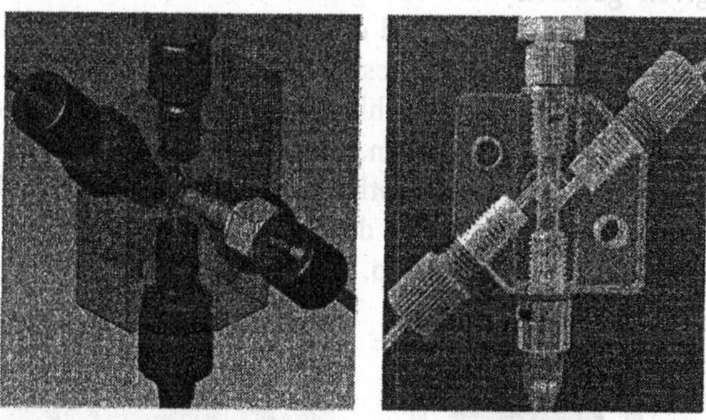

Figure 4. Sensor adapter optimized with regard to flow rates and dead volumes.

Through the use of standard fittings, the adapter can be rapidly integrated into a number of dispensing systems. The use of detachable sealed sensor elements enables these integrated sensors to be reused, thus avoiding unnecessary costs.

The benefit of using transparent material lies in the ability to monitor hidden and inaccessible spaces. However, before this monitoring tool can actually be implemented, further geometrical optimizations need to be carried out on the system and additional developments made to the sealing concept.

3.2 Checking process results

In order to be able to correctly assess the process results, both the dots and lines dispensed by the system need to be accurately measured and investigated. Gravimetric methods are only partially suitable as the mass decreases by the cube of the geometric dimensions. Thus, a drop of adhesive 1 mm³ in size weighs approx. 1 mg and a drop $100 \cdot 100 \cdot 20\ \mu m^3$ in size only 0.2 µg.

For reliable assessments regarding dispensing results in micro and nano assembly, primarily only optical methods are suitable as these are capable of measuring structures just a few micrometers in size. A high-performance vision system for in-line process control has been implemented at the institute using an SCC system (primarily developed for measuring particles on technical surfaces).

The system is capable of recording and measuring dot diameters smaller than 500 µm and line widths less than 500 µm. Any deviations from a desired pre-given geometry are recognized and reported. The measuring principle of an SCC camera is based on combined, flat, lateral illumination and / or incidental light from the test surface (Figure 5). In this way, an acceptable level of contrast can be achieved between the dots/lines dispensed and the rough surface they are on. With the aid of image-processing methods, they can be used to automatically record dots or lines. The system is able to reliably record and analyze dot diameters or line widths measuring approx. 8 µm (Grimme, R.; Klumpp, B., 1998).

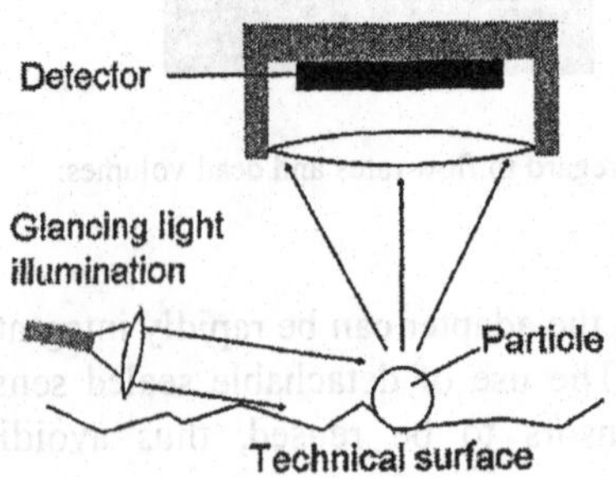

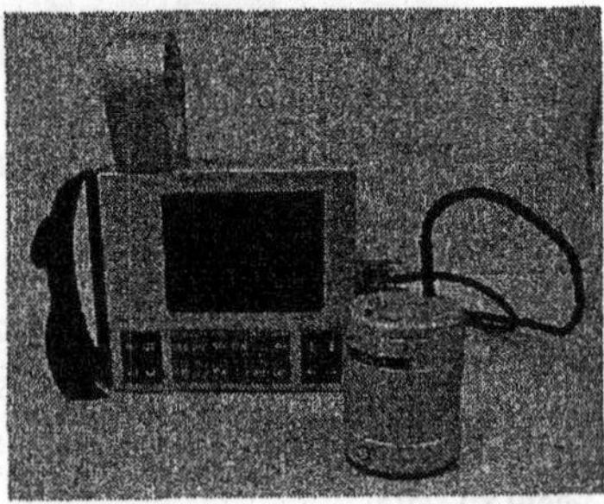

Figure 5. Prototype of a hand-held device for recognizing dots or lines on surfaces. Left: diagram of the principle of glancing light excitation inside a probe / Grimme 1998; Right: hand-held device with probe

4. PRESENTATION OF THE RESULTS

4.1 Measurement procedures / methods for monitoring process parameters

An overview of the results obtained is given below. To demonstrate investigations on system behavior, an under-supplied pump unit was used which was unable to prime enough medium because its lines were too narrow; as a result, cavitation effects occurred causing a strong, high-frequency pulsation of pressure values, as shown in Figure 6. A possible optimization solution would be to redesign the system by enlarging the cross-section of the lines, thus increasing process reliability.

With this and many other problems, it may be helpful to take the overall system or the process step and its environment into consideration in the investigations in order to find a solution.

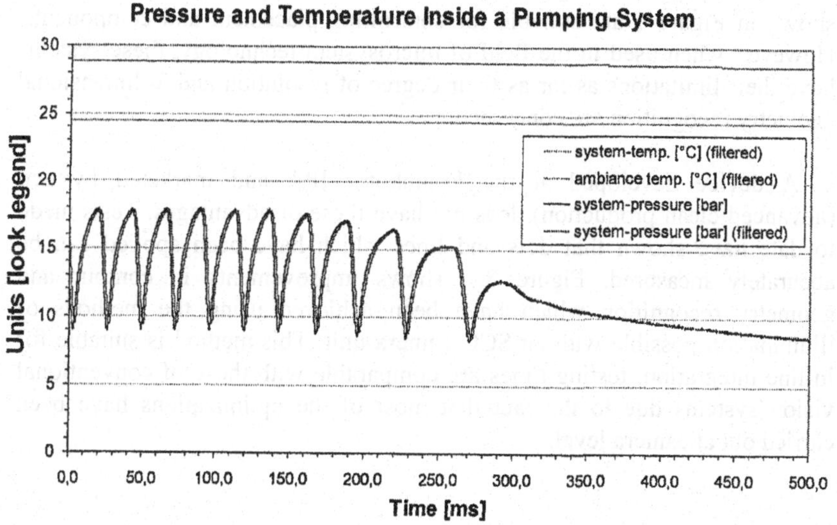

Figure 6. Pulsation inside a pumping unit due to cavitation effects.

4.2 Measurement procedures for controlling process results

The performance of the systems mentioned here was verified on dots and lines applied to silicone and ceramic substrates. In order to obtain statistical certainty, 10,000 dots or 100 lines were applied respectively and investigated using the various measuring methods.

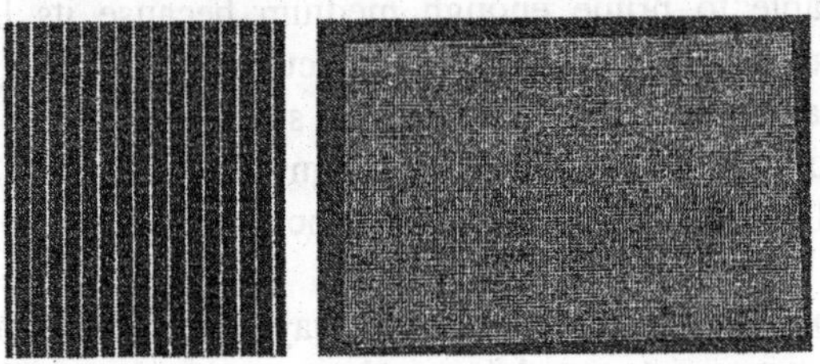

Figure 7. Results with a commercial or industrial vision system

Conventional vision systems used in industrial applications such as those shown in Figure 7 are suitable for controlling placement and components. However, when used in the field of microsystem technology, these systems have their limitations as far as their degree of resolution and 3-dimensional geometry recognition are concerned.

A device developed at the Fraunhofer IPA and marketed by acp (advanced clean production) does not have these disadvantages. Tests made to date have shown that dots and lines which have been applied can be accurately measured. Figure 8. shows improvements in contour and geometry recognition which have been achieved using the methods of illumination possible with an SCC camera unit. This method is suitable for in-line integration; testing times are comparable with those of conventional vision systems due to the fact that most of the optimizations have been carried out at camera level.

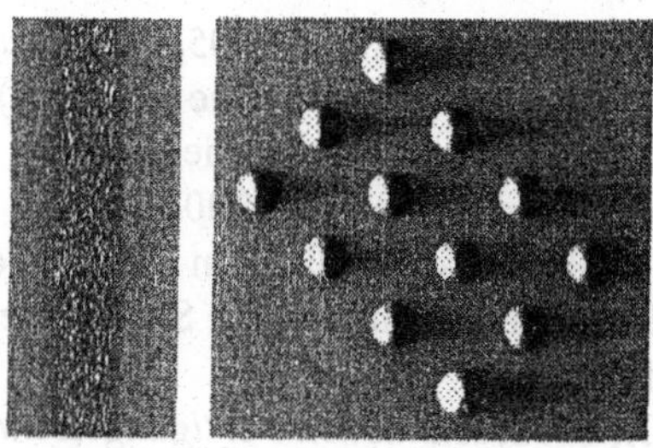

Figure 8. Results obtained with an SCC camera using only incidental light (left) and using glancing light combined with incidental light (right).

A laser scanning microscope (LSM) is used as a reference method / measuring system. Using the principle of light sections, this device is capable of depicting volume models of convex three-dimensional objects and of measuring them with an accuracy of approx. 1 µm. However, due to the long measuring times involved and the very small field of vision (200 x 200 µm²), this method is only suitable as a control or reference system. Figure 9 shows the measurement of volumes and of dot diameters and line widths of each of the dots/lines applied.

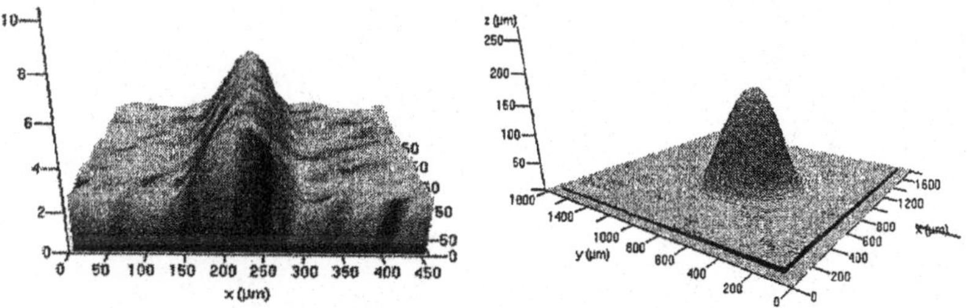

Figure 9. Test results obtained using an LSM.

REFERENCES

1. Gaugel, Tobias, 2003, Verfahren zum flexiblen Mikrodosieren von isotrop leitfähigen Klebstoffen, University of Stuttgart.
2. Grimme, R., Klumpp, B., 1998, Vorrichtung und Verfahren zur Überprüfung einer Oberfläche eines Gegenstandes, (10-1998), Patent: DE 197 16 264 A1.

3. Güttler , Stefan; Lenz, Thomas A., 2005, *Mikrodurchflusssensorik für höherviskose Fluide*, Springer, WT-Online (**03**-2005), pp. 162 et sqq.
4. Othman, Nabih 2004, Präzise Mikrodosierung von nieder-, mittel- und hochviskosen Medien, Adhäsion **48** (6-2004), pp. 22 et seqq.
5. Möller, Markus, 2002, Mikroapplikation von ungefüllten Klebstoffen zum Kleben in der Mikrosystemtechnik, Shaker Verlag, 2002, Aachen; also University of Aachen.
6. Hennemann, O. D., 1992, *Handbuch Fertigungstechnologie Kleben*, Carl Hanser Verlag, 1992, Bremen.
7. Luchs, R., 1998, Einsatzmöglichkeiten leitender Klebstoffe zur zuverlässigen Kontaktierung elektronischer Bauelemente in der SMT, Meisenbach Verlag, 1998, Bamberg; also University of Erlangen.

LASER SEALED PACKAGING FOR MICROSYSTEMS

Frank Seigneur
Ecole Polytechnique Fédérale de Lausanne
STI – IPR – LPM – Station 17
CH – 1015 Lausanne
frank.seigneur@epfl.ch

Jacques Jacot
Ecole Polytechnique Fédérale de Lausanne
STI – IPR – LPM – Station 17
CH – 1015 Lausanne
jacques.jacot@epfl.ch

Abstract Packaging is the last process of microsystem manufacturing. There are mainly two kinds of packages: plastic or metallic. The two main components of the package (base and cover) may either be glued or soldered. Each of these techniques has its advantages and drawbacks, and the choice should be driven by the functionality of the microsystem.

The advantage of gluing is that it is quite an easy production process. The drawback is that glue, like all polymers, is not hermetic on the long term. This is a problem when the package should protect the microsystem from oxygen or water vapour, an example being OLED displays (Organic Light Emitting Diodes).

The advantage of soldered metallic packages is that they are hermetic. The drawback is that the welding process takes place in an oven, and that the whole microsytem is heated to the fusion temperature of the joint. This is a problem for many microsystems, typically biomedical MEMS and OLEDS again.

The LPM-EPFL is working on the development of a two-part sealed packaging. One part of the package is metallic, the other part is made of glass.

The goal of the project is to soft solder the two parts of the package by the mean of a laser diode. The advantages of the laser soft soldered joint are:
– its water and air-tightness in regard to glue or a plastic joint,
– the possibility to heat only the solder joint, and not inside the package.

Keywords Microsystem assembly, packaging, hermetic sealing

1. INTRODUCTION

The goal of the study presented in this paper is to demonstrate the feasibility of a laser sealed packaging. First, a functional analysis of the packaging process is presented in section 1. The tools and setup are detailed in section 2. Two models developed to predict heating of the system during the packaging operation are presented in section 3. Section 4 presents the results of both models and experiments. Finally, future works are described in section 5.

2. FUNCTIONAL ANALYSIS OF THE PACKAGE AND THE PACKAGING PROCESS

Three main functions are searched for this packaging:

- tightness to O_2 and H_2O.
- heat control
- connection to the outside world

First of all, long term hermeticity is needed. It cannot be achieved with the use of glue or plastics, which are porous to gas. An interesting point is that a 2μm joint of aluminium is tighter than 1mm of plastic. The choice is thus the use of metals and ceramics for the package, and tin soldering for the welding seam. In our case, the base is ceramic, the cover is glass, and the solder is screen printed on the ceramics.

Another main function of the packaging process is to control the heating of the microsystem during the sealing operation. Many microsystems, typically bioMEMS and OLEDs must not be exposed to temperatures above 70°C. A common solution is to do the soldering operation in an oven, but then the whole microsystem is heated during this operation. That is why the use of a laser diode in combination with temperature sensors allows for control of the energy brought to the system during the sealing operation.

Finally, the sealed microsystem must be connected to the outside world. The type of connections depends on the application of the microsytem. In our case, screen printed electrical connectors have been tested.

The originality of the process is that the heating of the welding seam is achieved through the glass cover. The glass is transparent to the wavelength of the laser (940 nm). A thin layer of black dielectric allows the insulation of the contacts needed for the connection and also the conversion of the light to thermal energy in order to melt the welding seam.

3. TOOLS AND SETUP

The LPM-EPFL is equipped with a laser diode used for tin soldering of printed circuit boards. The advantage of a laser diode is first of all the possibility to easily control its power. The low power density is also very well adapted to the process. Another advantage, is that its cost is about the half of a Nd:YAG or CO_2 laser. The laser must only bring the tin to its fusion point (180°C), higher power would provoke vaporization of the material which is not needed for our application.

Another point is the possibility to perform continuous heating, which is an advantage in comparison to point to point soldering, regarding tightness of the welding seam. The package to be soldered is placed on an x-y table, allowing it to be moved under the laser spot. Several cameras are used to visually control the process during the soldering operation. The whole setup is placed inside a closed container for security reasons.

4. MODELS

An analytical thermal model of the welding seam has been developed. This model allows us to understand the influence of material choices on the quantity of energy needed to obtain the fusion of tin. The results of this model helped to chose materials in order to reduce the energy needed to heat the welding seam, and thus to reduce the heating of the microsystem. A second model was developed to determine the heating of the microsystem during the soldering operation. This model was compared to experimental results.

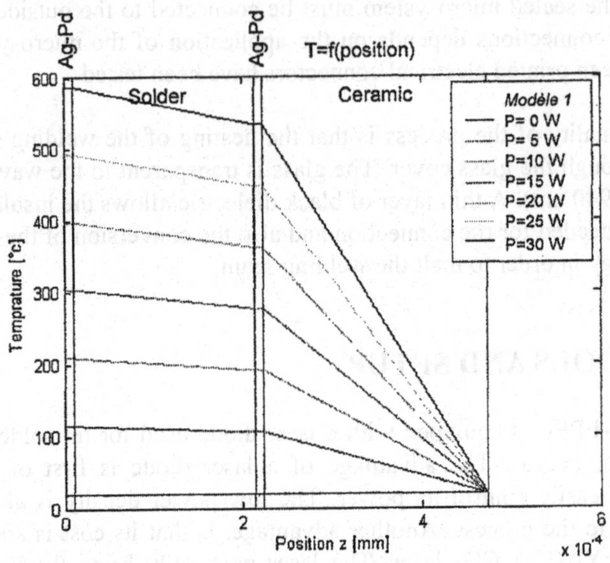

Figure 1. First model: temperature profile through the different layers for several powers of the laser.

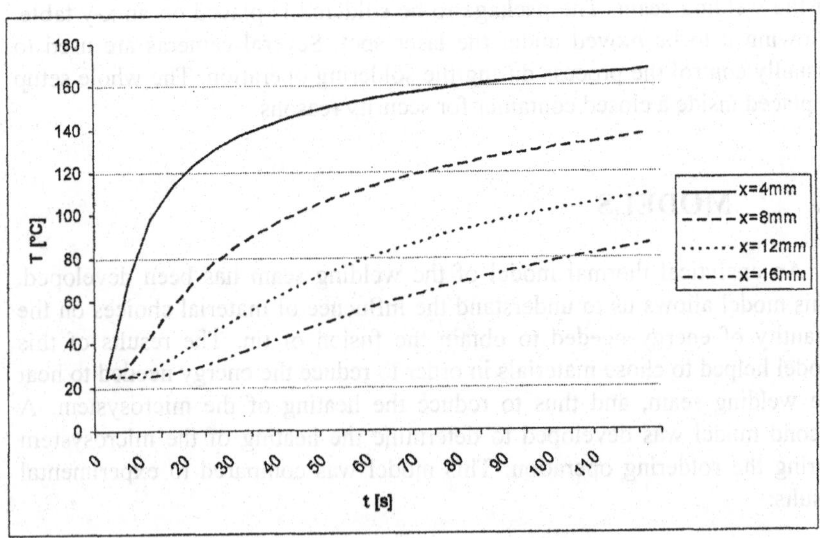

Figure 2. Second model: theoretical evolution of temperature inside the package for several distances from the laser spot.

The second model allows us to determine the temperature inside the package during the soldering operation (see Figure 2). The model is quite accurate for the first 10 seconds. For longer times, a gap can be observed between the predicted and measured temperatures. It is due to the fact that this model is one-dimensional. It is not an issue, because the soldering operation should not take longer than 10 seconds.

A prototype has been developed in order to validate the models. The different layers needed for the connections are made by screen printing. The welding seam is also screen-printed. This allows a good repetitivity in the geometry of the joint. The main drawback of this method is its lack of flexibility concerning the shape of the welding seam.

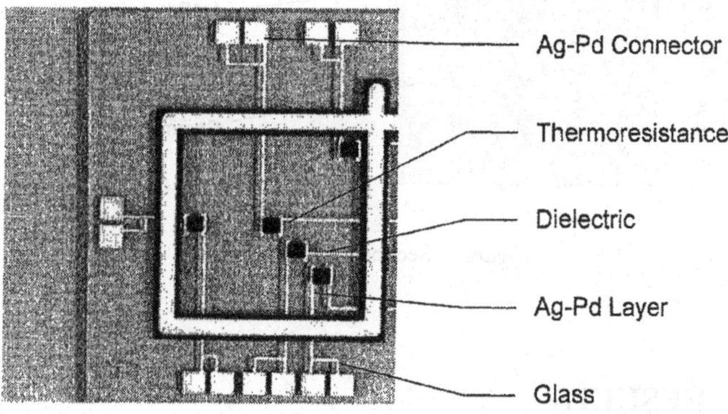

Figure 3. Layout of the prototype

Figure 3 shows the layout of the prototype. Note the thermoresistances used to measure the temperature inside the package during the soldering operation. These thermoresistances are needed for the development of the process.

Figure 4 shows the different layers and their functions. The first layer of Ag-Pd is used for the electric connections. They are insulated from the welding seam by a layer of dielectric.

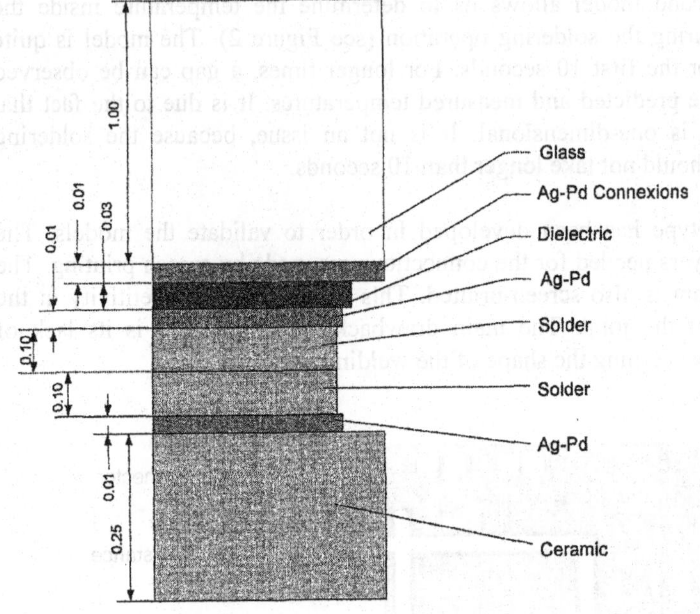

Figure 4. Section of the package

5. RESULTS

The first observation concerning the first model (see Figure 1) is the effect of the ceramic layer. It allows us to have a low thermal gradient in the solder. For example, the thermal gradient in the solder for a temperature of 100°C is about 1°C/mm. This allows us to have a homogenous solder joint. The second observation is that the power needed for a fusion temperature of 200°C is about 10W. This is higher than what has been observed practically. This may be due to the fact that the welding seam initial temperature is set to 20°C. It doesn't take into account the fact that the welding seam is being heated not only under the laser spot, but also due to its thermal conductivity.

The first tests showed that it was possible to heat the welding seam through the glass. However, it was critical concerning temperature, the glass part often broke depending on heating temperature and mechanical preloading on the package. The use of low fusion temperature soft solder might be a solution to improve this critical point.

The main difficulty is to obtain a good wettability of the tin on the glass part of the packaging. A lack of wettability was often observed during the soldering process. This wettability is obtained through metallization of glass, but this method still has to be improved.

Temperature inside the package was measured with temperature sensitive resistances. Figure 5 shows the temperature reached 4mm away from the welding spot. A temperature as low as 45°C was reached at the center of the package (9mm away from the laser spot).

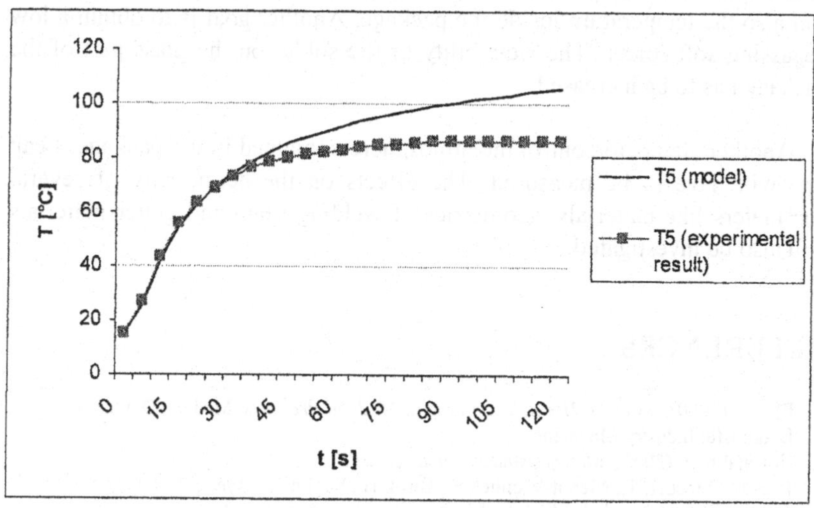

Figure 5. Second model: temperature inside the package for P=1W

Our first practical tests allowed us to compare the models to the experimental situation. The model concerning heating of the microsystem is well adapted for times shorter than 10 seconds. It was also possible to determine which parameters were significant regarding to the welding process: power, focusing and feed speed of the laser, but also composition and width of the welding seam. The tightness of the package has not been measured yet.

6. FUTURE WORKS

The feasibility of laser diode soldered packaging has been shown by soldering the two parts of the package together. The following steps need to be investigated in order to propose a robust solution.

First of all, the use of a special optics allowing us to follow the shape of the welding seam will be investigated. The goal of this optics is to allow soldering of more complex welding seams geometries and also to increase the scanning speed of the laser spot. The idea would be to heat the whole welding seam simultaneously by scanning it at high frequency. This would reduce constraints in the welding seam during the soldering process.

Development of soft solder will also have to be done. One objective is to reduce the fusion temperature in order to reduce the constraints in the glass, and also the temperature inside the package. Another goal is to obtain a low degassing soft solder. The wettability of the solder on the glass part of the package has to be increased.

Another step is to control the atmosphere contained in the package. Leak rate will have to be measured. The effects on the hermeticity of several parameters like materials, geometries of welding seam and getter materials will also be investigated.

REFERENCES

1. Ely, K. (2000). *Issues in Hermetic Sealing of Medical Products*. Medical Device & Diagnostic Industry Magazine.
2. Hoffmann, P. (2005). *Micro-Usinage par laser*. EPFL.
3. Jacq, C., Maeder, T., Menot-Vionnet, S., Birol, H., Saglini, I., Ryser, P. (2005). *Integrated thick-film hybrid microelectronics applied on different metal substrates*, EMPC, Belgium.
4. Nicollin, V. *Etude d'un packaging étanche et de son industrialisation*. Master project, EPFL.
5. Rey, C. (2005). *Etude d'un packaging étanche et de son industrialisation*. Semester project, EPFL.

MODELLING AND CHARACTERISATION OF AN ORTHO-PLANAR MICRO-VALVE

Olivier Smal[1], Bruno Dehez[1], Benoit Raucent[1], Michaël De Volder[2], Jan Peirs[2], Dominiek Reynaerts[2], Frederik Ceyssens[3], Johan Coosemans[3], Robert Puers[3]

[1]Université Catholique de Louvain, Dept. of Mechanical Engineering, Div. CEREM, Place du levant 2, 1348 Louvain la Neuve, Belgium ; [2]Katholieke Universiteit Leuven, Dept. of Mechanical Engineering, Div. PMA, Celestijnenlaan 300 B, 3001 Leuven, Belgium; [3]Katholieke Universiteit Leuven, Dept. of Electrical Engineering, Div. MICAS, Kasteelpark Arenberg 10, 3001 Leuven, Belgium.

Abstract: The main difficulties encountered in the development of microscale fluidic pumping systems stem from the fact that these systems tend to comprise highly three-dimensional parts, which are incompatible with traditional microproduction technologies. Regardless of the type of pumping principle, most of the hydraulic systems contain valves and in particular a one-way valve. This paper presents the design and modelling of an ortho-planar one-way microvalve. The main advantages of such a valve are that it is very compact and can be made from a single flat piece of material. An analytical model of the spring deflection has been developed and compared to FEM. A prototype with a bore of 1.5 mm has been build using a micro EDM (electro discharge machining) machine and also tested.

Key words: microvalve; pumping systems; ortho-planar spring.

1. INTRODUCTION

Advances in micro-mechanical and in medical areas generate a huge need for microfluidic systems. Recent studies have shown that there is a steadily growing market for Microsystems, and in particular for drug delivery systems. According to 1 the drug delivery market is estimated at US $20 billion. Micropumps have a large number of applications: dispensing of therapeutic agents into the body, for medical, but also cooling of microelectronic devices, miniature systems for chemical and biological analysis, micropropulsion, etc.

Microfluidic systems might also be used for actuation technologies in microdevices. Recent research revealed that fluidic microactuators might advantageously replace electrostatic and electromagnetic micro-actuators at microscale because they develop higher power and force densities as mentioned in 2 and 3.

Development of Micropumping devices began in the mid 1970's, now leading to many different principles as reported in 4. Several hundreds of papers reporting new micropumps or analysing micropump operation have been published.

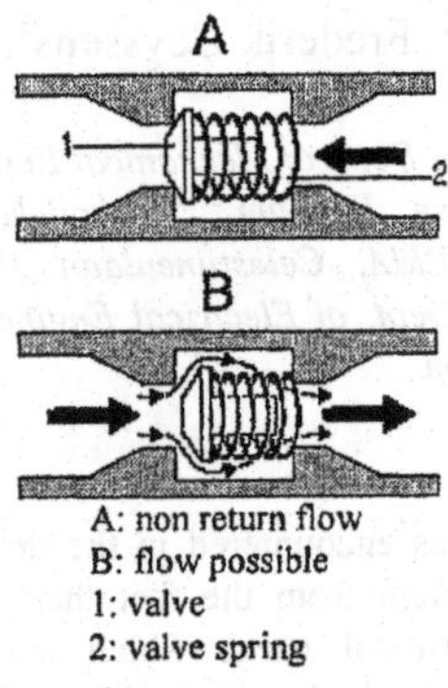

Figure 1. Non return valve principle (courtesy of Bosch)

Regardless of the pumping principle, most micropumps comprise active or passive microvalves. The most common valve is the check valve illustrated in Figure 1 5. It comes in a wide variety of shapes and materials but always with the same requirements: a low leakage in the reverse direction and a low pressure drop in the inlet flow direction. The spring stiffness should be carefully selected.

The valve presented in figure 1 is however very difficult to produce at micro scale. This paper presents the production and modelling of an ortho-planar one-way microvalve. The major advantage of such a valve is that it can be very compact and simple to build.

The second section focuses on the design principle. The third section describes an analytical model of the spring deflection while section 4 presents an FEM simulation. Section 5 presents the prototype fabrication and section 6 the prototype test and a comparison of models.

2. ORTHO-PLANAR MICROVALVE

Ortho-planar valves are valves based on ortho-planar springs. Such springs which can be either manufactured in or compressed down into a single plane 6. Examples of ortho-planar springs are the Belleville disc spring, the volute spring and the spider spring. Their major advantages are that they are very compact and can be made from a single flat piece of material. In this study, we will concentrate on spider springs, which are by definition manufactured in a plane. Several typical shapes are presented in Figure 2.

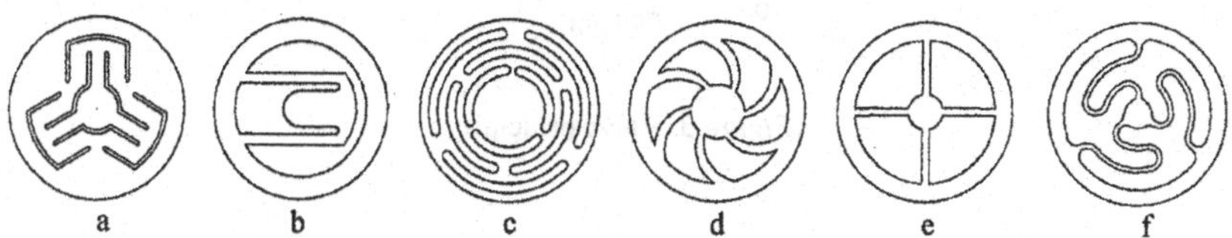

Figure 2. Spiders

The valve principle is presented in Figure 3. The central disc is connected to the outer disc by several beams. When pushing on the central disc in a direction perpendicular to the plate, the beams bend and apply a spring force to the central disc. If the plate is symmetrical then the central disc always stays parallel to the outer disc. Figure 3 shows the valve behaviour in its in-flow and back-flow state. When the fluid flows in (case a), the fluid pushes on the central disc and opens the valve. When the inlet pressure drops, the beams, acting like a spring, close the valve (case b). As a consequence, this valve acts as a one-way valve.

The various spiders illustrated in Figure 2 could theoretically be used in this configuration. However practical limitations should be considered: the fluid has to pass trough the spider cuttings. That is why shapes (a), (b) and (c) can be directly rejected because they offer too much resistance to the fluid flow. Spider (d), (e) and (f) are comparable in the principle, but as the beams are longer in (f), the spring stiffness will be smaller and the opening area larger. In the following we will concentrate on that case.

Figure 1 shows that to ensure that the valve is kept closed in its rest position, the spring has to be pre-stressed. This pre-stress defines the opening pressure of the valve. Normally, the opening pressure should be between 0.001 and 0.05 MPa 7. When the pressure is high enough, the opening should be as large as possible in order to reduce the pressure drop in the valve. It is therefore very important to have a good model of the spring.

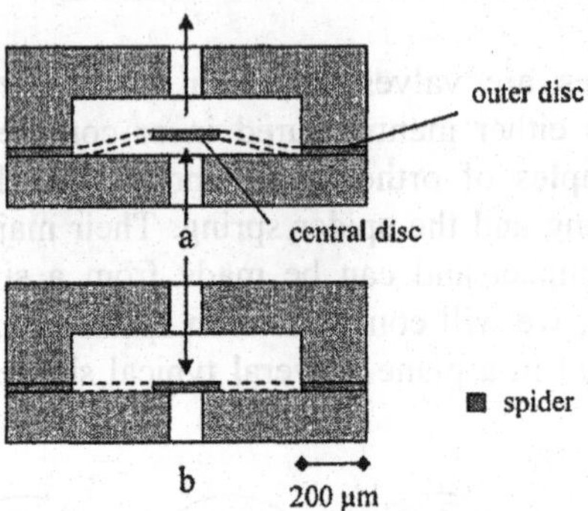

Figure 3. Valve principle

3. SPRING MODELLING

Spider have already been modelled and manufactured [6,8,9]. However, existing models consider only straight bar spring segments and take into account only the beam's bending, neglecting its torsion. Such a principle is therefore not suitable for curved beams as in case (f) presented in Figure 2. In this section, we propose a new model taking into account bending and torsion of a curved beam.

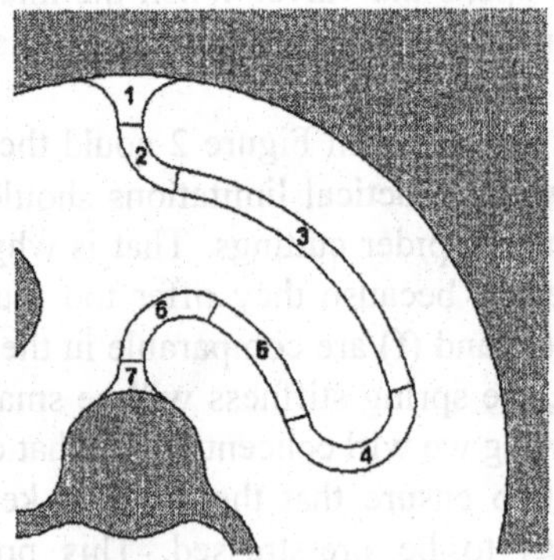

Figure 4. Beam decomposition

The analytical model presented in this section is a deformation model, the fluidic action being modelled by the simple action of the pressure on the central disc. Fluidic action on the beams are neglected

To model this structure, the valve is considered as a disc suspended on folded beams. The central disc is assumed to be non deformable, all the deformation occuring in the beams. Each beam can be considered as a spring connected in parallel. So the total force applied by the fluid on the central disc is equally divided between each beam The folded beam is divided into successive straight or curved standard elements, see Figure 4. The first and last element (1 and 7) are considered to be straight with constant cross section. Shear forces, moments, displacements and slopes are modelled for the two standard elements using Roark's formulae 10. The directions for forces and slopes are presented in Figure 5.

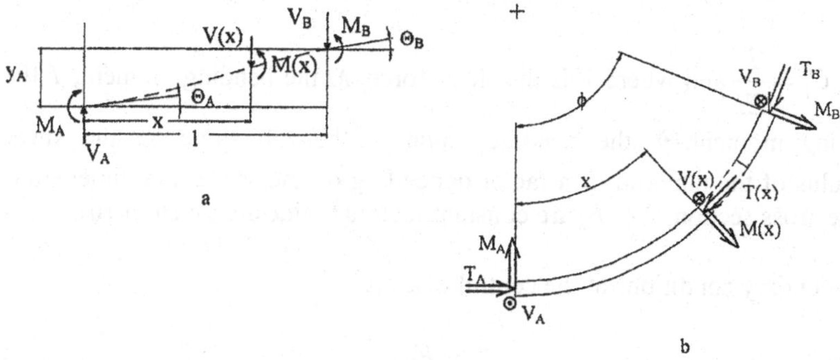

Figure 5. Forces and moment orientation, a: lateral view for a straight beam, b: top view for a curved beam.

Case (b) presents the conventions for elements 2, 5 and 6 in Figure 4. Formulas (1) and (2) present the results for this orientation. Formulas for the other orientation (as for elements 3 and 4 in Figure 4) can be easy derived and are not presented here. For a straight bar:

$$\begin{pmatrix} V(x) \\ M(x) \\ T(x) \\ y(x) \\ \Theta(x) \\ \psi(x) \end{pmatrix} = \begin{pmatrix} -1 & 0 & 0 & 0 & 0 & 0 \\ x & 1 & 0 & 0 & 0 & 0 \\ 0 & 0 & 1 & 0 & 0 & 0 \\ \dfrac{x^3}{6EI} & \dfrac{x^2}{2EI} & 0 & 1 & x & 0 \\ \dfrac{x^2}{2EI} & \dfrac{x}{EI} & 0 & 0 & 1 & 0 \\ 0 & 0 & \tan\left(\dfrac{xT_A}{KG}\right)\cdot\dfrac{1}{T_A} & 0 & 0 & 1 \end{pmatrix} \cdot \begin{pmatrix} V_A \\ M_A \\ T_A \\ y_A \\ \Theta_A \\ \psi_A \end{pmatrix} \quad (1)$$

and for a curved beam:

$$\begin{pmatrix} V(x) \\ M(x) \\ T(x) \\ y(x) \\ \Theta(x) \\ \psi(x) \end{pmatrix} = \begin{pmatrix} 1 & 0 & 0 & 0 & 0 & 0 \\ R\sin x & \cos x & -\sin x & 0 & 0 & 0 \\ R(1-\cos x) & \sin x & \cos x & 0 & 0 & 0 \\ C_1R^2F_3 & C_1RF_1 & C_1RF_2 & 1 & R\sin x & R(1-\cos x) \\ C_1RF_6 & C_1F_4 & C_1F_5 & 0 & \cos x & \sin x \\ C_1RF_9 & C_1F_7 & C_1F_8 & 0 & -\sin x & \cos x \end{pmatrix} \cdot \begin{pmatrix} V_A \\ M_A \\ T_A \\ y_A \\ \Theta_A \\ \psi_A \end{pmatrix} \quad (2)$$

with $C_1 = \dfrac{R}{EI}$ and where V is the shear force, M the bending moment, T the twisting moment, Θ the bending slope, ψ the roll slope, G the shear modulus of the material, K a factor depending on the shape and dimensions of the cross section; $F_1 \ldots F_9$ are constant factors (value are given in 10).

Boundary conditions at the central disc are:

$$V_{disc} = \frac{F_{fluid}}{j}$$
$$\Theta_{disc} = 0 \quad (3)$$
$$\psi_{disc} = 0$$

where F_{fluid} is the force applied by the fluid, $F_{fluid} = P \cdot S_{disc}$

j is the index of beam,

V_{disc} is the perpendicular force applied by the beam onto the central disc,

Θ_{disc} the bending slope on the central disc,

ψ_{disc} the roll slope on the central disc .

By combining relations (1) and (2), it is possible to evaluate results at any point on the beam starting with the effort at the beam extremity (on the outer disc side) as functions of M_N and T_N (bending and twisting moments on the central disc). Relation (3) can be used to evaluate M_N and T_N.

4. FEM MODELLING

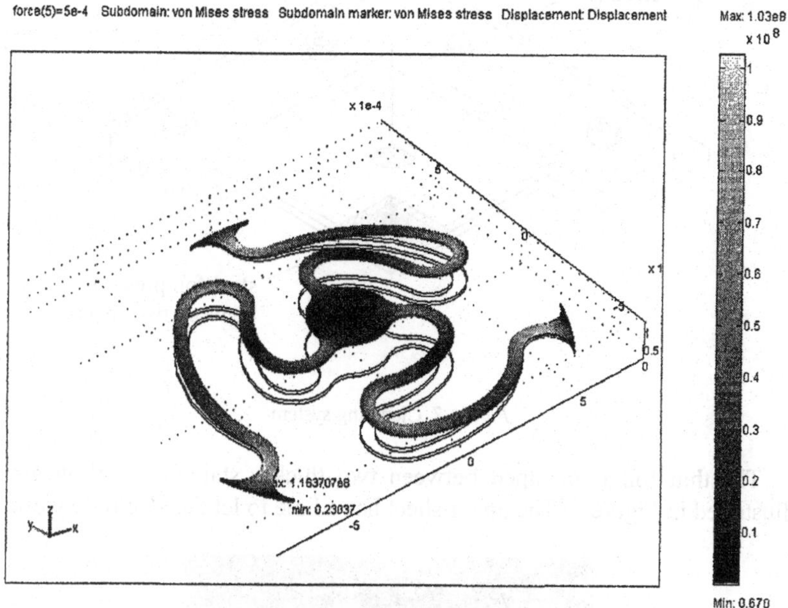

Figure 6. FEM simulation of the 3 beams spring

A FEM model has been developed. This is a 3D model, using a drawing of the valve computed with Autodesk Mechanical Desktop®. For a given force on the central disc, the values of the displacement and the Von Mises stresses are computed. As for the analytical model, this is only a deformation model, the fluidic part of the valve is not taken into account.

5. MANUFACTURING

This part presents only the manufacturing of the spring itself, the other components of the valve will be considered later. This small part is realised using a micro EDM (electro discharge machining) machine. Starting from a stainless steel micro-foil (thickness = 10 μm), the cuttings are made with a standard 150 μm diameter electrode.

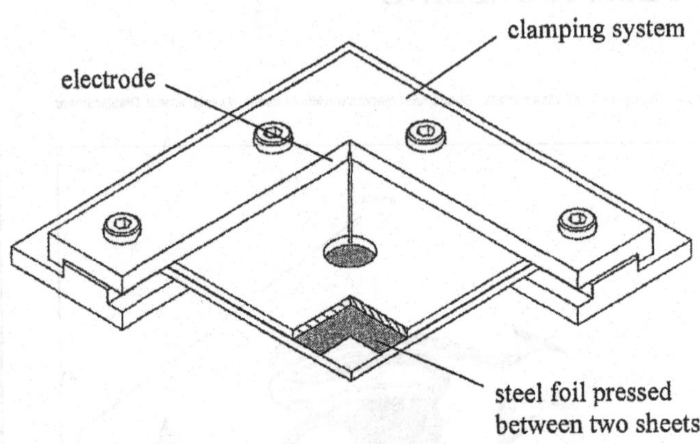

Figure 7. Clamping system

The thin foil is clamped between two thicker stainless steel sheets as illustrated in Figure 7. The upper sheet has a hole to let the electrode work.

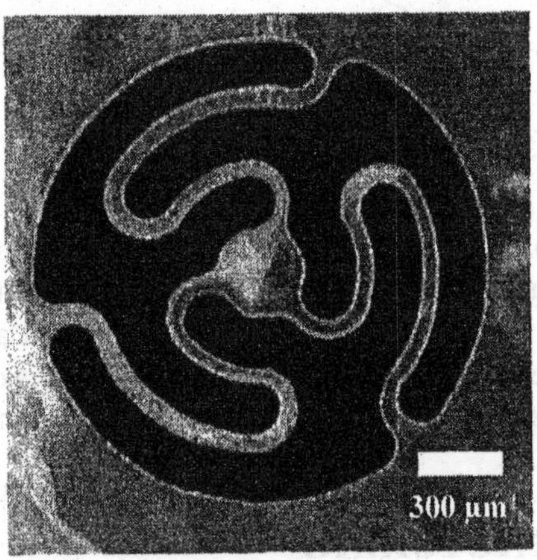

Figure 8. Valve spring

It will be possible to produce a smaller valve design if we work with a thinner electrode (up to 50 µm). On our EDM machine, there is a wire system to reduce the diameter of the electrode before machining. But for a first experiment we prefer to use the standard electrode without any modifications. Figure 8 shows the valve spring prototype.

6. RESULTS

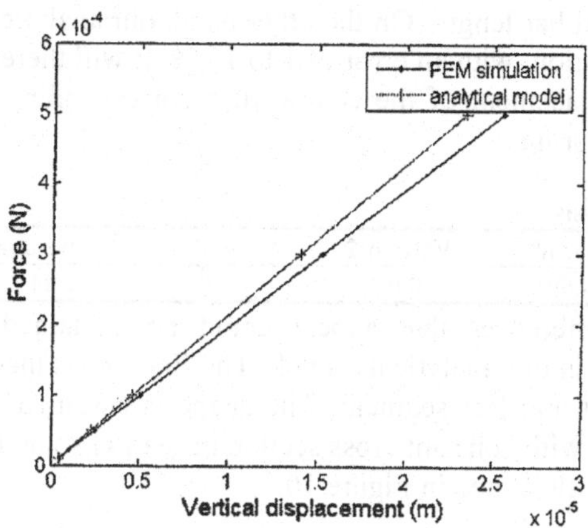

Figure 9. Spring force on the central disc

Figure 9 presents the evolution of the spring force as a function of the vertical displacement of the central disc for a 3-beam valve depicted in Figure 11. Good linearity can be observed. The FEM simulation was also conducted for the various shapes of Figure 11. The results for the stiffness are presented in Table 1 where they are compared with the calculations from the analytical model. The stiffnesses are evaluated with the values of the displacement and the applied force. The approximate model considers the beams as straight bars in the same way as models developed in 8. More precisely, for the case of Figure 4, if L_3 and L_5 are the lengths of segment 3 and 5, the stiffness is computed as being that of a straight beam of length $L_3 + L_5$ with one end fixed and the other end guided. This explains why, in general, the approximate stiffness is larger than that computed via an analytical model.

Table 1. Stiffnesses for the various valve designs (in N m-1) (*) = straight beam

Calculations	Valve n°1	Valve n°2	Valve n°3	Valve n°4	Valve n°5
Approximate model*	2.30	6.04	32.57	379.57	144.96
Analytical model	4.32	4.62	21.38	60.17	144.96
FEM	4.36	4.76	19.59	53.08	144.85

For valve 5, the 3 models yield similar results. For a curved beam however, the approximate model produces results between 50 % and 5 times removed from the FEM. The very large difference observed for valve n°4 is due to the fact that in this geometry, the lengths of segments 3 and 5 (see Figure 4) are very small, so the corresponding straight bar is extremely short, which produces a very high stiffness (the length is powered to three in the formulas). Table 2 shows the proportion of the considered length with respect to the total bar length. On the other hand, our analytical model gives a good approximation, with an error of 1 to 15 %. It will therefore be used to evaluate the displacement of the central disc corresponding to the desired pre-stress of the spring.

Table 2. models analysis

Calculations	Valve n°1	Valve n°2	Valve n°3	Valve n°4	Valve n°5
$(L_3+L_5)/L_{Total}$	0.69	0.61	0.51	0.34	1

The small differences that appear could be explained by the main hypotheses made in our analytical model. The first one is the approximation made on the first and last segment. The shape is assumed to be a simple rectangular beam with constant cross section instead of something somewhat more complex as illustrated in Figure 10.

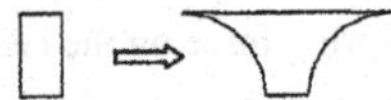

Figure 10. Shape approximation

The second one is the fact that we consider that the entire deformation occurs in the arms. In reality and in the FEM simulation, there is also a deformation in the central disc. The third one comes from the boundary conditions on our beam (see (3)). This hypothesis is a good approximation of what occurs in reality but not the exact truth.

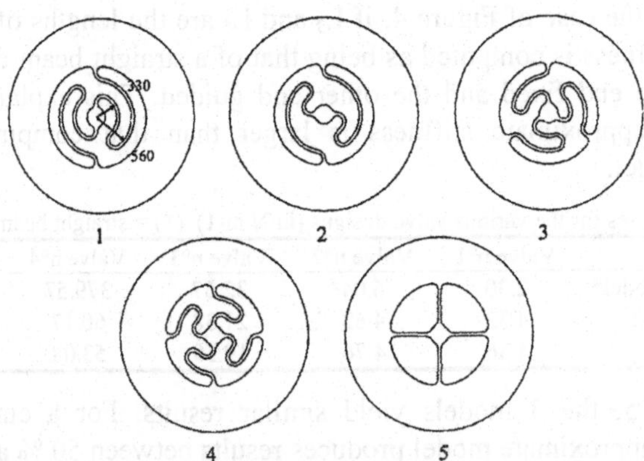

Figure 11. Thickness of the sheet = 10 μm; central disc diameter = 300 μm; beam width = 50 μm; spring diameter = 1.58 mm

Finally the stiffness of the spring presented in Figure 8 is measured using a very accurate microscope equipped with a probe. This probe allows to impose a vertical force from 10 up to 200 µN and to measure the induced displacement. Figure 12 presents the results of this test. It is important to note that the slope of the line (linear fit of the data) is quite close to the stiffness calculated and computed via FEM. We find 19.48 Nm^{-1} with the measurements, 19.59 Nm^{-1} with FEM and 21.38 Nm^{-1} by calculations.

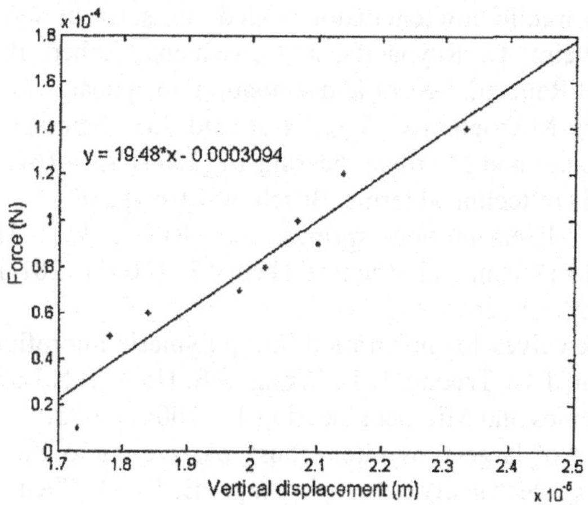

Figure 12. Stiffness measurement on spider n°3

In the near future, we will apply the same testing procedure with other spiders to validate our model for the various shapes.

7. CONCLUSION AND PERSPECTIVES

The analytical model proposed in this paper proved to be accurate enough for our application. In the near future we will use this model to optimize the valve design. Then we will perform an experimental comparison. Finally we will introduce fluid flow effect on the valve and build a prototype of the complete valve for hydraulic testing.

8. BIBLIOGRAPHY

1. Medical market for Microsystems, J. Malcolm Wilkinson, MSTnews n4 4,2002.
2. Production and Characterisation of a Hydraulic Microactuator, M. De Volder, J. Peirs , D. Reynaerts, J. Coosemans, R. Puers, O. Smal, B. Raucent, Journal of Micromechanics and Microengineering, to appear.
3. A novel hydraulic microactuator sealed by surface tension, M. De Volder, J. Peirs, D. Reynaerts, J. Coosemans, Robert Puers, Olivier Smal, Benoit Raucent, Sensor and Actuation, to appear.
4. A review of Micropumps, D.J. Laser and J.G. Santiago, Journal of Micromechanics and Microengineering 14 (2004) R35-R64.
5. Encyclopedia of technical terms, Bosch, www.ewbc.de.
6. Ortho-planar linear-motion springs, J.J. Parise, L.L. Howell, S.P. Magleby, Mechanism and Machine Theory 36 (2001) 1281-1299.
7. Festo, www3.festo.com
8. Micro check valves for integration into polymeric microfluidic devices, N.T. Nguyen, T.Q. Truong, K.K. Wong, S.S. Ho, C.L.N Low, Journal of Micromechanics and Microengineering 14 (2004) 69-75.
9. Development of large flow rate, robust, passive micro check valves for compact piezoelectrically actuated pumps, B. Li, Q. Chen, D.G. Lee, J. Woolman, G.P. Carman, Sensors and Actuators A 117 (2005) 325-330.
10. Roark's Formulas for Stress & Strain, 6th edition, W.C. Young, Mc Graw-Hill 1989.

MICROHANDLING AND ASSEMBLY: THE PROJECT ASSEMIC

Ana Almansa[1], Silvia Bou[2], Domnita Fratila[3]
[1,2,3] *Austrian Research Centres Seibersdorfs GmbH, 2444 Seibersdorf, Austria*

Abstract: Mechatronic competences represent a strong component in Microsystems Technologies, but very especially in Microhandling and –assembly, a field with challenging requirements. The Research and Training Network "Advanced Methods and Tools for Handling and Assembly in Microtechnology" (ASSEMIC) addresses this research field at a European scale. This paper presents aspects of the ASSEMIC project and some results achieved in its frame.

Key words: Training, Microassembly, Microtechnology, Microhandling

1. INTRODUCTION

The project "Advanced Methods and Tools for Handling and Assembly in Microtechnology" (ASSEMIC) was launched in January 2004, with the aim of addressing the broad and challenging research area of microhandling and assembly in the frame of a Marie Curie - Research and Training Network (RTN)[1].

With a consortium constituted by 14 participants from 10 different countries and a training and research schedule corresponding to more than 500 person months, the project ASSEMIC is among the largest Research and Training Networks funded by the European Commission. More than 30 foreign researchers, specifically appointed for the project, will be trained during its four years duration[2]. The members of ASSEMIC consortium are:

1. ISAS – TU Wien (Austria)
2. FSRM (Switzerland)
3. ARC Seibersdorf research GmbH (Austria)
4. IMT (Romania)
5. Politechnika Warszawska –WUT (Poland)

6. Uninova (Portugal)
7. University of Oldenburg (Germany)
8. Fundacion Robotiker (Spain)
9. FORTH (Greece)
10. Progenika Biopharma S.A. (Spain)
11. CCLRC – RAL (Great Britain)
12. Fraunhofer Gesellschaft – ILT (Germany)
13. Scuola Superiore Sant'Anna (Italy)
14. Nascatec (Germany).

The project is structured in several workpackages. A brief description of them content is given below[1]:

WP 1. Micropositioning: Positioning stages and elements with integrated sensors and feedback control, autonomous and mobile systems, microrobotics.

WP 2. Microhandling: Tools and methods for handling in different environments (normal room conditions, clean room, vacuum, fluids) and applications.

WP 3. Microassembly: Innovative tools, strategies and alternative approaches for efficient micro-assembly.

WP 4. Automation for industrial production: Including production chains, quality assurance, test and characterization issues, etc.

WP 5. Know-how management: Technology transfer and dissemination.

2. FIRST RESEARCH RESULTS

2.1 Micropositioning

The first workpackage is dedicated to technology for high accuracy positioning in microhandling applications. Different applications and potential uses have been considered in this workpackage such as: RAL's functionalized cantilever technology, which is applicable for novel Atomic Force Microscopy (AFM) probes, and optical position sensors developed by Uninova's.

Figure 1. Uninova's position sensor

2.2 Microhandling

The second workpackage deals with the development of technology for microhandling, which involves also a number of other further micromanipulation operations not directly related to assembly of MEMS, among others concerning biological and medical applications.

Examples of this are microsurgery, cell manipulation and detection and handling of biochemical macromolecules. Within the task devoted to the development of Advanced Microhandling tools, research has been done on different types of microgrippers, as well as special fabrication methods for such micro-grippers. Nascatec has reported an electrostatic microgripper and performed additional mechanical simulations by means of Finite Element Analysis.

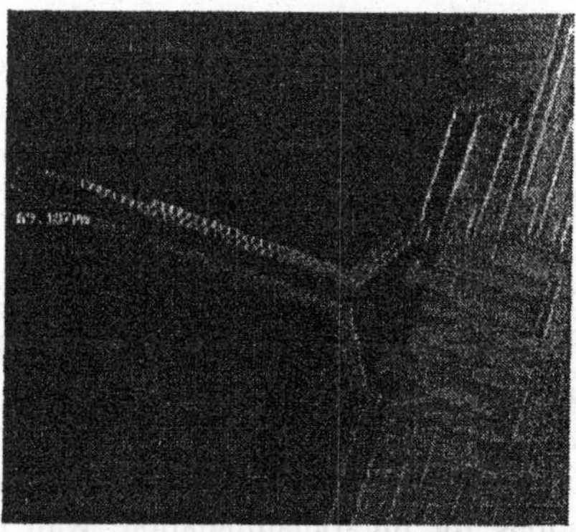

Figure 2. Electrostatic gripper (Nascatec)

While Scuola Superiore Sant'Anna has developed microgrippers by using a technology a technology for PST-actuated tools by using Shape Deposition Manufacturing (SDM), provided by previous collaboration with University of Uppsala in Sweden, Seibersdorf research has explored the combination of LIGA (Lithographie Galvanik Abformung) and PIM (Powder Injection Moulding) for the production of microgrippers.

Figure 3. Left: Tools for SDM processes (SSSA); Right: gripper photoresist structure on wafer (Seibersdorf)

Beyond the work on development of gripping tools, an extensive study has been performed by ARC Seibersdorf research, with contribution of Politechnika Warszawska –WUT, concerning strategies for handling of microcomponents. Concerning adhesion and material issues related to microhandling processes, Politechnika Warszawska–WUT has proposed two solutions for special intelligent coating with controllable adhesion. Furthermore, Scuola Superiore Sant'Anna performed some experiments to compare theoretical and real adhesion forces between sample and needle fingertip under different environmental conditions (normal and dry environment).

As regards microhandling applications, several application possibilities have been proposed and analyzed by the ASSEMIC participants, in order to test the tools and methods developed in the project. One of them, as reported by the University of Oldenburg, is the manipulation of TEM-lamella in the semiconductor industry.

The University of Oldenburg has also adapted and tested additional tools for manipulation of nanowires. They have done several experiments for gripping nanowires and for bonding them with the help of Electron Beam Deposition (EBD), with satisfactory results.

Another field of interest explored in Workpackage 2 comprises biological and medical applications for micromanipulation, with contributions from partners as Progenika and CCLRC (functionalization of biomedical cantilever based sensors), IMT-Bucharest (Surface Acoustic Wave micro-agitation device for DNA hybridization) and SSSA (mechanical and chemical measurement from vessels).

2.3 Microassembly

In a first approach to the development of Innovative Tools for assembly of microparts, results available at Fraunhofer Gesellschaft / Institute für Lasertechnik on integrated tools for gripper and joining of components, will serve as a basis for further work in the ASSEMIC project.

Another technology of interest for assembly of MEMS is micro-gluing. A study was done concerning adhesive used in Micro- opto- electromechanical systems (MOEMS) and microtechniques, as well as the different techniques of gluing and dispensing, including automated dispensing, UV-curing adhesives and those used for encapsulation[4].

Advanced processes for efficient microassembly, such as design for assembly, parallel assembly or self-assembly processes have been analysed and reported in a state of the art performed by Politechnika Warszawska – WUT and Seibersdorf.

Further, work related to the assembly, testing and improvement of the 4x4 and 8x8 cross connector switches for optical fibres was done in cooperation between FSRM and the University of Neuchatel. After the conception of an adequate assembly strategy, among other tasks, a driver PCB, a Visual Basic program for switching the 64 mirrors, and a new high voltage source were developed.

Figure 4. Side view of 4x4 packaged cross connection switch for optical fiber (FSRM/University of Neuchatel)

3. FURTHER WORK

This paper has presented main research achievements and activities performed in the ASSEMIC Network during the one and a half year. During the next years, further joint research is planned in the fields of microrobotics (miniaturized autonomous mobile platforms) and advanced control systems for intelligent manipulation, as well as human haptic interfaces. Research on biological and medical applications of micromanipulation will be extended with experiments for operation in special environments and conditions, such as vacuum, normal room conditions and fluids. Furthermore, work done in tools and processes for micro-assembly will be complemented with the analysis of alternative approaches to hybrid assembly, in comparison with monolythical integration. Finally, a new workpackage will deal with automation of micro-assembly for industrial production of MEMS and covering automated handling, test and characterization of assembled Microsystems.

ACKNOWLEDGMENT

This paper and the described work have been made possible thanks to the financial support provided by the Marie Curie Programme under the 6th Framework Programme of the European Commission through the Research and Training Network "Advanced Methods and Tools for Handling and Assembly in Microtechnology" (ASSEMIC). The work described in this paper has been performed by the 14 institutions participating in the ASSEMIC Network. The authors would like to thank permanent staff and appointed fellows working in this project at these institutions, for their excellent work and co-operation.

REFERENCES

1. European Commission, "Marie Curie Research and Training Networks (RTN) – Handbook" 2nd Edition, December 2003.
2. A. Almansa, S. Bou, W. Brenner, A. Locher: "A European Research and Training Network for Advanced Post-graduate Education", Proceedings of the IEEE International Conference on Industrial Technology; Hammamet; Tunesia, 2004. ISBN:0-7803-8663-9.
3. M. Fernandes, M. Vieira, I. Rodrigues, R. Martins. "Large area image sensing structures based on a-SiC: a dynamic characterization", Sensors and Actuators A, 113, (3), 2004, pp. 360-364.
4. S. Bou, A. Almansa, N. Balabanava, "Handling Processes in Microsystems Technology", submitted for publication at AIM 2005.

AUTHOR INDEX

Almansa, A. 327
Bartek, M. 133
Bert, J. 239
Bley, H. 121
Bossmann, M. 121
Bou, S. 327
Bourgeois, F. 267
Brecher, C. 11, 181
Burisch, A. 65, 83
Ceyssens, F. 315
Chaillet, N. 21
Chapius, Y. 53
Clévy, C. 21
Codourey, A. 193
Coosemans, J. 315
De Volder, M. 315
Degen, R. 109
Dehez, B. 315
Dembélé, S. 239
Fahlbusch, S. 21
Fleischer, J. 201
Fratila, D. 327
Freundt, M. 11
Heikkilä, R. 93
Heilala, J. 277
Helin, K. 277
Henneken, V. 155
Hesselbach, J. 65, 83, 101
Hoppe, G. 227
Hubert, A. 21
Jacot, J. 3, 267, 287, 307
Karjalainen, I. 75
Koerlemeijer, S. 3, 267, 287
Krahtov, L. 201
Kurniawan, I. 133
Lambert, P. 3
Lang, D. 33
Lange, S. 11, 181
Lefort-Piat, N. 239
Lempiäinen, J. 149
Lenz, T. A. 297
Lohse, N. 215
Lutz, P. 43
Michler, J. 21
Montonen, J. 277
Mussard, Y. 193
Olea, G. 251
Othman, N. M. 297
Oulevey, M. 287
Paris, M. 43
Peirs, J. 315
Perrard, C. 43
Perroud, S. 193
Peschke, C. 11
Prusi, T. 93
Puers, R. 315
Raatz, A. 65, 83, 101
Ratchev, S. 53, 215
Rathmann, S. 101
Raucent, B. 251, 315
Reynaerts, D. 315
Salmi, T. 149
Schäfer, C. 215
Schöttler, K. 101
Seigneur, F. 3, 307
Slatter, R. 65, 109
Smal, O. 315
Smale, D. 167
Soetebier, S. 65
Takamasu, K. 251
Tichem, M. 33, 133, 155
Tuokko, R. 75, 93
Turitto, M. 53
Ususitalo, J. 93
Väätäinen, O. 277
Volkmann, T. 201
Weinzierl, M. 181
Wrege, J. 65, 83, 101

KEYWORD INDEX

2dof Actuators 251
Active Alignment 11
Adhesive Bonding 297
ADS 227
Agent-Based 215
Assemblability Analysis 149
Assembly 11, 181, 227
Assembly Applications 109
Assembly Force 93
Assembly Planning 121
Assembly Process Decomposition 215
Assembly Process Testing 93
Automation 21, 201
Capillary Forces 3
Classification Scheme 43
Communication 167, 227
Compliant Mechanism 83
Contactless 53
Control 227
Cost of Ownership Analysis 277
Cost-Modeling 287
Design 167
Design for Assembly 121, 149
Desktop Assembly 251
Desktop Factory 65
DFA 149
Dispensing 297
Distributed Manipulation 53
Electrostatic 33
Equipment Configuration 215
EtherCAT 227
Eupass 227
Evolvability 227
Feeding 43
Fibre-Chip Coupling 155
Flexibility 21
Flexible Micro Parts 11
Flexure Hinges 83
Fluid Self-Alignment 155
Grip Force 33
Gripper System 11

Gripper-Integrable Milti-Axes System 11
Handling 11
Hermetic Sealing 307
High Precision Robotics 83
Hybrid Microsystems 133
In Process 101
Laser Scanning 101
Man-Machine Cooperation 267
Mechanical Stop 155
MEMS 133
MEMS Technology 155
Micro Actuator 109
Micro Assemble 297
Micro Assembly 75
Micro Components 181
Micro Gripper 11
Micro Harmonic Drive 109
Micro Robot 65
Microassembly 3, 21, 53, 133, 327
Micro-Assembly 149, 201, 267, 287
Microfactories 239
Microfactory 21, 43, 193
Microfeeding 53
Micro-Gears 65
Micro-Gripping 33
Microhandling 327
Micromanipulation 21
Micromanipulation Cell 21
Micro-Mechatronical Products 201
Microsystem Assembly 307
Microtechnology 327
Microvalve 315
Miniature Production Line 193
Miniature Robot 193
Modeling 33
Modular Assembly System Design 277
Modular Assembly Systems 215
Modular Design 251
Mosaicing 239
Multi-Degrees-of-Freedom 251
Ontology 215
Ortho-Planar Spring 315
Overall Equipment Efficiency 277

Packaging 133, 307
Parallel Mechanisms 251
Parallel Robot 193
Parallelograms 251
Parallel-SCARA 65
Passive Alignment 181
Planar Motors 251
Planning Systematics 201
Pneumatic 53
Positioning 251
Positioning Errors 93
Precision 251
Precision Assembly 101, 267
Process Control 297
Process Monitoring 297
Product-Internal Assembly Functions 155
Pseudo-elastic SMA 83
Publusher / Subscriber 227
Pumping Systems 315
Scanning Electron Microscope 21
Semi-Automatic Assembly 267
Sensor Integration 297
Servo Gripper 75
Silicon Optical Bench (SiOB) 155
Simultaneous Engineering 121
Skills 227
Submilimetric Application 3
Supervision 239
Surface Tension 3
Tolerance Budgeting 155
Tool Changer 21
Training 327
Ultra Precision Machining 181
UPnP 227
Vision System 297
Visualisation 167
Work Cycle Time 93